DATE DUE

MAINTENANCE WELDING

MAINTENANCE WELDING

Edgar Graham

PRENTICE-HALL, INC. *Englewood Cliffs, New Jersey 07632*

Library of Congress Cataloging in Publication Data

Graham, Edgar.
  Maintenance welding.

  Bibliography: p.
  Includes index.
  1. Welding. 2. Industrial equipment—Maintenance and repair. I. Title.
  TS227.G655 1985   671.5′2   84-11439
  ISBN 0-13-545468-9

Editorial/production supervision and
  interior design: *Mary Carnis*
Cover design: *Wanda Lubelska*
Manufacturing buyer: *Anthony Caruso*

© 1985 by Prentice-Hall, Inc., Englewood Cliffs, New Jersey 07632

All rights reserved. No part of this book may be
reproduced, in any form or by any means,
without permission in writing from the publisher.

Printed in the United States of America

10 9 8 7 6 5 4 3 2 1

ISBN   0-13-545468-9   01

PRENTICE-HALL INTERNATIONAL, INC., *London*
PRENTICE-HALL OF AUSTRALIA PTY. LIMITED, *Sydney*
EDITORA PRENTICE-HALL DO BRASIL, LTDA., *Rio de Janiero*
PRENTICE-HALL CANADA INC., *Toronto*
PRENTICE-HALL OF INDIA PRIVATE LIMITED, *New Delhi*
PRENTICE-HALL OF JAPAN, INC., *Tokyo*
PRENTICE-HALL OF SOUTHEAST ASIA PTE. LTD., *Singapore*
WHITEHALL BOOKS LIMITED, *Wellington, New Zealand*

To George Redden, Angelo (Tony) Cabrera, and Richard (Dick) Barnes of the San Francisco Community College District; to Joe Marden, president of the Marco Company; to all the steelmill, shipyard, pipeline, oil-refinery, construction, and job-shop welders who shared their good days and their bad days with the author; and to the inquisitive students who enlivened the author's classroom with question on question in the Socratic mode.

CONTENTS

PREFACE xv

1 SHIELDED METAL-ARC WELDING 1

Mastering the Arc 2
Understanding the Arc 2
Understanding the Weld Puddle 6
Understanding the Machine 8
  Troubleshooting the Welding Circuit 13
  Setting the Machine 13
The SMAW Technique 14
  Flat-Position Welding (Down-Hand) 15
  Horizontal Welding 16
  Vertical-Up Welding 16
  Vertical-Up Fillet 17
  Overhead Welding 17
Electrode Classification 17

2 OXYACETYLENE WELDING 21

In the Beginning 21
The Torch Today 22
Basic Processes 22

Fusion Welding 22
Brazing 23
Braze Welding 25
Torch Brazing 26
Sweating 29

Surface Alloying 29

Understanding the Torch 30

Welding-Tip Size 30
Lighting the Torch 31

Hot-Spray Metallizing 34

3 TUNGSTEN INSERT-GAS WELDING 37

Understanding Tungsten Inert-Gas Welding 37

Fact and Fable 37

Shielding Gases 39

Gas Flow Rate 41

Choice of Electrode 42

Setting the Current 43

Selecting the Filler Rod 44

4 METAL INERT-GAS WELDING 45

Understanding the Arc 45

Shielding Gases 47

Arc Transfer 48

Welding Techniques 49

5 METALLURGY FOR WELDERS 51

What Is Metallurgy? 52

Pure Metals 53

Transformation of Pure Metals and Alloys 56

Classes of Alloys 58

Class I Alloys 58
Class II Alloys 60
Class III Alloys 63

Iron–Carbon alloys 65

Heat Treatment of Steel 69

Understanding Austenite 69
Control of Steel Properties 70
From Austenite to Pearlite 71

Heat-Treatment Processes 73
Summary 75

6 IRON–CARBON STEELS 77

Low-Carbon Steels 77
Medium-Carbon Steels 79
Controlling Hardness 82
High-Carbon Steels 83
Stick-Welding Carbon Steels 84
Gas-Welding Carbon Steels 86
TIG-Welding Carbon Steels 86
MIG-Welding Carbon Steels 87
Wrought Iron 90

7 TOOL-AND-DIE STEELS 92

Slow (Natural) Cool-Down of High-Carbon Steels 92
Hardness or Toughness? 98
> *Tempering* 98
> *Martempering* 100
> *Austempering* 100

Types of Tool-and-Die Steels 100
> *Water-Hardening Steels* 101
> *Oil-Hardening Steels* 101
> *Air-Hardening Steels* 102
> *High-Speed Steels* 102
> *Hot-Working Die Steels* 103
> *Low-Alloy Tool Steels* 103
> *Mavericks and Renegades* 103

Maintenance Welder's View of Tool-and-Die Steels 104
> *Preventing Cracked Welds* 104
> *Preheating Temperature* 105

Welding Tool-and-Die Steels 106
Silver-Brazing Tool-and-Die Steels 107

8 CAST IRON 108

Understanding Cast Iron 109
> *Gray Cast Iron* 109
> *Pearlitic Gray Iron* 110
> *Addition of Various Metals to Gray Iron* 110

White Cast Iron   112

Malleable Cast Iron   113

Nodular Iron   113

Welding Cast Iron   114

   Transformations of 3% and 5% Cast Irons   114
Joint Preparation   116
Heat Treatment   116
Welding Gray Cast Iron   117
Welding White Iron   120
Welding Malleable Iron   120
Welding Nodular Iron   121
Using Nickel Electrodes   121
"Cold"-Welding Cast Iron   122

Brazing Cast Iron   124

Soldering Cast Iron   124

Thermit-Welding Cast Iron   124

Summary   127

9 STAINLESS STEEL   128

Understanding Stainless Steel   128

AISI Classification   129

   2xx Series—Austenitic   129
3xx Series—Austenitic   129
4xx Series—Martensitic   131
4xx Series—Ferritic   131

Welding Stainless Steel   131

   Stick-Welding Stainless Steel   131
Gas-Welding Stainless Steel   132
TIG-Welding Stainless Steel   133
MIG-Welding Stainless Steel   134

Brazing Stainless Steel   134

Soldering Stainless Steel   135

SMAW Electrodes for Stainless Steel   135

10 OTHER METALS   138

Nickel-Based Alloys   138

   Welding Techniques   139
Characteristics of Selected Alloys   142
Brazing Nickel-Based Alloys   145
Cutting Nickel-Based Alloys   148
Gas-Welding Nickel-Based Alloys   149

Copper-Based Alloys   150
Tough Pitch Copper   150
Deoxidized Copper   150
Aluminum-Copper Alloys (Aluminum Bronze)   151
Beryllium-Copper Alloys   151
Copper-Zinc Alloys (Brasses)   152
Copper-Zinc-Lead Alloys   152
Copper-Zinc-Tin Alloys   153
Copper-Tin Alloys (Phosphor Bronze)   153
Copper-Silicon Alloys   153

Nickel-Silver Alloys   154

Cupronickels   154

Manganese Bronzes   154

Other Nonferrous Metals   154

Low-Melting Metals   155

Precious Metals   156

Dissimilar Metals   156

11 ALUMINUM AND ITS ALLOYS   159

Understanding Aluminum   159
Wrought Aluminum Alloys   160
Aluminum-Copper Alloys   160
Heat Treatment of Aluminum   161

Welding Aluminum   163
Gas-Welding Aluminum   164
TIG-Welding Aluminum   165
Stick-Welding Aluminum   170

Soldering Aluminum   171

Brazing Aluminum   172

Characteristics of Aluminum Weldments   173

12 AUSTENITIC MANGANESE STEEL   175

Understanding Manganese Steel   175
Characteristics That Affect Welding   176
Welding Procedures and Recommendations   177

13 WELDING TECHNIQUES   178

Joint Design   179
Stress and Strain   180
Welding Symbols   189

Metal Working 199
> *Oxyacetylene Cutting* *199*
> *Oxygen-Arc Cutting* *208*
> *Arc Cutting* *208*
> *Air-Arc Cutting* *209*
> *Flame Straightening* *211*

Fabricated Shop Facilities 213

14 PIPE WELDING   216

Taking the Pipe Test 217

Mathematics of Pipe Fitting 219
> *Fractions* *220*
> *Square Root* *221*
> *Right Triangles* *223*
> *The Circle* *224*
> *Capacities of Common Tanks* *226*

Pipe Fitting 228
> *Joining Pipe Sections* *228*
> *Flanges* *230*
> *Laying Out Angles Using a Square* *235*
> *Laying Out Angles Using a Protractor* *237*
> *Laying Out Miter Cuts for Pipe Turns* *237*
> *Branches and Headers* *240*
> *Fabricating Fittings from 90° Pipe Turns* *243*
> *Welding Offsets* *246*
> *Installing a Steam Line* *247*
> *Miscellaneous Fittings* *248*
> *Reading Isometric Pipe Sketches* *249*

Ancillary Skills 251
> *Being Able to Fit In on the Job* *251*
> *The First Day* *252*
> *Precautions* *253*

Appendix: Trigonometry Table 254

15 WEARFACING   258

Understanding Wearfacing 258

Types of Wear 259

Wearfacing Materials 262

Use of Wearfacing in Industry 264

> *Steelmills* *264*
> *Earthmoving* *265*

Logging and Lumber 269
Mining 271
Well Drilling 272
Automotive 272

APPENDICES 273

A GLOSSARY 274

B SYMBOLS AND ABBREVIATIONS 289

C UNIT CONVERSION FORMULAS 295

D TABLES 300

E BIBLIOGRAPHY 327

INDEX 329

PREFACE

This book is written by a welder who throughout his welding career has always seemed to find himself welding an unfamiliar metal under the worst conditions with inadequate or inferior equipment.

That first paragraph may well be considered the life story of a maintenance welder: the poor chap who never knows from one day to the next what metal he will encounter and what welding process he will have to use. Will he first silver-braze a complex assembly requiring two applications at two different temperatures, jump to a preheat and brazeweld of a heavy cast-iron section, air-arc a large casting, return to the delicate touch of a TIG weld on a composite die, and finish his day with a hot-spray overlay on the worn teeth of a sprocket gear?

Such a day is quite typical in the life of the maintenance welder or job-shop welder. Would such a welder need a welders' guide in much the same way as an amateur chef would need a cookbook? The writer addresses that question especially to the industrial mechanic, the heavy-duty mechanic, the machinist, the stationary engineer, and all others who find that welding is an ancillary skill required in their trades.

The book has been designed as a reference manual more than as a course of study. It is recommended, however, that the purchaser browse through the entire book in order to get acquainted with its content. He should read the table of contents, the index, and the glossary; and should study Chapter 5, "Metallurgy for Welders," briefly but often. Chapters 6 and 7, "Iron–Carbon Steels" and "Tool-and-Die Steels," should be studied together with the metallurgy section.

The major thrust of the book is toward a basic understanding of metals, predicated on the author's belief that armed with the basics, the mechanic can rise to the challenge, whatever the challenge may be.

ACKNOWLEDGMENTS

I wish to thank the following companies for their help and cooperation in the preparation of this book: Lincoln Electric Company; Hobart Brothers; Crane Valves; Kaiser Aluminum; Reynolds Aluminum; Tube-Turns, Inc.; Marco Weld Products; Huntington Alloys, Inc.; High Technology Materials Division, Cabot Corporation; Stoody Company; Arcair Company; and Victor Equipment Company.

Special acknowledgment is owed to the American Welding Society and to T. B. Jefferson, whose many publications the author has used, perused, and sometimes abused over a welding career of forty years, be it welding, selling, or teaching.

Edgar Graham

1

SHIELDED METAL-ARC WELDING

Shielded metal-arc welding (SMAW), which during the past 40 years has been destined for oblivion according to the pessimists who have said, "The machines will put us out of business," still commands upward of 70% of the welding trade; and it promises to maintain a substantial lead in the industry, due to its adaptability to more welding situations than can be handled by any other single welding process.

SMAW, sometimes called *stick welding* to distinguish it from other arc welding processes, such as gas metallic and flux-core arc welding, is versatile and dependable. It is readily available and requires a minimum of setup time. Regardless of the position of the weld, in spite of adverse welding conditions, and despite the condition of the weldment, such as paint, rust, grease and grime, adhered concrete, galvanized coating, poor fit-up, and many other challenges to the welder's ability, welds of good fusion and reasonable appearance can be obtained if the welder suits the electrode type and joint preparation to the job at hand.

Despite the predictions of the doomsday oracles, stick welding has grown by leaps and bounds: from the thousands of welders in the barewire days, through the welder's early experiments with tape-covered wire, or bare rods dipped in gelatin and rolled in sawdust, until the present day, when millions of welders use hundreds of different welding electrodes.

Aluminum, bronze, copper, Monel, Inconel, and nickel electrodes are used today by experienced welders whose places of employment cannot or will not expend their capital for tungsten or metal inert-gas or other, more sophisticated processes and are used where it is impractical to set up the more prestigious equipment.

Stick welding of the hundreds of alloys of the chrome, nickel, manganese, molybdenum, and vanadium steels, and other combinations of several alloying ingredients, metallic and nonmetallic, maintains a respectable pace with the other welding processes. Stick welding will be around as long as there are welders to weld and metals to join.

MASTERING THE ARC

A student of any mental-manual contest, be it work or play, will do well to begin by understanding his adversary, the equipment he will use, and the objective of the game.

In all of the manually operated arc welding processes, the welder's tool is the electrode itself. He must master it as the player masters a bat or racquet. The welder, however, must rely entirely on dexterity. He cannot beat his opponent into submission by a trilogy of skill, strength, and aggression. He must, instead, approximate more the delicate hand of the artist. He must control the molten metal from an electrode in much the same manner as the artist controls the flow of paint, widening and narrowing and thickening or thinning the paint as he shapes the result according to the picture he carries in his mind.

But the welder is not as free to express himself as is the painter. At best, welding may be likened to painting a portrait; at worst, he must *paint by the numbers*. He is controlled either by what is proper for the job and what is customary in the trade, or by the demands of his superiors. But regardless of the case, he must have a mental image of the finished product.

This imagery comes from experience. The mechanic-welder may look at pictures of welds or may look at samples of welds, or may be told what to carry in his mind as he welds, but nothing can equal the imagery that comes from practice and failure, then practice and fair approximation, and ultimately complete control of every cubic millimeter of every bead and every bead of a multiple-pass weld.

UNDERSTANDING THE ARC

No doubt the mechanic has seen the arc that is produced when a positively charged cloud shorts out to earth or shorts out to another cloud that is less positively charged or negative to it. The arc is nature's means of reestablishing order in an unstable electrical condition. Nature's intolerance of electrical imbalances has proved of great utility to human beings and their machines.

If the mechanic has observed what happens when he inadvertently shorts out two wires of his house current, he knows that an arc is produced momentarily, and if a fuse does not blow, the two wires will continue arcing and melt apart, or fuse together and then melt.

Understanding the Arc

Gas-engine-driven arc welder. (Courtesy of Lincoln Electric Company.)

He knows that if he shorts out his car battery, something somewhere must burn out, and that if this does not happen, his battery will burn up.

He knows, certainly, that if he connects a light bulb across the two wires, the light bulb will not be damaged, unless, of course, he tries to light a 12-volt (V) bulb in a 110-V circuit. The filament of the light bulb is designed to *drop* the voltage and allow only enough current to pass through to dissipate the amount of power printed on the glass envelope, such as 100-watt (W), 50-W, and so on.

Perhaps the mechanic knows that his light circuit is rated at 110 V. How, then, does the light bulb, subjected to a force of 110 V, use only the number of watts printed on its face? The same voltage is applied to the 100-W bulb as to the 50-W bulb.

George Simon Ohm, a German physicist, first put it into words and is credited with the law of nature bearing his name, *Ohm's Law*. He was first to say that 1 volt will push 1 ampere (A) through a unit of resistance bearing his name: 1 ohm (Ω). Count Alessandro Volta and André-Marie Ampère are the other two gentlemen whose names appear in the formula.

When a circuit is closed (turned on), the total current available at the switchbox attempts to get through the line (around the closed circuit). At the first surge, a very large current travels through the filament of the bulb. But as the filament becomes hotter by trying to dissipate all the power, it increases its resistance to the current until only enough current can get through to dissipate the amount of wattage printed on the envelope according to the formula: voltage times current equals power (watts).

The carbon-arc searchlight functions similarly in that the light is produced by a voltage trying to force the current through an air space (or vacuum) between an anode and a cathode. The power is dissipated in the arc.

A German physicist, Gustav Kirchhoff, first wrote down another law of nature (called *Kirchhoff's Law*): that "all the voltage drops around a closed circuit must add up to zero." Thus, according to Kirchhoff, we would have 110 V on one side of the arc (or filament) and no voltage on the other, assuming that the voltage drop in the wiring is negligible.

Another law of nature is that power will distribute its dissipation in proportion to the resistance offered against the completion of its rounds. Thus, when the welder closes the circuit by touching an electrode to a weldment, he holds in his hand all the power that the welding machine is set to deliver; and as he establishes an arc, all of that power must be used up in the circuit. If power dissipates according to the resistance in the circuit, it must dissipate in the arc, the point of highest resistance.

The welder can demonstrate this resistance phenomenon by touching the electrode to the weldment and keeping the electrode in contact with the weldment until the electrode becomes so hot that it melts. (Go easy on the machine; use small-diameter wire.) Since the cable to the electrode is copper and of relatively large diameter, the power loss before reaching the electrode is negligible; and since the weldment is the end of the line, all of the power is dissipated between the electrode holder and the weldment. The electrode does not increase its resistance sufficiently to restrict the flow of current as does the filament of the light bulb.

When the welder raises the electrode and establishes an arc, the highest resistance is in the arc, and the power is dissipated there, except for a relatively small amount of power that is dissipated in the electrode itself. The power loss in the electrode varies according to the electrical conductivity of its material composition: as the welder will notice as he progresses from E-6010 to E-7018 and then on to stainless steel electrodes. These power losses can be observed, however, by the experienced welder when he sees that the electrode burns a little hotter and faster as it becomes shorter, thus reducing the resistance to the current. In cable 1 in both circuits of Figure 1.1, the power loss is the same

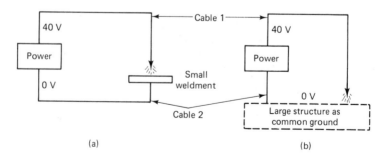

Figure 1.1 Typical welding circuits.

(assuming that they are of equal diameter). In cable 2, the loss is greater in circuit (a) according to the difference in cable length.

In theory (Ohm's Law) the ground should be as close to the arc as is conveniently possible; but since the weldment must be connected to the welding machine in order to complete the circuit, the ground cable is usually connected to the part of the weldment closest to the machine. Thus the welder may be working with a very long electrode cable and a very short ground cable, as shown in Figure 1.1b, on very large weldments such as ships, bridges, buildings, and so on; and with cables of equal lengths on very small weldments, such as a detail assembly on an ungrounded workbench, or on a concrete slab, or on the earth itself.

As can be seen in Figure 1.1b, although the electrical conductivity of steel is much lower than the welder's copper cable, the magnitude of the weldment is usually so great that there is no voltage drop or power loss between the *work* side of the arc and the machine.

To return to Ohm's Law (1 volt, 1 ampere, 1 ohm), the welder can quickly see that the conductivity of the cable can be very important. If, as is customary, the welding machine has a load voltage of 40 V, and the welder needs 5000 W to burn the electrode, the resistance of the cable must be well below 1 Ω. Sometime in his career, the welder will be on a job requiring several hundred feet of cable and find that he must reset the welding machine far beyond the recommended setting in order to burn the electrode satisfactorily.

As stated previously, the arc gap is the greatest resistance in the circuit. Thus if an air gap is resistant to the flow of current, shortening and lengthening the arc must affect the arc considerably: the shorter the arc, the less the voltage drop.

Is a welder, then, expected to hold an arc at precisely 4, 5, or 6 millimeters (mm)? No. That would be impossible for the steadiest hand in the business. So, fortunately for welders, the machine is designed to raise its voltage output in accordance with the varying resistance in the arc gap. For this reason we call the machine a *constant-current* type, as opposed to the *constant-voltage* type found in automatic-wire-feed types, such as metal inert-gas and flux-core.

Transfer and deposition of molten metal across the arc is affected by a multiplicity of factors: surface tension, capillary attraction, the jet stream of the arc, a magnetic pinch effect, and the universal gravitational pull that the acceleration of a larger mass exerts on an accelerating smaller mass (this gravitational force being separate from earth's gravity, which tends to help the transfer in down-hand welding, to deter the overhead transfer, and to cause overhang—sagging—in the horizontal transfer).*

Although different electrodes may dictate a change in technique, as a general rule, the short arc is preferred (less heat input to the parent metal, less

---

*See the Glossary for definitions of "surface tension," "capillary attraction," and "magnetic pinch effect," as well as for definitions of many other terms used in the field of welding.

spatter, etc.) One of the functions of the coating on an electrode is to shield the weld puddle from contaminating air during transfer across the arc and until the deposit cools. A long arc reduces the effectiveness of the shielding. The E-7018 electrode, for example, requires a very short arc. This electrode is a low-hydrogen type for high-quality welds, and if the welder's arc is too long, he has a low-hydrogen electrode (if properly cared for) but may not have a low-hydrogen deposit because hydrogen from the ambient air has infiltrated the shielding effect of the electrode coating. The individual characteristics of electrodes are covered in ensuing chapters together with the metals for which they are particularly recommended.

UNDERSTANDING THE WELD PUDDLE

The welding of steel, which constitutes far and away the greater percentage of all metal joining, is similar to the steelmaking process itself. Although the welder cannot be expected to be the melder on the furnace floor and the metallurgist in the laboratory, he can better understand his successes and his failures if he understands what is happening as he melts a filler metal, remelts an area of the parent metal, controls their mixing together, and cools their admixture to normal (room) temperature.

Perhaps the reader has heard that "the weld is stronger than the steel itself." This claim is justified only if the welder has done his work well. Steelmakers tend to make their product as economically as is consistent with the end product desired. The open-hearth process, for example, is more economical than the electric furnace process. However, the *electric steels* are the finest steels. Thus the welder, because he is literally making an electric steel in his weld puddle, produces a weld metal that is superior to open-hearth steel.

Gas-engine-driven generator providing an ac power source and ac/dc welding circuits. (Courtesy of Lincoln Electric Company.)

The welder produces this electric steel with an electric arc in a manner similar to that employed in an electric furnace, which does it with carbon electrodes. When we know that steel melts at or below 1535° Celsius (C) [2800° Fahrenheit (F)], depending on its carbon content, and that the heat of the arc is upward of 2800 °C (5075 °F) [as high as 5000 °C (9000 °F) at the core of the puddle], we see immediately that the steel melt is superheated if it is permitted to *cook* for a period of time after melting. This superheating and cooking produce a boiling action in the melt which removes impurities and produces a nonmetallic slag blanket that controls the cooling of the melt and prevents recontamination by the ambient air. The observant welder can see the boiling action in the weld puddle and the formation of the slag blanket.

We may further compare welding and steelmaking by comparing the pouring of the furnace steel into ingot molds and the welder's filler metal into the welded joint. Both are poured into a confined space and just as ingot molds are sometimes preheated to prevent the molten metal from cooling too rapidly, so a weldment is preheated to slow down the cooling. (The effects of cooling of a carbon-steel weld are shown in Figure 6.1.)

The actual fusing of the parts being welded is accomplished by the intense heat of the arc melting both the filler metal and both sides of the joint, and creating an in-solution melt that begins to freeze at approximately 1500 °C (2730 °F) and is completely solidified into one amalgamated mass at approximately 1200 °C (2190 °F) (both figures are relatively dependent on the carbon content). This homogeneity, however, will not last, because many changes occur as the solid further cools below the solidification temperature (see Figure 5.10).

As the solidification of the weld metal begins, the observant welder will note that the solid metal will assume a shape that closely approximates the shape of the weld puddle. Therefore, if the welder wants the solid metal to assume a certain shape, he must manipulate the weld puddle in such a manner that the molten metal closely approximates the desired solid metal. This manipulative skill is what was referred to in the earlier comparison of a welder with an artist.

The mechanic–welder will soon discover that the molten metal is more difficult to shape when welding in one position than it is in another. Welding in the flat position, for the inexperienced welder, appears to be the least troublesome because the molten metal from the filler electrode will fall on the parent metal and tend to flatten out from its own weight. Thus the welder's faulty manipulation of the puddle has less dire results. In vertical-up welding, the welder is stacking molten metal drop on drop, which necessitates better control of the puddle. In overhead welding, the welder is asking molten metal to hang suspended from the parent metal, and he must immediately see that a closer arc, a smaller or thinner puddle, and a quicker freeze are necessary.

An understanding of two phenomena in nature will speed up the mechanic's acquisition of skill in handling the weld puddle. These phenomena are surface tension and capillary action. These phenomena are generally neglected in discussing arc welding, but they are necessary to the art.

Capillary action is the result of *capillary attraction,* which is the tendency of a liquid to flow between two closely abutted surfaces and to spread out when coming in contact with a solid surface. The welder can demonstrate this phenomenon by filling a spoon with water, placing the tip of the spoon against a wall, tilting the spoon slightly, and observing the liquid spreading out on the wall in all directions, including upward. The success of this demonstration will depend on the surface tension of the wall's material composition. This diffusion of the liquid indicates the absorbency of the wall. It is, in actuality, capillary attraction at work: The liquid is seeking out the interstices of the closely adhering particles of the composition. If the wall material is steel, the water will be seeking the closely packed grain boundaries.

Surface tension, a result of *molecular attraction,* is the tendency of a liquid to contract to its smallest possible mass. This phenomenon can be demonstrated by spilling water on a freshly waxed car and observing how it beads up. Striking one of the beads with a finger will cause the bead to break up and reform into smaller beads. The liquid's tendency to spread out is thwarted by the surface tension of the waxed surface, and its tendency to contract into its smallest possible mass causes it to form a globule.

These phenomena are covered more extensively in Chapter 2. See, in particular, Figure 2.1.

As the welder deposits the first bead in the flat position, he will increase his skill more rapidly and will better prepare himself for the challenge of vertical and overhead welding if he will begin to see both of these phenomena in the liquefying and solidifying weld puddle. He should observe the shape of the weld puddle: it is dish-shaped. The arc is digging into the parent metal and melting it to depths varying with the type of electrode and the type of current—alternating current or direct current and *straight* or *reverse polarity*. The penetrating (digging) action of the arc is at the center of the puddle and forces molten metal out and away from the center. The edges of the dish should be observed and controlled; the edge determines all of the factors that shape a neat-appearing weld.

UNDERSTANDING THE MACHINE

The arc was first recorded by Sir Humphry Davy. He discovered the arc in 1801 and by 1809 had managed to harness it for short periods of time. His interest was in using it as an arc light. The first recorded fusion welding with the arc was the welding of lead by Auguste de Meritens in 1881, using a carbon electrode.

Since these and several subsequently improved instances of arc welding used batteries, the first welding machines were of the direct-current (dc) type. The welder needed only a dc source of power and a form of resistance in the circuit with which to drop the voltage to deliver only the current he wanted. Industrial plants, such as steel mills, had a source of power ready and capable

Understanding the Machine

in their traveling overhead cranes. The welder had but to connect his variable resistance bank to the crane's traveling electrodes by means of a copper cable equipped with a metal hook. If the welding wire he was using worked better on one polarity or the other, he reversed the connecting cables. With this same resistance bank, a municipal transportation company welder could hang one cable from the trolley wire and ground the other to one of the streetcar's rails.

Of the two most prevalent welding machines in use today, the motor-generator type was the earlier of the two. The generator may be designed to provide alternating current (ac), but in the great preponderance of welding machines, the generator provides dc. The motors driving the generators are preponderantly ac, with a relative scarcity of dc types. On job sites where electricity is not readily available, gasoline or diesel engines drive the generators.

Although the well-designed motor–generator welding machine is strong and durable, it does require some consideration by the welder in its care and maintanance. If he sees an arrow accompanied by the word *rotation*, he should check the start of rotation as he presses the start button; and if the rotation is wrong, he should change the connecting leads at the power source. Additionally, brushes wear out, bearings run dry, commutators become worn. Worn brushes should be replaced, bearings cleaned and regreased, burned commutators should be cleaned with fine sandpaper and wiped clean, and if commutators are badly burned or worn, they should be turned in a lathe and reslotted.

In smaller shops or plants, much of the responsibility for the proper care of equipment rests with the welder. Larger firms, of course, have their own maintenance electricians, but they do appreciate impending serious damage being brought to their attention.

A 250-A ac/dc rectifier-type welder.
(Courtesy of Lincoln Electric Company.)

In the case of the gasoline and diesel engine types, aside from proper engine and generator maintenance, the welder should be acquainted with the automatic idler control. The machine should be set so that the engine will rev-up quickly as the welder strikes the arc and will not idle down for several seconds after he breaks his arc. These several seconds should give him time to change electrodes.

The ac-dc transformer-rectifier type of welding machine is the most versatile because it provides a quick change from dc to ac as the need arises. Additionally, in contrast with other types of machines, it requires less maintenance, is much lighter in weight relative to its power output, is less noisy, and is easily adaptable to gas-tungsten-arc welding by the addition of a high-frequency unit.

The internal structure of the transformer-rectifier machine is relatively simple. There are no moving parts except a ventilating fan. A step-down transformer lowers the line voltage from 220 or 440 V to a much safer open-circuit voltage of 80 V, which drops again to an even safer voltage when the circuit is closed. A line voltage of 220 or 440 V is quite lethal, and the welder must be aware of the danger involved in coming into contact with the primary side of the transformer.

Although considered safe under most conditions, the 40-V closed-circuit potential is only relatively safe. As any welder who has worked in the field on a rainy day can attest, 40 V can make him have to fight to get a fresh electrode into the holder. A better idea is to wear rubber gloves inside wet or even very damp gloves. (*Caution:* Do not pick up two electrode holders, one in each hand. If one machine is set electrode positive and the other machine electrode negative, the circuit voltage is doubled.)

The schematic diagram of Figure 1.2 shows the internal circuitry of the typical ac-dc machine. Assuming an input voltage of 220 V and a machine rated at 300 A, at maximum load the welder would require a power of 40 V times 300 A or 12,000 W [12 kilowatts (kW)]. This means, discounting losses in the transformer itself, that he would need an input power of 12,000 W or 220 V times 54.54 A. In ac welding, the machine would function without additional parts, but this would mean that the welder could use only an electrode designed to weld at 300 A—a very large electrode indeed. According to the natural law—1 volt, 1 ampere, 1 ohm—he will need a variable resistance in the circuit to control the flow of current as required by electrodes of different diameters.

One method immediately apparent to the eye (Figure 1.2) is varying the proximity of the secondary coil of the transformer to the primary coil. The voltage step-down is accomplished by the ratio of the number of turns in each coil. The current is picked up by the secondary by induction and can be controlled by moving the two coils closer together or farther apart; the greater the distance between them, the less the current transferred. This method of control is called the *movable-core* type.

Other methods of control are (1) magnetic-shunt control, wherein a movable magnetic iron shunt moves between fixed primary and secondary coils;

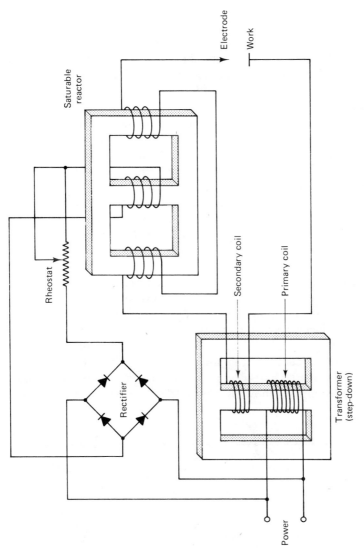

Figure 1.2 Typical ac/dc machine (as bypasses rectifier).

(2) reactor control, wherein a variable reactor is connected between the secondary coil and the welding load; and (3) reactor-rheostat control, which is similar to the variable reactor type, but the reactor is stationary and is controlled by a rheostat. The reactor–rheostat type is easily adaptable to remote control, such as a foot pedal.

When we switch to dc, the rectifier is connected into the circuit and converts the ac to dc. In the interest of having a smooth current, the rectifier should be a full-wave rectifier. Our arc will note the difference between a current of 60 pulsations per second and 120 per second, just as our eye will see the difference between a motion picture of 24 frames per second and one at 12 frames per second.

The ac (Figure 1.3) from the transformer changes polarity 60 times per second. Our electrode will be alternately negative and positive; and the current across the arc will change direction 60 times per second. Many of our electrodes and many of the metals we weld will tolerate this reversal of current, and for some, the ac has certain advantages, such as more rapid deposition rates and more efficient use of power. Other metals, however, respond adversely to ac and are better suited to dc, with both straight and reverse polarities having their distinctive advantages. The oxides of some metals, for example, tend to produce a rectification effect and extinguish the arc on the reverse cycle (see Chapter 3).

If we wish to weld with dc, we must convert the ac accordingly. When we switch to dc, we connect the rectifier into the circuit, and the diodes will pass current in one direction only. In effect, we block out one-half of the current, as shown in Figure 1.4, and have an interrupted current that flows and stops each $\frac{1}{60}$ of a second. This converter is called a *half-wave rectifier*. By the addition of more diodes (as in Figure 1.2) we can fill in the gaps, so to speak, by passing the other 60 half-waves in the desired direction, thus doubling the current and the pulse, as shown in Figure 1.5. We can further stabilize this pul-

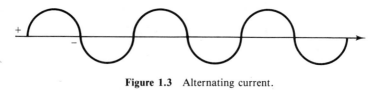

Figure 1.3 Alternating current.

Figure 1.4 Direct current, half-wave rectification.

Figure 1.5 Direct current, full-wave rectification.

Understanding the Machine

sating current by adding other refinements to the circuit, such as iron-core chokes to smooth out the ripple and capacitors to compensate for core losses.

Troubleshooting the Welding Circuit

With rare exceptions, the welder will not be expected to repair the internal parts of a machine, but he will be expected to diagnose and correct malfunctions in the welding-load circuit and, perhaps, the input circuit.

1. Start with the machine itself. Is it running? If the ambient noise level and vibration level are very high, press the off button, wait a few seconds, place your hand on the machine, and press the *on* button. In the case of the motor-generator type, you should hear or feel the start of rotation; and with the transformer type, you should hear or feel the "thump" as the laminated plates of the transformer slam tightly together when the initial surge of current sets up the magnetic field in the transformer core.
2. If the machine does not start, check the main power switch. If the switch is open, do not close it before determining that it is safe to start any other equipment operating from the same power source.
3. If the machine is running, check the load circuit in this order:
   (a) Is the weldment grounded?
   (b) Is the machine delivering power to its output terminals?
   (c) Check the cables, electrode holder, and ground clamp for breaks.
   (d) If electrode cable is several hundred feet, trace the cable back to the machine for bad connectors, or severed or damaged cable.

Setting the Machine

Just as there are many ways of controlling the welding load circuit, the respective knobs and dials on machine panels vary considerably. But if the mechanic understands that each machine has a load limit and that each electrode has a limited useful current range, he can immediately set the machine relatively close to the heat best suited to the electrode, to the welding position, and to the condition of the weldment.

Doing it by the numbers:

1. If the machine has two control knobs, one will be the range setting and the other will be the particular setting (similar to *rough* and *fine* tuning of high-quality radio receivers).
    Assume a machine with range settings of 50-100, 100-200, 200-350, and 350-500 A, and a fine control of 1 to 10. Set the range control at 50-100 A. The fine control at zero will then deliver 50 A and at 10 will deliver 100 A. Thus each gradient of the fine control adds 5 A to the minimum 50 A on the range control. Setting the range control on 100-200 A

means that each gradient on the fine control will add 10 A to the minimum 100 A on the range control. For example, with the range control at 100–200 A and the fine control at 6, we should have 160 A.
2. With machines having only *high* and *low* ranges, merely divide the numbers on the dial into the range and get the amperage value of each gradient on the scale and add it to the minimum stated on the range.
3. With machines having only one dial and no range settings, divide the numbers on the dial into the rated maximum load and obtain the amperage for each gradient on the scale. Machines having only one knob and one dial are usually marked in amperage and need no explanation.

Doing it by experience. Setting the machine as detailed above assumes that the machine is functioning properly, is accurately calibrated, and that the dials are still readable. Unfortunately, machines are not always in the best working order, and the welder must get acquainted with the machine assigned to him or the one that goes with the job for that day.

Unreadable dials and faulty machines are no great challenge to the experienced welder. If he cannot read the dials, he sets the machine in what he guesses to be the middle, strikes an arc to see what he has got, and adjusts the machine accordingly—rarely are more than two resettings required.

To cite one common occurrence, assume an unreadable machine with one crank. Crank the machine all the way to the right, then count the number of turns of the crank as you crank the machine all the way to the left. Crank the machine to the middle and strike an arc. A couple of turns to the right or left will tell the welder the machine's maximum output and the amperage represented by each turn of the crank.

THE SMAW TECHNIQUE

As mentioned in the opening paragraphs, the welder will be called on to weld many different metals using many different electrodes. As a general rule, the core material of the electrode will approximate the composition of the material to be welded—steel electrodes for steel, bronze for bronze, aluminum for aluminum. When welding dissimilar metals, the welder may have to select an electrode composed of an alloy of metals that are compatible with the two metals to be welded. In all cases, however, the welding technique will vary only slightly.

SMAW electrodes, as the title implies, must be provided with some form of shielding for the weld puddle. (Some bare-wire welding is still performed, but such welding is confined to a few applications in wearfacing.)

The shielding is provided by a coating that was designed initially to stabilize the arc, provide a shielding effect for the weld metal during its critical heat range, produce a purging effect that would float out the impurities, maintain a fluid state until the gases could escape, and provide a protective slag blanket as the weld deposit cooled. The function of the coatings has since been broadened to other purposes, such as speeding up the deposition rate and introducing special alloying ingredients into the weld deposit.

The coating of the electrode determines to a great extent the method of weld metal distribution. The same manipulative technique cannot be used for sparsely coated electrodes as will be used for electrodes having heavy coatings. The characteristics and welding techniques for the individual electrodes are covered in subsequent chapters, together with the metals for which they are designed.

In keeping with the general practice in vocational schools, we shall introduce the mechanic to the sparsely coated electrode because it requires the striking of an arc, a maintenance of a particular length of arc, and gives the welder a better vision of what is taking place in the weld puddle. For this reason, we have chosen the experienced welder's old favorite, the E-6010.

Flat-Position Welding (Down-Hand)

1. Select a piece of steel of any size and shape, preferably clean at this stage of the welder's development, and preferably about 1 centimeter (cm) in thickness.
2. Select E-6010 electrodes, 4 mm (5/32 in.) in size, and set the current at 110–120 A.
3. Strike an arc by tapping or scratching the weldment gently, bearing in mind that you must quickly withdraw the electrode from the weldment several millimeters (about 3 to 4). If the electrode is not withdrawn quickly enough, it will stick—break it loose with a twist of the wrist. If the electrode is withdrawn too far, it will extinguish—strike again. Practice this striking procedure until your hand is educated to the nature of the arc.
4. Run short beads in a straight line.
    (a) Establish the arc.
    (b) Observe the puddle form.
    (c) Tilt the electrode about 10° in the line of travel.
    (d) Try to see the side and trailing edges of the puddle.
    (e) Try to distinguish the metal from the slag—they are discernible.
    (f) Keep the arc gap as uniform as possible.*
    (g) Observe the piling up of the bead and try to keep its width, depth, and *ripple* uniform.
5. Practice restarts.
    (a) Break the arc by moving it slightly ahead of the bead, then upward, instead of lifting the electrode abruptly. This practice will prepare you for some specialized procedures in future chapters.

*The proper arc with the E-6010 will produce a crisp crackling sound. Although this sound has been variously described as a "frying sound," "like bacon in a skillet," and others, it has a distinctive sound all its own. To educate the ear and eye, vary the arc length and observe the results. *Ear*: Too long, and the sound softens to a hiss and blow; too short, and it will choke and sputter. *Eye*: The proper length will transfer metal across the arc in a steady spray of fine globules, with an absence of spatter.

(b) Practice the T-type restart; that is, restrike the arc several millimeters ahead of the solidified bead, stabilize the arc, and return it to the travel line deposited last. Try to see the trailing edge of the puddle blend with the solidified metal smoothly.
6. Run practice beads using a very slight circular motion. The motion should be relatively slow and, above all, smooth; and the circle should be no greater in diameter than twice the diameter of the electrode. Practice this circular motion both clockwise and counterclockwise.

Horizontal Welding

1. Position a steel plate perpendicular to the horizontal plane and run horizontal beads 10 to 12 cm in length, one on top of the other. Follow all procedures for flat-position welding, but also try tilting the electrode both directions in the vertical plane.
2. Run beads of varying thicknesses, using a very small circular weave of not more than 6 to 9 mm: 6 mm (¼ in.) with the electrode tilted upward and 9 mm (⅜ in.) with it tilted downward. Observe the natural tendencies of weld metal distribution. Wider weaves, called *lacing*, are possible, but their merit scarcely offsets the high degree of skill required.
3. Run stringer beads, one on top of the other, blending each new bead with the solidified previous bead so as to present a relatively smooth surface overlay (see Figure 1.7).

Vertical-Up Welding

1. Position a steel plate perpendicular to the horizontal plane.
2. Starting at the bottom, establish the arc and hold steady until a puddle of molten metal fuses with the weldment.
3. Move the arc upward (less than 1 cm) and allow the puddle to solidify.
4. At the very moment that you observe the freeze of the molten metal, while the deposit is still red, move the arc downward to form a new puddle atop the solidified metal. Repeat and repeat, stacking the metal drop on drop. (*Note*: The temptation of many inexperienced welders is to raise the electrode too far. The whip should be timed so that the new puddle is formed on the solidified metal while the solid is still very soft.)
5. Use the inverted-tee (T) technique to flatten the face of the deposit somewhat, that is, move upward with the arc in the center of the intended deposit and when returning to form the puddle, cross the T with a very slight left-to-right or right-to-left motion.
6. Weave beads about 1 to 1.5 cm in width, using a very flat Z pattern. Do not whip in and out of the puddle; this weave must be with a steady rhythm, and the arc must remain in the puddle.

7. Another pattern with experienced welders is the figure 8. Perform as in step 6. The figure 8 will generally be flattened considerably. The box weave is also useful in vertical-up wide weaves.

Vertical-Up Fillet

1. Assemble and tack two steel plates as a T (see Chapter 13).
2. Weld vertical-up in the joint using the inverted-T method. Flatten the face of the bead in such a manner that both plates are fused and the same amount of metal fuses on each side.
3. Weld a larger bead, using the J-reverse J (ᒎ) technique: that is, whip upward, return to the puddle, move toward the other plate, whip up, return to the puddle via that plate, and move toward the other plate. Some welders may refer to this manipulation as a U, and it would be a modified U if metal were deposited on the upward whip. The J-reverse J helps eliminate undercutting by always pushing toward the side metal that suffered the digging action of the arc.
4. Experiment with larger beads using the flat Z and figure 8 weaves.

Note: When whipping upward with the ᒎ technique, do not pull away from the joint; move the arc into the joint. When using the flat Z and figure 8 weave, do not whip; keep the arc in the puddle.

Overhead Welding

For the experienced welder, it may be true that overhead welding and flat-position welding are identical, but there are three very great differences: (1) the welder is in a much more awkward position, (2) the welder's vision of the puddle is impaired, and (3) arc length and weld distribution are more critical. Therefore, observe all the rules for flat welding and keep a tight arc. (Do not try to make the arc blow the metal upward.) Carefully observe the flow-out of the weld pool against the sides of the joint to avoid undercutting.

ELECTRODE CLASSIFICATION

All of the welding with the E-6010 electrode discussed previously applies to the welding of mild steel. Other electrodes for mild steel and low-alloy steel welding are described in Chapter 6.

The AWS classification of electrodes is detailed in Table 1.1. Reading right to left, the final digit specifies the coating, the penultimate number signifies the welding positions recommended, and the remaining digits represent the tensile strength in pounds per square inch; for example, in E-7018, the 8 represents a low-hydrogen iron-powder coating, the 1 represents all-position welding, 70 means 70,000 psi, and E indicates electrode.

TABLE 1.1 AWS Classification of Electrodes

AWS Classification	Recommended Current	Penetration	Composition of Coating	Usability	Tensile Strength (psi)
E-6010	DCEP	Deep	Cellulose–sodium	All positions	60,000
E-6011	ac, DCEP	Deep	Cellulose–potassium	All positions	60,000
E-6012	ac, DCEN	Medium	Rutile–sodium	All positions	60,000
E-6013	ac, dc	Light	Rutile–potassium	All positions	60,000
E-7014	ac, dc	Light	Rutile–iron powder	All positions	70,000
E-7024	ac, dc	Light	Rutile–iron powder	Flat, horizontal	70,000
E-7015	DCEP	Medium	Low-hydrogen sodium	All positions	70,000
E-7016	ac, DCEP	Medium	Low-hydrogen potassium	All positions	70,000
E-7018	ac, DCEP	Medium	Low-hydrogen iron powder	All positions	70,000
E-7028	ac, DCEP	Medium	Low-hydrogen iron powder	Flat, horizontal	70,000

In Figure 1.6 we see the appearance of the E-6010 electrode when used on a joint-welded vertical-up. Note the keyholing of the root bead and the slight overwelding of the cap or cover pass. The welder in this instance is displaying journeyman quality. This same technique applies to the V-grooved butt joint (see Chapter 13).

Figure 1.7 shows the appearance of the E-6010 when running horizontal stringer beads with the parent metal positioned vertically. The welder in the figure is displaying relatively mediocre skill.

A typical horizontal open corner joint welded with the E-6010 electrode is shown in Figure 1.8. Note the keyholing of the root bead, the planning of successive beads, and the very slight overweld at the edges. The appearance of the stringer bead technique will be approximately the same for this joint and

Figure 1.6 Joint welded vertical-up.

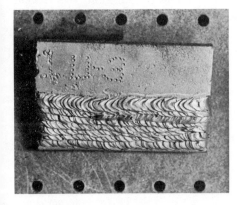

Figure 1.7 Horizontal stringer beads with the parent metal positioned vertically.

Figure 1.8 Typical horizontal open corner joint.

the V-grooved butt joint regardless of the welding position. The welder was of journeyman caliber.

Figure 1.9 shows a multiple-pass fillet weld of six beads and a lacing cover welded in the horizontal position using an E-6010 electrode. Note that the planning of the beads has produced a weld deposit of equal legs. Although the weld is approaching journeyman caliber, the stringer beads should blend more smoothly and the lacing is a little erratic.

Figure 1.9 Horizontal fillet: six-bead, three beads showing, with a lacing cover.

Finally, Figure 1.10 illustrates a welding technique for obtaining the equivalent of a back weld when the back side of the joint is inaccessible. This technique requires a relatively high degree of skill, especially when welding chrome–nickel pipe and in the high-nickel alloys. The motion of the electrode is a very slight forward-and-backward sawing motion, keeping the arc blow on the back side of the opening as shown at point A. Electrodes E-6010 and E-6011 may be whipped up and away, but electrodes of the higher alloys, especially the stainless steels and high-nickel alloys, must not be whipped; they can, however, be moved very slightly in the same manner.

For this technique, except for E-6010 and E-6011, the gap or root opening must be wide enough to give access to the entire electrode, coating and all; and attention must be given to the fact that, due to the shrinking of the weld deposit, the gap will close rapidly unless restrained.

Note: Before attempting this backwelding technique on the higher alloys, study the welding techniques of those alloys in succeeding chapters.

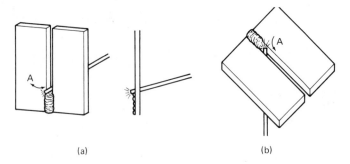

Figure 1.10 Obtaining the equivalent of a back weld.

2

OXYACETYLENE WELDING

IN THE BEGINNING

The history of gas welding as we know it today is a relatively short one. If we can ignore the Greek-Roman-Egyptian method of fusing metals from the heat of alcohol-air and oil-air, gas welding had its beginning with the discovery that oxygen produced the hottest flame obtainable with gas. It is probably due to the fact that the first oxygen was produced by the electrolysis of water that hydrogen, a by-product of that electrolysis, was one of the first gases used in gas welding (circa 1880).

Other gases, such as coal gas, were combined with oxygen, but these were laid aside when about 1895 a Frenchman named Henri Louis Le Chatelier announced that the combustion of oxygen and acetylene produced a hotter flame than that of any other gas.

The first practical oxyacetylene welding torch was also produced by a Frenchman, Eugène Bourbonville. In 1906, he brought this technique to America, where it was first used in maintenance and repair welding.

In the early part of the twentieth century, through World War I and beyond, the oxyacetylene torch was the major welding process for fabrication, construction, maintenance, and repairs. It enjoyed this popularity because of its versatility in welding all of the major metals. Its decline in popularity resulted from the development of faster and, in many cases, superior methods, such as improvements in the arc welding processes, the development of shielded electrodes in the early 1900s, and the introduction of the TIG process near the mid-century.

THE TORCH TODAY

Although the oxyacetylene process may be thought by many to be in a moribund state, it is alive and well in the maintenance welding field and in some production processes. It is a prime tool of the stationary engineer, the industrial mechanic, the machinist, and a necessary tool for other trades, such as electrical and plumbing. It is still the preferred method for brazing, metallizing, and soldering, and for the application of many of the carbides in wearfacing.

This versatile welding process can be added to the maintenance mechanic's equipment by the purchase of a few welding nozzles that are generally available as attachments to the common oxyacetylene cutting torch at a cost less than the mechanic's one-day take-home pay.

BASIC PROCESSES

There are five basic methods of metal joining with the oxyacetylene torch:

1. *Fusion welding:* The metals to be joined are melted by the welding flame and fused into one amalgamated mass, with or without the addition of a filler metal.
2. *Brazing:* The parent metal is heated to a proper bonding heat, and a filler metal of a lower melting point is distributed around and through the joint by capillary attraction.
3. *Braze welding:* The parent metal is heated as in brazing (generally, at a somewhat lower heat) and a filler metal is distributed along the joint in a bead of proper size and shape.
4. *Surface alloying:* A filler metal is applied as in brazing and braze welding, and both filler metal and parent metal are atomically admixed.
5. *Sweating:* The surface of the parent metal is brought to the melting point, and a filler rod is fused with the "sweating" surface. This method is called "sweating" to distinguish it from the usual fusion welding.

Fusion Welding

As a general rule, fusion welding is done with a neutral flame (see Figure 2.6). Fusion welding may be accomplished with or without a filler rod. Without a filler, the flame melts the two edges of the joint, and the melt flows together and solidifies. This method is generally restricted to corner and edge welds. Very little practice is needed before producing welds of excellent appearance. With a filler rod, both sides or edges of the parent metal are melted, the filler rod is melted as needed, and the molten puddle solidifies into an amalgamated mass, the size, shape, and appearance of the weld being dependent on what is desirable and the welder's skill in manipulation of the torch and rod. Generally, the filler metal matches or very closely approximates the compositional structure of the parent metal.

Forehand welding. Forehand welding is the preferred welding procedure in most cases of fusion welding and in all cases of braze welding. In this method of torch and rod manipulation, the torch is between the filler rod and the advancing bead, the torch being angled slightly ahead of the bead and away from the puddle. Although proper fusion must be observed more closely, this method is preferred, especially for its better weld appearance.

Backhand welding. Backhand welding is advantageous only in obtaining deeper penetration and greater speed. With this technique, the filler rod is dipped between the torch and the advancing puddle, with the torch angled into the advancing puddle. This weld cannot equal the forehand weld in appearance—it will be much rougher and more uneven. If speed is essential, a slightly carburinizing flame will reduce the melting point of the metals and speed up the process. If penetration and appearance are factors to be considered, many welders will run a root pass backhand and fill with the forehand method. Proper preparation of the joint will eliminate a need for deep penetration.

Brazing

A welder of the author's acquaintance some few years ago read—or did not read well enough—an article in a magazine about "brazing aluminum." Unfortunately, the welder believed that brazing was using a bronze filler metal to join other metals. Equally unfortunately, he did not think of the great difference in the melting points. He did not know, or ignored, the fact that the filler metal must melt at a lower temperature than the melting temperature of metals to be joined. The marriage he envisioned had about as much chance of consummation as would the shotgun wedding of a dog and a cat. But out of failure, he learned that brazing is a process that is adaptable to many metals and many different filler metals.

To adapt the process readily to any metal-joining problem that arises, the welder must first understand two natural phenomena: *capillary action* and *surface tension*, which were described briefly in Chapter 1. Capillary action is more graphically described by the siphoning effect encountered by installers of metal roofs. Water descending over the edge of a metal sheet and onto a second metal sheet begins to creep between the two surfaces, and upon reaching the top of the underlying sheet has formed a siphon that flows a steady stream of water, as shown in Figure 2.1. In (a), the water first tends to bridge over the edge,

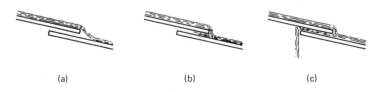

(a)   (b)   (c)

Figure 2.1 Siphon effect.

but the flow-out or wet-out tendency overcomes the bridging tendency and wets the edge of the top sheet. In (b), water fills in the bridged area, and the creep tendency begins. In (c), capillary action is still present but is not necessary to the siphoning action. The capillary creep in Figure 2.1 would be compared with the capillary flow necessary for the brazing of the pipe or tubing joints of Figures 2.2 to 2.5.

Understanding surface tension. Surface tension is the result of molecular attraction. Molecular attraction is the tendency of a liquid to contract into its smallest possible mass. Surface tension may be illustrated by observing the action of water on a freshly waxed automobile. Every car owner judges a wax job by how the water "beads up." Few, however, have taken the trouble to strike a bead with their finger and to observe how the bead will break up and reform into smaller beads. If the experimenter will now recall that the water did not bead up as much, if at all, on the unwaxed car, he may deduce that the surface tension of the waxed surface is greater than that of the unwaxed surface. The surface tension is resisting the capillary action of the water. The surface tension of the water itself causes it to ball up instead of flattening out. Thus, when a liquid and a solid meet, we have two tendencies in nature that oppose each other.

The behavior of a liquid metal contacting a solid metal is the same. The surface tension of the solid resists the capillary action of the liquid. The liquid is trying to obey both laws of nature: (1) it is trying to flow out on the solid surface and infiltrate the molecular interstices of that solid, and (2) it is contracting to its smallest possible mass as it solidifies. In brazing, the welder must break down the surface tension of both metals and control the capillary action.

Understanding heat. Everyone knows that heat, in large enough quantities, will convert a solid to a liquid. However, little thought is given to the fact that heat breaks down the surface tension of a solid or a liquid. Every experienced swimmer knows that staying afloat in warm water consumes more energy than is required to stay afloat in cold water. It is not necessary that the swimmer know that this phenomenon is due to the greater surface tension of cold water, but the welder must know that the heat of a brazing joint is very important to a successful weld. The statement that "the filler metal follows the heat" or "the heat sucks the metal" should be used only in a figurative sense. The molten metal is not attracted to the heat as a bug is attracted to light. The metal follows the heat only because the heat is preparing a continuous path for the metal to follow by natural capillary action. For capillary action to work, the heat must be uniform. The molten metal will not flow from one heated area across an unheated section toward a torch heating yet another area.

Understanding flux. Contrary to common belief and disregarding appearances, flux is not for the purpose of cleaning the metal. Flux may aid in cleaning by breaking down the surface tension of the adhered material and floating it away, but the primary purpose of flux is to break down the surface tension of the metal itself. The metal to be brazed should be cleaned (by whatever means: grinding, filing, machining, scrubbing) and then fluxed prior to and during application of the filler metal. Some silvers are said to—and will—wet out on clean copper, but if this does not happen, don't fight—flux!

What is flux? It can be anything that breaks the surface tension of another substance. Although the physician calls it an enzyme, we may say that the saliva in one's mouth is a flux, in that it breaks down the surface tension of food. Some welder's wives may be more knowledgable about flux than their welder-husbands. They know which are the best detergents for their dishes and which are the least harmful to their hands; and they know which are best for getting out certain stains. The welder may show his knowledge by pointing out that the suds floating on top of the water are no help at all to the fluxing action below, but his wife may be hard to convince. He may be able to convince her that she needs more flux for hard water than for soft water, and more flux for cold water than for hot water, but she has her own way of knowing when the water is right—she simply adds flux until she is satisfied with the "silky" feel as she pinches the water between her thumb and fingers.

The welder should remember that after he has cleaned the metal he is about to braze, the metal immediately begins to oxidize; that is, the oxygen in the air combines with the metal to form oxides that are designed by nature to resist further oxidation. Unfortunately, this thin film of oxides resists the welder's attempt to bond the filler metal to the parent metal. He should remember, too, that as a metal becomes hotter, its resistance to oxidation is lessened. The flux is therefore called on to do several things.

1. Provide a protective coating that prevents further oxidation as the metal is heated
2. Break down and float away the residual oxidation that formed immediately upon cleaning and the oxides that form during the brazing process
3. Reduce the surface tension of the parent metal and thus enhance the capillary action of the molten filler metal
4. Readily displace itself as the molten filler flows into the joint or out onto the surface
5. Reach its functionable liquidity at a temperature at or slightly below the proper bonding temperature
6. Protect the deposited metal from oxidation until it has cooled
7. Be easily removable after welding by the least destructive means possible

Braze Welding

The distinctions between brazing, braze welding, and soldering have been an attempt by the welding trades to separate three methods of joining metals by using a filler metal with a melting point below the melting point of the metals joined. Between soldering and brazing, the distinction is made by temperature: soldering below 425 °C and brazing above 425 °C (800 °F). Between brazing and braze welding, the distinction is made that brazing is "thin flow" and braze welding is "beading up."

Where brazing and soldering depend on capillary action to carry the filler metal along, around, and through the joint, braze welding, to a degree, resembles the stacking up or beading up of fusion welding with stick or torch, and capillary action occurs only in the small area covered by each drop of filler metal.* The most discernible distinction is in joint preparation. Where in brazing, the joint is kept as tight as possible, in braze welding, the joint is veed out and filled to the depth of weld desired, or it is built up as a fillet. Where in brazing the joint is heated broadly, and the filler is permitted to flow freely, in braze welding, the filler metal is controlled drop by drop. Where brazing requires relatively little manual dexterity, braze welding requires a manual dexterity equal to that of stick welding, oxy-gas welding, or tungsten inert-gas (TIG) welding.

The manipulation of the torch (TIG, carbon, gas) in braze welding and the melting off of the filler rod depends on the size and shape of the desired bead. The torch movement may be circular, back and forth, or moved steadily in the direction of the bead. The filler rod is dipped into the molten puddle of metal. In any case, the heat must "tin" the parent metal slightly ahead of the deposit, keep the puddle molten, and melt off a portion of the filler rod.

Good bonding is indicated by a slight flattening out of the newly melted drop of molten filler toward the sides of the bead and forward of it. As in all welding, the welder must have a mental image of the size and shape of weld he wants and then do what is necessary to obtain it. (If a heavy deposit is desired, welding "uphill" will stack up the bead more easily than will welding flat.)

A successful weld of neat appearance is dependent on several factors.

1. *Type of filler rod:* fast freeze or slow freeze (narrow or wide plastic range).
2. *Amount and concentration of heat:* too hot and the filler spreads too widely, too cold and the bead is rounded with overhang (or "nonfeathered" at the edges). This nonfusion at the edges of a bead, this "balling up," also indicates improper bonding.
3. *Steady travel of torch.*
4. *Constant rhythm of "dipping"* (the dipping of the filler rod into the molten puddle.)

Torch Brazing

As noted earlier, before attempting to braze by torch, the welder must understand (1) capillary action, (2) surface tension, (3) flux, and (4) the function of heat.

*Many books make the distinction that capillary attraction is not involved. Capillary attraction may not be so noticeable, but its wetting-out characteristic is necessary if the filler metal is to be properly bonded.

Basic Processes

In brazing, the joint is made as tight as possible. The tolerance for high-grade silvers should be within the range 0.002 to 0.005 in. Other brazing fillers may require a looser fitting in accordance with their thinness of flow and their "wetting" characteristic.

The selection of the filler metal depends on several factors:

1. The metal to be brazed (filler must be compatible)
2. The type of joint
3. The strength desired
4. The conditions to which the weldment is subjected during its service life
5. Cost

Metals that can be brazed include the following:

1. *Aluminum and aluminum alloys:*
   (a) Aluminum alloys high in magnesium content are difficult to braze because of their poor wetting characteristics.
   (b) Torch-brazed aluminum cannot be anodized.
   (c) Corrosion is likely unless flux is removed by thorough scrubbing.
2. *Magnesium and its alloys:* These are highly susceptible to corrosion. Otherwise, they react similarly to aluminum.
3. *Copper:* There are no special problems with copper. Most silver, copper, and copper–zinc groups may be used.
4. *Low-carbon and low-alloy steels:* There are no special problems with these steels. Silver, copper, and copper–zinc groups may be used.
5. *Stainless steels:* Although many books say that you can use silver, copper, and copper–zinc groups, copper should be avoided due to electrolytic action. Best are the nickel groups, such as nickel–silver, high-grade silver, and Nichrome.
6. *High-carbon and high-speed steels:* High-carbon steel brazes well with nickel–silver, and nickel–silver–menganese. High-speed steels (HSSs) should not be brazed unless done prior to or during the hardening process. [High-speed steels may be *welded* by arc (stick) with a balanced austenitic–ferritic electrode of the very high alloy type if done rapidly with an absolute minimum of melt-down of the steel. The resultant narrow heat-affected zone is, of course, no longer high-speed steel. The extent of the "softened" area is visible to the naked eye immediately after welding.]
7. *Nickel and high-nickel alloys:* Use nickel and silver groups. Avoid sulfur and metals of low melting points such as lead, bismuth, and antimony.
8. *Cast iron:* Try to avoid brazing cast iron; braze-weld it.

The general procedure for torch brazing is as follows:

1. Clean the proposed brazing area thoroughly.
2. Apply flux to all areas of the joint. If the joint is a tube and sleeve, for instance, brush flux inside the sleeve and around the end of the tube.
3. Heat broadly until the flux appears liquid and transparent.
4. Apply filler metal at the point selected.
5. Move heat over the joint as necessary to distribute the filler metal through the joint by natural capillary action.
6. Remove the flux residue. (Flux residue will usually *pop* off if water is sprinkled on the weldment while still hot.)

Typical braze joints:

Sleeve and Tubing (Figure 2.2). The procedure is as follows:

1. Clean thoroughly.
2. Flux the inside of the sleeve and the outside of the tubing.
3. Insert the tubing into the sleeve.
4. Heat the entire area until the flux liquefies.
5. Apply filler metal at A.
6. Continue a circular heating pattern, moving the torch slowly toward point B.
7. *Note:* The "absorption" of the filler into the joint can be observed better if the weldment is positioned vertically.
8. Repeat the procedure on the other end of the sleeve.

Bell and Spigot (Figures 2.3 & 2.4). The procedure for the bell and spigot is similar to that for sleeves and tubing.

Thick to Thin Application (Figure 2.5):

1. Flux the brazing area.
2. Heat the entire assembly broadly; concentrate heating on heavier section; avoid overheating thin section.

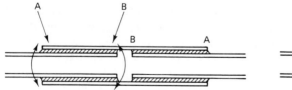

Figure 2.2 Sleeve and tubing.

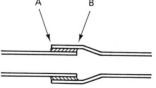

Figure 2.3 Bell and spigot.

Surface Alloying

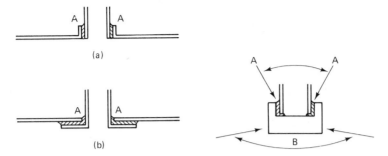

Figure 2.4 Flanged (upset) and Lap Joints   **Figure 2.5** Thin-to-thick Application

3. Apply filler at A, completely around the joint.
4. Heat around at B until the filler is observed to sink into the joint.

Sweating

"Sweating" may be thought of as somewhere between brazing and fusion welding. In this technique, steel is heated to a white heat [1200 °C (2200 °F)] with an excess acetylene flame. At this white heat, the unconsumed carbon of the excess acetylene combines with the surface layer of the steel and reduces its melting point well below the normal melting point of steel. This phenomenon can easily be understood by recalling that carbon reduces the melting point of iron to as low as the eutectic at 4.3% carbon. [See the liquidus of the iron–carbon equilibrium diagram (Figure 5.10).] Properly heated, the surface layer of the parent metal should melt at approximately 1200 °C (2200 °F). No attempt should be made to puddle the filler rod. If the surface is properly heated, the melt will flow out and spread like solder and bond with the parent metal with a union that is stronger than the alloys themselves. The sweating technique is especially advantageous in the application of many hardfacings (such as the tungsten carbides), where very thin overlays are desirable. The maintenance welder may, with discretion, find many uses for variations of this procedure in applications other than hardfacing.

SURFACE ALLOYING

Surface alloying is more a natural phenomenon than it is a welding technique: The parent metal is heated to the proper bonding temperature as in brazing, and the filler metal is applied as in brazing; but instead of bonding by freezing some of the filler metal within the expanded, then contracted, parent metal, the filler metal bonds by alloying with the surface layer of molecules by atomically admixing with those surface molecules. That is, the surface molecules of the parent metal and some of the molecules of the filler metal exchange atoms and create a third alloy between the deposited metal and the parent metal. This phenomenon occurs due to the great affinity the filler metal and the parent metal have for each other. The difference between ordinary braze welding and surface alloying can be best observed by comparing a reheat of the welded joint: In ordinary brazing the parent metal and the deposited metal can be separated by merely heating the joint slightly above the original bonding temperature, whereas the surface-alloyed joint cannot be separated so easily—it appears to have been fusion welded.

UNDERSTANDING THE TORCH

The oxyacetylene torch is basically an apparatus for the admixture of oxygen and acetylene in the proper proportions to provide a combustion that produces a flame with a heat ranging from 3200 to 3480 °C (5800 to 6300 °F), the mean temperature being 3370 °C (6100 °F). The neutral flame (an exact 1:1 ratio of the gases) should be the mean temperature, and excesses of the gases will increase or decrease that temperature to a low of 3200 °C (5790 °F) (excess acetylene) and a high of 3480 °C (6260 °F) (excess oxygen). The neutral flame is so called because all of both gases are combusted.

The neutral flame is the most widely used flame, but all three of the flames have their applications. Uses of flames other than the neutral flame will be discussed under the recommended welding procedures for the various metals. The welder should know at this point, however, that an excess of oxygen tends to make a porous weld due to oxidation (hence the flame is often called an "oxidizing flame"), and that an excess of acetylene introduces carbon into the weld (hence it is often called a "carburizing flame").

If the welder is blessed with high-quality equipment maintained in its best condition, the flame, once set, should remain stable. These ideal conditions, unfortunately, are the exception instead of the rule. Such problems as creeping regulators, loose torch valves, defective or dirty mixing chambers, and a host of others are always around to aggravate the welder. He must, therefore, keep a sharp eye on his torch flame. Excess acetylene is readily apparent from the feather appearing beyond the flame cone (Figure 2.6) and excess oxygen is noticeable by a shrinking of the inner cone, a change of color to bluish purple, and a slight hissing sound. Figure 2.7 shows an accepted means of designating excess acetylene. $2\times$, for example, means that the feather extends twice as far as did the original neutral cone; $3\times$, three times as far. Excess oxygen generally causes the welder more aggravation than does excess acetylene.

Welding-Tip Size

Suiting the tip size to the job is of paramount importance. Attempting to weld a joint with too small a tip ranges from the impossible to an unnecessary waste of time. Table 2.1 should be consulted until the welder has gained enough ex-

Understanding the Torch 31

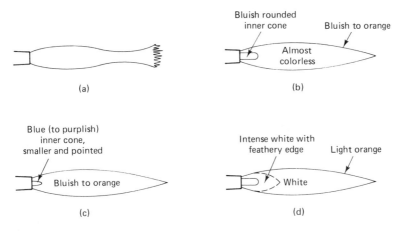

Figure 2.6 Acetylene flames: (a) pure; (b) neutral; (c) oxidizing; (d) carburizing.

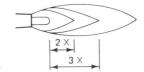

Figure 2.7 Excess acetylene designation.

perience to dispense with it. The table is based on the welding of steel. Some metals will require a variation in the tip size due to their very low melting points (such as lead, pewter, zinc die-cast) or because of their high thermal conductivity (such as copper, bronzes, aluminum). Each tip has its specific oxyacetylene range, and a lot of problems may be avoided by closely following the recommended gauge settings for the tip. Generally, the gauges can be set at midrange and then readjusted up or down as the welding conditions dictate. The tip size and gauge settings should be such that the welder has a wide control at the valves of his torch. He should be able to adjust the flame with a gloved hand without danger of the flame popping out.

Lighting the Torch

Always use a "striker," a flint-friction lighter, preferably the approved safety-lock type. Open the acetylene valve halfway; the acetylene will burn with a smoky yellow flame and may give off clots of "soot." If these clots appear, quickly open the valve wider until the clots disappear. If the valve is opened too far, the flame will show a gap between the nozzle and the flame. Adjust the valve until the flame just retouches the orifice. Open the oxygen valve. If the gauges have been properly preset, the acetylene flame should then accommodate the oxygen without popping out. When turning off the torch, turn off the acetylene first. An absolute law in most places of employment is: *Never lay down a lighted torch.*

TABLE 2.1 Welding-Tip Size

Thickness of Steel[a] (in.)	Tip Size
Up to $1/32$	000
$1/64 - 3/64$	00
$1/32 - 5/64$	0
$3/64 - 3/32$	1
$1/16 - 1/8$	2
$1/8 - 3/16$	3
$3/16 - 1/4$	4
$1/4 - 1/2$	5
$1/2 - 3/4$	6
$3/4 - 1\,1/4$	7
$1\,1/4 - 2$	8
$2 - 2\,1/2$	9
$2\,1/2 - 3$	10
$3 - 3\,1/2$	11
$3\,1/2 - 4$	12

[a]Applies also to cast iron.

The gauge settings given in Table 2.2 are for oxyacetylene operation of several torches and should suffice for most standard brands. The settings between maximum and minimum will be determined by the type of metal and its thickness. For best results:

1. Open the oxygen bottle "all the way."
2. Open the acetylene bottle $1/2$ to $3/4$ turn, and if the bottle is of the wrench-type valve, leave the wrench on the valve (for emergency turnoff).
3. With the torch valve half-open, set the gauges to the desired flow of gas or oxygen. This will give the operator wider, more flexible control at the torch valves. (Who wants hair-trigger control!)
4. Keep the settings toward the "low" side of the range, but if the torch "pops" out occasionally, increase the acetylene. However, if the torch pops out over the entire range, check for leaks, dirty tip, "creeping" valves, and so on. Also check the tip size—welders sometimes enlarge the tip orifices by "sawing" too vigorously with the tip cleaner. Such tips should be discarded.
5. Remember that the heavier the metal, the larger the tip needed.
6. For additional or more comprehensive information, write the manufacturer for product literature.

Figures 2.8 and 2.9 show a light-duty torch and a relatively complete oxyacetylene welding and cutting outfit.

Understanding the Torch

TABLE 2.2 Gauge Settings for Easy Control of Torch Valves

	Oxygen		Acetylene	
	Max.	Min.	Max.	Min.
Welding-tip size				
000	5	3	5	3
00	5	3	5	3
0	5	3	5	3
1	5	3	5	3
2	5	3	5	3
3	7	4	6	3
4	10	5	7	4
5	12	6	8	5
6	14	7	9	6
Cutting-tip size				
000	25	20	5	3
00	25	20	5	3
0	35	25	5	3
1	35	30	5	3
2	40	35	6	3
3	45	40	7	3
4	50	40	12	5
Rosebud size				
2	8	5	5	3
4	10	6	7	4
8	20	10	12	8

*Settings are psi; for KPa, multiply by 6.895.

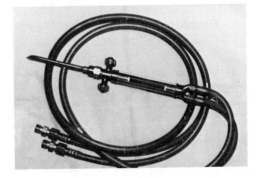

Figure 2.8 Oxyacetylene welding torch for light duty, tip sizes 000 to 2, with $\frac{3}{16}$-in. twin-hose whip and adapters to the $\frac{3}{8}$-in. hose.

Backfire. If the torch flame goes out with a loud pop, the torch valves should be closed and the equipment checked for the following defects or faulty welding technique:

1. The seat of the tip may be nicked or have dirt on it.

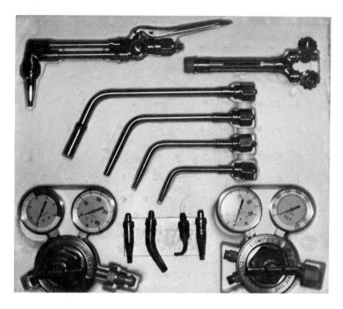

Figure 2.9 Relatively complete oxyacetylene welding and cutting outfit. The curved tip is used for general scarfing and gouging, and the 90° tip is especially adapted to rivet washing.

2. The inside of the tip may have carbon or metallic deposits.
3. The torch connections (nozzle or tip) may be loose.
4. There may be insufficient gas pressure for the size of tip used.
5. The tip has become overheated.
6. The operator is touching the tip to the puddle; this will smother the flame. Keep the inner cone out of the puddle.

Flashback. Evidenced by a squealing or hissing sound. The flame has burned back inside the torch. Close the torch valves immediately and check the equipment. Check the gas pressures. Repeated flashbacks indicate that there is a serious defect in the equipment, and it should not be used. If you are unfamiliar with the repair of torches and regulators, the equipment should be referred to the oxyacetylene supply company or to the manufacturer.

HOT-SPRAY METALLIZING

The hand-held hot-spray metallizing torch has found wide acceptance in maintenance welding, especially in wearfacing and the restoration of worn areas. The descriptive procedures are for the author's personal torch shown in Figure 2.10 but should adapt adequately to other torches. We have called this torch a *hot-spray* torch to distinguish it from the *cold-spray* method. Where the cold-spray method bonds below 250°C (500°F), the hot-spray torch welds at 650

Hot-Spray Metalizing

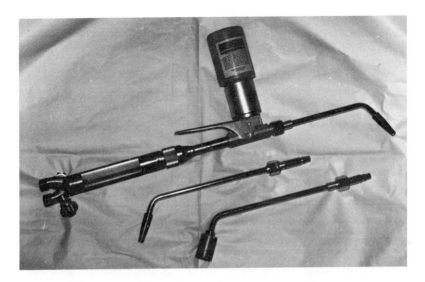

Figure 2.10 Typical hot-spray torch with two single-orifice spray nozzles and a multi-orifice spray.

to 760 °C (1200 to 1400 °F). Where cold spray relies on some mechanical strength, such as slots, dadoes, or serrations, the hot-spray method is applied and welds the same as the ordinary braze-welding method with torch and filler rod.

Using the same techniques and observing the same precautions as in braze welding, the experienced welder will have no problem using the hot-spray torch. The procedure for best results is as follows:

1. Clean the weld area thoroughly. Do not assume that a shiny surface is welding clean. This polish is often a thin film of oxides that resists good bonding.
2. Although slots, dadoes, grooves, and so on, are not necessary, the better practice is to grind the surface, bearing in mind, of course, that grinding cast iron tends to smear the free graphite, which tends to inhibit good bonding.
3. Set the oxygen at 138-172 kPa (20 to 25 psi) and acetylene at 83 kPa (12 psi). Adjust to a neutral flame. Depress the spray lever to test the flame, as in flame cutting.
4. Attach the powder container to the torch. Do not depress the spray lever until the metal has reached the welding temperature.
5. Heat a small area as in ordinary braze welding until a red spot appears. Raise the torch and the red spot should disappear. This temperature is called *black heat*; it is the bonding temperature.
6. Lower the torch until the red spot reappears and depress the spray lever, spraying a light coating of powder on the weld area. Release the spray

lever and heat the area until the powder is seen to melt into a thin film of molten metal. This is the same as *tinning* when doing ordinary braze welding.
7. When the powder has melted and has wet out, depress the spray lever again and begin the braze-welding process.
8. The rate of travel will depend on the amount or thickness of the weld deposit desired. The tinning should always precede the buildup by at least 5 mm ($\frac{3}{16}$ in.).
9. As a general rule, the lever should not be depressed constantly but should be opened and closed intermittently (pulsing). This technique has much the same effect as does the dipping of a filler rod into the molten puddle.
10. Visualize that more of the powder is carried on the outer periphery of the flame cone. Therefore, if working on a relatively sharp edge or corner, the flame cone should not be centered on the sharp edge but should be angled to one side. This does mean a waste of upward of 50% of the metal, but centering the torch on the sharp edge will mean a loss of up to 90% and an overheating and possible meltdown of the parent metal.
11. The multiorifice nozzle will find excellent use for those thin overlays of broader surfaces, especially those expensive tungsten carbides. Overlays as fine as 25.4 micrometers (microns) (0.001 in.) are possible.

3

TUNGSTEN INERT-GAS WELDING

UNDERSTANDING TUNGSTEN INERT-GAS WELDING

The shortest definition of tungsten inert-gas (TIG) welding would be "arc welding with a tungsten electrode within an inert-gas envelope." But if the reader is experienced with the TIG process, he does not need a definition; and if he has not become acquainted with TIG, the definition is worthless.

To describe the process in detail, we shall assume that the mechanic is familiar with stick welding and oxyacetylene welding. In TIG welding, the welding machine is or can be of the same type as that used in stick welding, that is, one delivering a constant-current varying-voltage supply to the arc. The TIG electrode, however, is not melted. Although it is called a nonconsumable electrode, it does waste away (as does a carbon electrode).

TIG is similar to oxyacetylene welding in that (1) the electrode and its holder are called a torch, (2) the arc performs and is manipulated similarly to the oxyacetylene flame, and (3) the deposit is formed by feeding a filler rod into the molten puddle. The electrode does not melt because the melting point of tungsten is upward of 3370 °C (6100 °F) and the electric arc delivers only 2760 °C (5000 °F). Of course, a carbon electrode could be used, but then we would have another process called CIG (carbon inert-gas) welding.

As well as being an acronym for "tungsten inert-gas," TIG is a short form for the engineer's acronym GTAW, for "gas tungsten-arc welding."

Fact and Fable

Heliarc is TIG, but all TIG is not Heliarc. Heliarc is a well-known, well-respected trade name. The welder cannot "Heliarc" with an Airco machine; he must "heli-weld." Hobart Welding Company dubs it Cyber-Tig. Lincoln Electric Company simply adds TIG to their trade name Idealarc.

The welder does not "ball the point" because he is welding aluminum, but because he is using alternating current; and he does not sharpen the point because he is welding stainless steel, but because he is using dc straight polarity.

"Inert gas" means exactly what it says: It is inactive. It is incapable of a chemical reaction with another substance. The welder who has used a little excess acetylene when gas welding in order to get a shielding effect can appreciate having a scrupulously clean nonreactive gas shield around the molten puddle which he is trying to protect from the active (noninert) contaminants in the air.

Argon gas is heavier than air and its liters-per-minute (cubic-feet-per-hour) [1pm/cfh)] flow is less than one-third that of helium, which is lighter than air. Helium gives a hotter, more penetrating arc; argon gives a more stable arc.

The electrode need not be recessed in the cone (or cup). It is allowed to protrude as the welder sees fit. The "stick-out" may be adjusted to the length needed for good vision of the weld and as necessary to reach into corners, holes, and other obstructed areas. The 1pm (cfh) must then be increased accordingly. Typically, the stick-out should be 1/2 for fillets, 1/4 to 1/2 in. for V butt welds, and 1/8 in. for edge welds.

Direct-current reverse polarity can be used, but is rarely used. It is sometimes used when welding magnesium and metals of very thin section. The size of electrode must be somewhat larger and will waste away somewhat faster than with dc straight polarity (DCSP). If high-frequency stabilized alternating current (ACHF) is available, the welder will prefer the ACHF because of easier arc starting and, especially, there is less pickup of weld metal on the electrode. (On reverse polarity, the current is flowing from the work to the electrode; hence the electrode instead of the weldment suffers the atomic bombardment, the result being that the electrode becomes overheated and contaminated by material from the weld puddle.) Table 3.1 summarizes the recommended currents for various metals.

Water-cooled equipment is not an absolute. It is required, however, when using high heats on heavy sections when welding time is prolonged.

Although TIG has been enshrined by many welders as the "miracle weld," it has not solved the problems of stress, strain, and distortion, nor has it eliminated the need for heat treatments. A fusion weld, by whatever means, exceeds the transformation range of the metal fused. For example, a fusion weld by the TIG process will convert malleable iron to white iron in the same way as will gas or stick; and the welding of hardened aluminum will put the weld area in an annealed condition. TIG may cause greater distortion than stick or MIG because its slowness of welding speed results in greater heat input per unit of weld measurement. TIG does concentrate the heat-affected zone more narrowly than does oxyacetylene welding.

TABLE 3.1 Recommended Currents[a]

Material	ACHF	DCSP	DCRP
Aluminum, wrought	Good	N.R.[b]	N.R.
Aluminum, cast	Good	N.R.[b]	N.R.
Brass	Good	Good	N.R.
Cast iron	Fair	Good	N.R.
Copper, deoxidized	N.R.	Good	N.R.
Copper, silicon	N.R.	Good	N.R.
Magnesium, up to 1/8 in.	Good	N.R.	Fair
Magnesium, 3/16 + in.	Good	N.R.	N.R.[b]
Magnesium, cast	Good	N.R.	Fair
Steel, low-carbon	N.R.	Good	N.R.
Steel, high-carbon (thin)	Good	Good	N.R.
Steel, high-carbon (0.30 + in.)	Fair	Good	N.R.
Stainless, thin	Good	Good	N.R.
Stainless (0.30 + in.)	Fair	Good	N.R.
Hastelloy	Fair	Good	N.R.

[a] N.R., not recommended.
[b] (0.30 + un.)

High-frequency stabilization is not absolutely necessary when using dc. The TIG process may be used with any dc supply simply by substituting the TIG torch for the stick-electrode holder and adding a properly controlled gas line. This unsophisticated equipment does require touch-starting of the arc. It cannot be used on ac because the arc will extinguish itself on the negative cycle due to a rectification phenomenon. (See "TIG-Welding Aluminum," Chapter 11.) Because of this rectification phenomenon, the machine should be set on "Continuous" for ac, but may be set on "Start" if using dc. Some less sophisticated equipment does not have automatic controls. These units are usually fingertip controlled, and the welder must keep the control depressed constantly when using ac. If using dc, he may release the control after the arc is established and depress it again and again for restarts. If, however, the control also controls the gas flow, the welder must keep the control depressed as long as the arc is functioning.

SHIELDING GASES

The majority of maintenance welders will usually be restricted to one gas, and for the most part their choice will be argon. The larger plants, however, will have occasion to use other gases and will often mix the gases to obtain advantages peculiar to each.

High-frequency-stablizing unit which can be used with the usual welding equipment for TIG welding, such as with the 250-A ac/dc machine shown in Chapter 1. (Courtesy of Lincoln Electric Company.)

It is ironic that air is the enemy we protect our weld from, yet is the source of most of our shielding gases. By volume, air is 78% nitrogen, 21% oxygen, 0.94% argon, and 0.06% carbon dioxide and other gases. Our atmosphere also contains water vapor, its content varying with the weather. Water is a compound of hydrogen and oxygen (H_2O). The greatest enemies of a good weld are oxygen, nitrogen, and hydrogen.

Oxygen. This gas is very active. It combines with other elements in the weld pool to form oxides and gases. Free oxygen combines with iron to form iron oxide (the same iron oxide that we call "rust" and the slag that results from our burning of steel with the oxyacetylene torch). Oxygen will also combine with the carbon in steel to form carbon monoxide, which, if trapped in the weld deposit, causes porosity.

Nitrogen. Nitrogen is a more serious problem, but only because there is so much of it in air (almost four times the amount of oxygen). Nitrogen causes hardness in and around the weld and often leads to cracking. Large amounts cause porosity.

Hydrogen. When welding carbon steels, hydrogen causes an unstable arc. The welder can usually work the hydrogen out of the weld pool. The solidifying weld tends to reject it also, but if it is trapped in the weld, the results are cracks in the weld, fisheyes, and underbead cracking.

Argon–helium. These gases are inert; that is, they will not combine with other elements. Helium is a by-product of the natural-gas and petroleum industries. Argon is a by-product of the liquefaction of air during the production of oxygen. Both gases are nonflammable. Helium is lighter than air; thus a greater flow is required. Argon is heavier than air, and its lpm (cfh) flow can be one-third that of helium. Helium is less expensive than argon. Argon is the preferred gas by most welders and in most TIG applications. It produces a quieter, smoother arc, and better cleaning action with alternating current on aluminum and magnesium. Helium produces a hotter arc and deeper penetration, and is preferred on thick sections of steel and for metals with a high thermal conductivity. It is preferred for automatic high-production-rate welding. Argon permits greater changes in arc length without breaking the arc, and starts are more consistent with ac. The gases can be mixed to compromise their differences. (*Caution:* Use only plastic hoses for these gases. Helium will diffuse through rubber and fabric hoses and let in air.)

Argon and helium with hydrogen. The addition of hydrogen increases arc voltage. The increase is greater when mixed with helium. Twenty percent hydrogen added to argon increases the arc potential to that obtainable with helium alone for the same cfh flow. Use the argon–hydrogen for stainless steels only, no other application.

Nitrogen. Nitrogen may be used to weld deoxidized copper and as a backup shield for stainless steel to produce brighter weld roots.

GAS FLOW RATE

Gas flow rates are given in liters per minute (cubic feet per hour) [1pm (cfh)]. The rate is indicated by a flow meter, in combination with its companion regulator, or is used with a standard regulator (see Figure 4.1).

Gases are expensive. Too great a 1pm (cfh) flow not only wastes gas but can also cause porosity in the weld. Thus the welder is on the horns of a dilemma: Too much gas and the gas cannot escape fast enough as the puddle solidifies, and too little gas lets in contaminants from the ambient air. As if this were not enough of a problem, the welder must be aware of any agitated or disturbed air blowing the gas away from the weld pool. The welder's best indicator of proper flow is the electrode tip itself: Proper gas flow maintains a shiny, slightly rounded tip; improper gas flow causes a burning, scaling erosion of the electrode.

The proper flow is dependent on the following: (1) type of gas, (2) type and size of electrode, (3) type and amount of current, (4) size and shape of the cup, (5) type of weld joint, (6) size of the weld puddle, (7) welding speed, (8) welding position, (9) length of stick-out, (10) angle of the torch, and (11) air currents blowing across the weld area. With so many variables our reader can sympathize with our reluctance to include a table reflecting 1pm (cfh) flow! For what it is worth, the author once read in a manufacturer's operational manual that "the gas flow rate should be high enough to protect the weld pool and the electrode tip."

CHOICE OF ELECTRODE

Tungsten electrodes are almost universally used. Of all the elements, tungsten's melting point is second highest, second only to carbon. Carbon melts at 3500+ °C (6330 °F), tungsten slightly above 3400 °C (6150 °F), both well above the 2760 °C (5000 °F) heat of the electric arc. Carbon electrodes are sometimes used, but their high rate of consumption and their fragility more than offset their lower price. If carbon electrodes are used, the process is called "carbon inert-gas" (CIG).

Although called a nonconsumable electrode, tungsten does waste away, but at a much slower rate than carbon. Tungsten electrodes are expensive, and the welder should care for them properly. They are not as fragile as the carbon electrodes, but they will break relatively easily. Other causes of excessive wasting are (1) too little 1pm (cfh) flow, (2) touch starting, and (3) inadvertent touching of the weldment and/or the filler rod.

Following are the common types of electrodes:

1. *Pure tungsten:* The pure tungsten electrode is generally used only with ac welding.

2. *Zirconium:* This type is also excellent for ac welding. It has a high resistance to contamination, it will carry more current size for size, and it provides a more stable arc.

3. *Thoriated tungsten:* These electrodes are more expensive than pure tungsten but should be used for dc straight polarity, especially if touch starting

Fully self-contained ac/dc rectifier-type welder for TIG welding.

is used. Their lower rate of consumption probably offsets their higher cost. Their arc stability is excellent, especially at the lower current settings.

4. *"Striped" electrode:* This is a thoriated electrode, but the 2% thorium is not distributed throughout the tungsten but is inserted as a separate entity, as a wedge down the full length of the electrode.

Electrodes are available in diameters (in inches) from 0.010 to 0.25 and in lengths 3, 6, 7, 18, and 24, and 6 in. being the preferred length for manual operation.

Current capacity. The current-carrying capacity depends on the type of electrode, type of gas shielding, cooling afforded by the holder, length of stick-out, and polarity used.

Welding procedures. The techniques peculiar to TIG welding are covered in subsequent chapters which detail the welding of specific metals. See especially the discussion of the TIG welding of aluminum and its alloys in Chapter 11.)

SETTING THE CURRENT

Just as stick-out and gas flow are varied according to varying welding conditions, so are the settings of the current varied. Although the experienced welder may be able to set the heat by guess and immediately know whether he was right or wrong, it is a better practice for the welder–mechanic to make a trial run on a practice piece of metal. Table 3.2 is offered as a general guide but as the reader can quickly see, we have given upward of 30% variation. Note that the size of the cup will have a bearing on the lpm (cfh) flow. Current settings are for the flat position, and in the higher ranges will be somewhat less for vertical and overhead welding. The gas flow rate can vary considerably. (See "Gas Flow Rate," this chapter.)

TABLE 3.2 ACHF Welding of Aluminum

Thickness of Metal (in.)	Current (A)	Electrode Diameter (in.)	Filler Rod (in.)	Cup Size (in.)	Argon [lpm (cfh)]
1/16	60–90	1/16	1/16	1/4–3/8	7 (15)
1/8	125–160	3/32	3/32	3/8	8 (17)
3/16	190–240	1/8	1/8	1/2	9.5 (20)
1/4	250–330	1/8–3/16	1/8–3/16	1/2–3/4	12 (25)
3/8	310–390	3/16–1/4	3/16–1/4	3/4	14 (30)
1/2	375–450	3/16–1/4	3/16–1/4	3/4	15 (32)

SELECTING THE FILLER ROD

As shown in Table 3.2, the filler rod increases in size with the thickness of the weldment. Since the maintenance welder rarely has a complete inventory of welding rods that would cover the entire table, he should select the filler rod of the next largest or next smallest size. It should not take him long to recognize the utter futility of trying to weld 1/2-in.-thick plate with a 1/16-in.-diameter rod with a 275A current. Types of filler metals are covered in subsequent chapters dealing with specific metals.

4

METAL INERT-GAS WELDING

Definitively, metal inert-gas (MIG) welding is *semiautomatic arc welding with a consumable electrode within a gas envelope*. Here, the welder's terminology, usually so descriptive, is not quite as proper as the engineer's term, *gas metallic-arc welding* (GMAW), since noninert gases, such as carbon dioxide and oxygen, have expanded the general utility of the process.

We can compare MIG with TIG only in the fact that both use a gas shield for the fusing area. Their differences are: (1) TIG uses a constant-current varying-voltage supply for the arc, whereas the supply for MIG is constant-voltage varying current; and (2) the TIG electrode is nonconsumable, whereas the MIG electrode is not only consumable but can be set for extremely high rates of consumption.

UNDERSTANDING THE ARC

As we learned in studying the arc in stick welding, the melt that takes place in the arc is the result of the consumption of electrical energy. In stick welding, we set the current at a constant value, and our somewhat erratic manual control varied the voltage in accord with the resistance in the arc, which varied with the length of the arc. We increased the electrical energy to the arc (in order to burn a larger electrode or to burn it at a faster rate) by resetting the machine to deliver a higher current to the arc.

The MIG-welding machine (Figure 4.1) is designed to deliver the amount of energy necessary to consume the electrode, regardless of the size of the electrode and speed at which it is fed to the arc. Our control of the arc is restricted

Figure 4.1 Gas metallic-arc welder featuring easy mobility around plant facilities. (Courtesy of Lincoln Electric Company.)

to our presetting of the constant-voltage value and our setting of the rate of wire feed; and, of course, variation of the stick-out.

Unlike stick welding, the operator of MIG does not strike an arc by touching the weldment momentarily and withdrawing the electrode a fraction of an inch; instead, he pulls the trigger of his welding gun, and the wire feeder pushes or pulls the wire forward until it strikes the weldment. Upon touching the weldment, the electrode closes the circuit, and current flows. Momentarily, there is zero voltage in the arc, and the open-circuit voltage sends a great surge of current to the short-circuited tip of the electrode. The electrode burns off, producing a gap. The width of this gap is the *arc length*. The design of the machine is such that the current will then stabilize at the value which, coupled with the arc voltage value, produces the electrical power necessary to consume the electrode and maintain the arc.

In *shielded metal arc welding* (stick welding) we learned that the resistance of the arc varied with the arc length; and because the machine was a constant-current type, the voltage had to increase or decrease in order to keep the current constant; all in accord with Ohm's Law of *1 volt, 1 ampere, 1 ohm*. By interpolating Ohm's Law, we can see that if the voltage is forced to remain constant, the current must vary with the resistance offered by the arc length.

If we assume an operating voltage of 30 V and a wire speed that must consume an average of 150 A, 30 V divided by 150 A would give us an average arc resistance of 0.2Ω. Thus, if the arc resistance is increased to 0.25Ω, our current will drop to 120 A in order to keep the voltage constant at 30 V, and if the arc resistance drops to 0.15Ω, our current must increase to 200 A. We are dealing with such minute changes in resistance that we can observe fluctuations in current by lengthening and shortening the electrode extension (stickout) and watching the ammeter on the panel of our welding machine.

From the discourse above, we can deduce that the electrode's burn-off rate is determined by the current, and the arc length is determined by the constant-voltage setting. The welder who has noticed in stick welding that his manual control of arc length had a profound effect on the arc action will now notice that his voltage setting will profoundly affect the arc action in MIG welding. The interaction of the 1 volt, 1 ampere, 1 ohm law is the same except that, once set, the interaction is controlled automatically by the machine (except for variations in stickout, of course). The welder's control is one of selection; that is, he chooses the voltage and the wire feed rate, and these two choices determine the arc action (which is the subject of the next several sections). But before entering on the discussion of the many arc actions encountered in MIG welding, the welder should bear in mind that the type of shielding gas can also greatly affect the action of the arc.

ARC TRANSFER

Spray transfer. The melting electrode is transferred across the arc in very small droplets of metal, which the welder should observe as never exceeding the diameter of the wire being used. The welder may compare the MIG spray transfer as being similar to the arc action of stick electrodes, which he describes as having a *lively arc action* or *good purging action* or *good overhead rod*. Spray transfer is generally used with inert gases and dc reverse polarity. It requires higher heats than globular transfer and may prove troublesome on sections thinner than 3.2 mm ($\frac{1}{8}$ in.). It is almost spatter free, and is excellent for fast fill and for the vertical and overhead positions. The welder should learn to discern between the spray and the globular by sight, sound, and distribution of weld metal.

Globular transfer. At the lower currents, the transfer across the arc tends to become globular; that is, the melt-off forms first as a globule on the tip of the electrode and becomes noticeably large before separating from the electrode for transfer. The stick welder may observe it as a *lazy* or *sluggish* arc. Other characteristics are excessive spatter and reluctance to flat-out on overhead welds. Globular transfer is often used on thin-gauge steel with carbon dioxide (CO_2) shielding.

Short arc. We have chosen the welder's term for this MIG process because of its simplicity and its prevalence in the trade. A more complete name is around, but it is rarely used: *short-circuit welding*. The process does short circuit and is also called *fine wire* welding. In reality there is no transfer across the arc. The process features a high rate of feed and low current, permitting the electrode to come in contact with the weldment before the heat can melt the electrode or before a globule can form. The high heat generated by the short-circuit literally explodes the melting tip, and a gap occurs between the electrode and the weldment, reestablishing the arc. Short-arc welding is excellent for thin gauges, thin to thick, and for poor fit-ups because this type of transfer is less penetrating and the melt is less fluid. Excessive current will result in a globular transfer. Short-arc began with carbon dioxide shielding but soon expanded to using pure argon, argon–oxygen, and argon–CO_2, as well as the pure carbon dioxide. Argon–oxygen is excellent for vertical-down welding of stainless steel linings, thin to thick, in the petroleum industry.

Buried arc. We have used the term *buried arc* instead of *submerged arc* in order not to confuse the MIG submerged arc with the submerged arc using the slag blanket. As the term "buried arc" would indicate, there is no transfer across the arc. The arc digs a hole in the weldment, and the melting electrode deposits itself within the molten pool. Stabilizers and deoxidizers have been added to the electrode as a core, resulting in a flux core carbon-dioxide shielded process.

Pulsed arc. This process features low heat input and the desirable spray transfer that is normally unobtainable at the lower heats (see globular versus spray transfer above). Two power sources are required: one provides a steady background current to heat the electrode and weldment, and a second source supplies a high-density pulsing current in bursts of heat 60 times per second. The process was developed especially for thin sections of aluminum and magnesium, which required low heat and inert-gas shielding.

SHIELDING GASES

The great preponderence of MIG welding is done with argon, helium, and carbon dioxide. They are used in their pure state and in mixtures. Other gases are sometimes mixed with argon, helium, and carbon dioxide to obtain specific benefits. A knowledge of their characteristics will help the welder understand the variations in arc behavior so that he may put their purposes to best possible use.

Argon. Argon provides low thermal conductivity, stable arc, reduced spatter, deep penetration at the center of the bead, and shallow penetration at edges of bead. It is rarely used with carbon steels due to its lack of uniform penetration and its tendency to undercutting. Argon is excellent for use with thin-gauge material; aluminum, magnesium, copper, and their alloys; and for out-of-position work.

Helium. Greater arc heat results from using helium; penetration is more shallow but more uniform. It is excellent for heavy sections and metals with high heat conductivity, and for aluminum, magnesium, and copper alloys.

Carbon dioxide. Carbon dioxide is much cheaper than other gases because it is extracted from the waste gases given off by the burning of natural gas, coke, and fuel oil. It gives a uniformly deep penetration and relatively wide flat bead contour. The arc is unstable and spatter is a problem. Spatter can be reduced by adding stabilizers as a core in the electrode. (*Caution:* The heat of the arc breaks the gas down to its constituent parts, and a dangerously high amount of carbon monoxide may be produced. Therefore, if using carbon dioxide shielding, weld only in a well-ventilated area.)

Argon–helium. These two gases may be mixed in any ratio desired by the welder. The gases are stored in separate bottles and are intermixed by regulators, and the mixture then controlled by a flow meter. An estimate of the effect on arc action of a given mixture ratio may be extrapolated from their individual characteristics as detailed in the preceding paragraphs.

Argon–CO_2. This mixture reduces the spatter that is characteristic of pure carbon dioxide and broadens the narrow penetration area that is characteristic of argon. It is excellent for low-carbon steels, low-alloy steels, and some stainless steels. The filler wire should contain stabilizers to counteract the ill effects of the oxygen. The addition of carbon dioxide helps promote a spray-type transfer. The mixture is also desirable for its reduction in costs, carbon dioxide being approximately one-fifth the price of argon.

Argon–oxygen. When added to argon, oxygen, in amounts not exceeding 5%, widens the penetration area and reduces undercutting when welding steel. It is excellent for aluminum alloys, carbon and low-alloy steels, and stainless steels, and for relatively wide weaves in vertical-down welding of stainless steel liners. (See "Short Arc," this chapter.)

WELDING TECHNIQUES

In manipulation of the gun and distribution of filler metal, the transition from stick to MIG should pose no problem. Some adjustment of technique may be necessary when welding butted joints that have a root opening. The electrode must be aimed into the puddle as it crosses the root opening, lest the wire squirt through the opening without melting. This squirt-through is evident from the short lengths of unconsumed wire standing on the back side of the joint. Welders call these stubs *tree stumps*; a great number may be dubbed a *forest*. They are also called *whiskers*. Particular care should be taken to avoid squirt-through when welding butted pipe joints. A forest created inside a pipe by a MIG weld is as undesirable as the *grapes* hung there by the careless stick welder. The use of consumable backup rings virtually eliminates squirt-through. Due to the deeper penetration of MIG, root faces may be wider and root openings may be narrower than in stick welding.

Spatter. Arc spatter can adhere to the nozzle to such an extent as to interfere seriously with the smooth flow of gas. Turbulent or erratic gas flow causes an erratic arc. A pair of needle-nose pliers (preferably with a hooked nose and side cutters) is a handy gadget for stripping spatter from inside the nozzle and for trimming back the electrode for restarts. Nozzle *dip* will reduce the amount of accumulated spatter. If not available, ordinary grease is better than no dip at all.

Nozzle angle. The *nozzle angle* of MIG produces the same penetration and bead characteristics as does the electrode angle in stick welding. A pulling angle concentrates the heat in the puddle, and results in deeper penetration and higher bead contour; a pushing angle forces some molten metal ahead of the bead and results in a more shallow penetration and a wider, flatter bead. As in stick welding, an angle of no more than 10° out of the perpendicular is recommended. The loss of shielding effect by excessive tilting of the MIG nozzle is more serious than is the excessive tilt of the stick electrode.

Wire feed. Although the machine is designed to increase the current as needed to burn off the electrode and maintain the arc, the wire feed rate may exceed the machine's output. Excessive wire feed will cause the bead to pile up, producing a lumpy, convex contour. In extreme cases, the operator may feel the electrode thump against the weldment. Such a failure to burn off has been called stubbing and or *roping*. *Birdnesting* occurs when the wire backcoils between the cable and the drive roll. It may be caused by improper alignment, friction within the feed cable, or by the operator allowing a kink or sharp bend in the feed cable while welding.

Travel speed. The travel speed of MIG welding affects the weld in the same way as it does in stick welding. Slow travel begets deeper penetration, a larger bead, and more heat input; fast travel produces the opposites. Suit the travel to the burn-off rate.

Arc length. Although the machine controls the arc length automatically, the operator can alter it by resetting the voltage control. He can also vary the arc length slightly by varying the stick-out. The proper arc length produces a crisp, crackling, frying sound that is similar to that related to the proper arc length of an E-6010 stick.

Suit the filler wire composition to the weldment, the wire size and feed to the size of desired bead, the voltage to the proper arc action, stick-out to the proper arc sound, travel to the burn-off rate, and weld, weld, weld. As the old timer advised the neophyte: "There ain't nothin' to it. You're just weldin' with a stick that won't quit."

5

METALLURGY FOR WELDERS

The question, "Does a welder have to know metallurgy?" is too often asked disparagingly. To those questioners, the answer is "A welder is a metallurgist." Certainly, he *specializes in metals*; and although others may be more elaborately involved with metals, none are more intimately associated. He is, in a sense, a *melder* (from the word *meld*, which is a "blend of melt and weld"). He is a steelmaker similar to the melder on the furnace floor. The melder in a steel mill melts and blends his product in a furnace, sends periodic samplings to the laboratory for analysis, adds alloying ingredients as needed to bring his heat to specifications, and when the proper analysis is achieved, pours his meld into an ingot mold for solidification. The welder must recognize the composition of the weldment, choose the proper alloying elements in the electrode or filler rod, and blend the weldment and the filler metal in the mold that he has prepared by veeing out or otherwise arranging the metal to be welded. He has a further responsibility. The mold does not shape the meld as does the ingot mold. He must make the meld fit the mold. Where the steelman's ingot then goes to the soaking pits to be brought to the very top of the solidification temperature and is thereafter rigidly controlled during the cool-down, the welder's meld is most often left to the mercy of the ambient air and the rapid cooling effect of the mold. The end product, the structure of the solidified meld, is generally the result of an educated guess; and the greater the welder's knowledge of the product, the closer is the guess.

WHAT IS METALLURGY?

Metallurgy is the "science and technology of metals." Science, of course, means a *knowing*; therefore, a scientist is one who knows. Technology, as best befits our purpose, is *applied science*. May we, then, say that a metallurgist is one who knows metals and knows what to do with them? Would that description fit the welder? Perhaps not all welders. Those welders who weld where, when, and how they are told may fall somewhat short of the mark, but the maintenance welder in the smaller shop and many from the larger shops who are often sent out on their own must make decisions; and in order to make decisions, they must know something about metals and know the *how* of welding them.

The understanding of metals begins with that smallest particle that is peculiar to an individual metal: that particle which, in conjunction with millions and billions of its siblings, identifies the whole as *iron* or *nickel, silver,* or *gold*.

How much simpler would have been the welder's task if we had been satisfied with this pure iron or pure nickel. But we were not satisfied. Iron was too soft and weak, nickel was too expensive, and lead melted at too low a temperature. All of these metals had special properties that people wanted to put into that one miracle metal that had everything, so we began mixing them. Even that best of all metals, gold, was too soft, so we mixed gold with other elements, reducing gold from 24 carat to 18 carat, to 14 carat, and down to 10 carat, in keeping with the desired hardness and, perhaps, a certain target price.

Unhappy with the performance of iron, people added carbon to iron to produce hundreds of iron–carbon alloys, which were named "steels." We then added nickel, chromium, molybdenum, manganese, copper, and other metallic elements, and further, added nonmetallic elements, such as sulfur, silicon, cerium, selenium, and others.

Human beings somewhat belatedly extracted aluminum from bauxite and for a time were satisfied with its purity and its whiteness; but running true to form, they alloyed it with copper and other elements. They extracted magnesium from several sources, principally seawater, and although it had useful properties, they were immediately dissatisfied with it and mixed it with aluminum, manganese, silicon, and tin.

There is no doubt that our metallurgists will go on with new mixtures and will discover new elements. Will the welder's knowledge gained today be obsolete tomorrow? Of course not, iron is iron, nickel is nickel, and tin is tin. Metals retain many of their characteristics when they are the principal constituent, and impart some of their character to their foster parent when they are a minor constituent. They do, sometimes, have a profound effect on the structure of the whole even if their percentage is low. For example, when 12 to 14% manganese is added to steel, it prevents the steel from transforming its gamma iron to alpha iron as it would normally tend to do.

To add to the problems they have created for the welder, metallurgists have hot-worked, cold-worked, heated, and quenched metals to further change their structures. They have created metals of such unstable structures that a touch of the welder's arc can split them asunder as easily as a sudden jolt can explode unstable TNT. Despite this seeming conspiracy against welders, metallurgists do keep us in mind. The weldability of a newly conceived metal is of great concern to them, because a metal that cannot be welded has a greatly restricted utility.

PURE METALS

When a metal freezes (changes from liquid to solid), it crystalizes. During the process of solidification, the atoms distribute themselves in an orderly arrangement, thus forming a definite pattern in space. We call this arrangement a *space lattice*. The space lattice is made up of a series of points in space that form a geometrical structure of atom groups.

There are a number of different space-lattice groups. One very common space lattice is the cubic pattern (Figure 5.1a–c) of metals such as copper, iron, aluminum, lead, nickel, molybdenum, and tungsten. Some metals form the *body-centered cubic* (bcc) lattice, some the *face-centered cubic* (fcc) lattice (Figure 5.1b and c, respectively). Another relatively common pattern is the hexagonal space lattice (Figure 5.1d), sometimes called *close packed* because of its number of atoms (17). This pattern is found in zinc, magnesium, cobalt, cadmium, antimony, and bismuth.

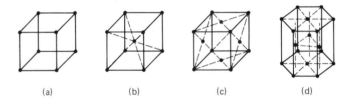

Figure 5.1 Space-lattice patterns: (a) cubic lattice; (b) body-centered cubic lattice; (c) face-centered cubic lattice; (d) hexagonal space lattice (close-packed).

The character of metals is dependent to a great degree on the type of space lattice formed during solidification. Metals with a face-centered lattice have a ductile, plastic, workable crystal state, whereas metals with a hexagonal lattice tend to lose their plasticity rapidly upon shaping. Antimony and bismuth are very brittle.

Some metals, such as iron, our greatest concern, change from one lattice to another during natural cool-down in the solid state. This change is called an *allotropic change*, and it can have a profound effect on the character of iron. When we know that iron, at a definite temperature, changes from a face-centered lattice to a body-centered lattice, we can see from the patterns in Figure 5.1b and c that five atoms are squeezed out. Where do these atoms go? These atoms will be found to be very useful to us when we study the iron–carbon alloys.

The crystal diagram in Figure 5.2 has been illustrated in two dimensions. Crystals, however, grow omnidirectionally, like a grain of sand. If we may so analogize, the atom groups fill in the areas A by aligning themselves along the axes as flesh is draped around the animal skeleton.

When the liquid metal (*melt*, in the pure form; *meld*, if alloyed) cools to the freezing point, groups of atoms orient themselves into one of the common crystal patterns. These groups are nuclei, each nucleus being a potential crystal and able to grow to form a crystal visible to the naked eye.

At the top of the solidification range, these nuclei spring into being throughout the entire melt. Were it not for this simultaneous birth of many nuclei, the melt might well become a single giant crystal. Once into being, each nucleus sends out shoots similar to the branches of a tree. These *shoots* are called *axes of solidification* (see Figure 5.2) and form the skeleton of a crystal in much the same way as frost patterns occur. Atoms then attach themselves to the axes of the growing crystal in progressive layers, finally filling up these axes and forming a completed solid (A in Figure 5.2). We call the crystals *grains* because of their variety of shapes. Since so many nuclei are trying to grow, it is inevitable that there is some competition for space, and some nuclei will not have as much room to grow as do their neighboring nuclei, the result being crystals of all sizes and shapes. At this point, pertinent to later study, we should fix in our mind that time and space are the essence of the size of a crystal. The more time and space in which to grow, the larger the crystal or, as generally stated, the coarser the grain. From these phenomena, we can see that a crystal may vary in size and may take different *external* shapes, but the atomic arrangement is always the same. We call the parallel lines between the rows of atom groups *planes* of solidification, but since it is along these planes that slipping or cleavage occurs, we more often call them *slip lines* and or *cleavage lines:*

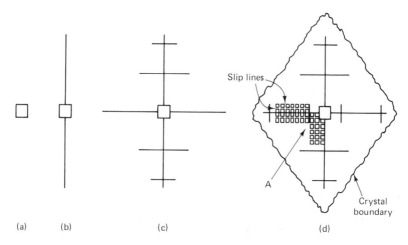

Figure 5.2 Skeleton of a crystal: (a) nucleus; (b) first axis; (c) second stage; (d) third stage.

slip lines in the case of deformation, cleavage lines in the case of rupture.

Pure metals never (at least hardly ever) break along the crystal boundaries. This phenomenon is due, in part, to the irregular shape of the crystal. From Figure 5.2, a comparison of the irregularity of the crystal boundary and the precise shaping of the atom groups will show that the atom groups would slide along each other much more easily. It is also believed that some atoms could not make up their minds which crystal to join up with and are free atoms between the boundaries, and act as a kind of abradant key.

Due to the law of disorder, no nucleus will form with its planes or atomic alignments the same as those of any other nucleus. Thus the planes of one crystal will never line up with the planes of another crystal. And since slipping or cleaving takes place along these planes (never at crystal boundaries), any slipping or cleaving must run into a brick wall and change direction at the boundary of the next crystal, and run an erratic, zigzag course through the material under stress. Therefore, if a metal slips or cleaves along the planes of the atom groups and must change direction at the boundary of each sequential crystal, we can deduce that a fine-grained metal should resist fracture or deformation more than would a course-grained metal, because the slipping or cleaving would have to change direction more times per unit of linear travel.

While on the subject of atomic arrangement, which will be pertinent throughout our study of physical metallurgy, we have attempted in Figure 5.3 to show the elasticity, the elastic limit, and the point of no return: *set* or *permanent deformation*. In Figure 5.3a, we see the orderly arrangement of atom groups typical to a very small segment of a crystal. When a stress is applied,

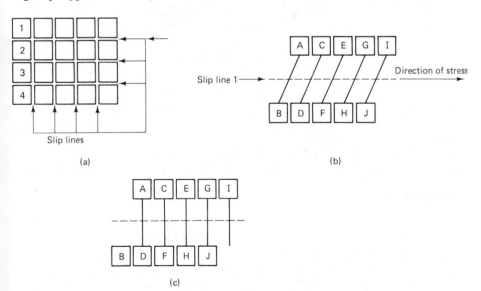

Figure 5.3 Slippage of atomic tiers within a crystal: (a) orderly arrangement of atoms; (b) shift within the elastic limit; (c) shift beyond the elastic limit.

the atom groups will shift. If the stress is within the metal's elastic limit, the atom groups will shift as in (b): group A lining up with group D, group C with group F, and so on along the tiers; but they can return to their original positions when the stress is removed. If the stress exceeds the elastic limit, the metal is *set* as in (c), and the atom groups cannot return. The metal is permanently deformed.

The displacement along the parallel lines may be analogized and seen by comparing it with the shearing effect when we accidentally drop a ream of paper and the sheets slither out over the floor like a deck of cards fanned out on the table by the dealer in a gambling casino. Our deck, of course, is the stack of atomic tiers in Figure 5.3a. Figure 5.3b represents only two cards, but in (b) we can easily visualize them being pushed back, one atop the other. In (c) the cards must remain as they are.

When a metal becomes set or permanently deformed, its tensile strength and hardness are increased. Stiffness is increased only slightly or remains unchanged. Plasticity is decreased. If continually deformed, the metal becomes brittle. An extreme case of continued deformation is when we bend a metal wire back and forth until it breaks. This work-hardening phenomenon, however, has beneficial applications also, as we shall see in later study. Hadfield steel, for example, seems to thrive on abusive treatment and is excellent for withstanding the pounding of locomotive and freight-train wheels over frogs and crossovers of railroads.

The reason metals become harder and stronger is not of paramount importance to the welder, but for the curious welder, there are several theories that he may investigate. One is *amorphous cement theory*, wherein the constant rubbing along the slip lines creates a kind of binder, perhaps akin to the galling effect when two metals wear against each other. A second theory is *slip interference theory*, wherein the crystals are said to fragmentize, and the fragments then interfere with the slipping just as cigarette ashes and other debris picked up by the cards will interfere with a fast deal when playing poker.

Of more importance to welders is how to undo this work hardening when it becomes too great or is not wanted at all. Fortunately, welders can put the metal back into its original condition simply by reheating the metal and starting all over again. This process is called *annealing*. And if we wish to retain the hardness but reduce the instability of the metal, we give it a less drastic treatment, called *stress relieving*. Both treatments will be covered more extensively later in the book.

TRANSFORMATION OF PURE METALS AND ALLOYS

Principal characteristics of pure metal transformations are as follows:

1. Pure metals freeze at a constant temperature; that is, the temperature of the total mass remains the same until the entire mass has solidified. Cooldown then resumes.

Transformation of Pure Metals and Alloys

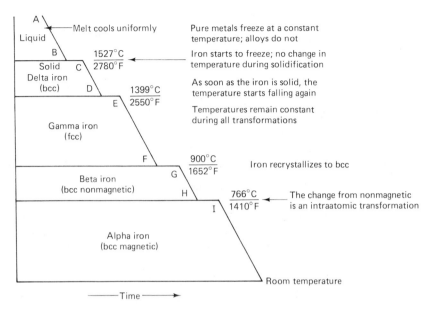

Figure 5.4 Cool-down of pure iron.

2. All transformations occur at a constant temperature.
3. Structures of pure metals may be altered only by cold working (work hardening) and subsequent annealing.

Characteristics 1 and 2 may be seen in the cool-down of pure iron shown in Figure 5.4. The horizontal lines illustrate the time in which the temperature remains constant while the transformations take place. The descending oblique lines illustrate the cool-down between the transformations, thus:

A-B: Liquid cools to 1527 °C (2780 °F).

B-C: Temperature remains at 1527 °C until entire melt has solidified. Cool-down resumes.

C-D: The solid (delta iron) cools to 1399 °C (2550 °F) and remains constant.

D-E: Delta iron (bcc) changes atomic structure to gamma iron (fcc). Cool-down resumes.

E-F: Gamma iron cools to 900 °C (1652 °F).

F-G: Gamma iron changes to beta iron (bcc). Cool-down resumes.

G-H: Beta iron cools to 766 °C (1410 °F).

H-I: Beta iron (bcc, nonmagnetic) changes to alpha iron (bcc, magnetic).

Delta and beta irons are of relative unimportance to the welder (unless he should encounter a structure called *beta martensite*). Gamma and alpha irons, on the other hand, are of great importance in the understanding of the alloying process and the heat treatment of steels. The transformations in structures described above are allotropic changes, as diamonds and graphite are allotropes of carbon.

Unlike pure metals, alloys do not solidify or transform at constant temperatures but do so during the cooling-down period. These gradual transformations make possible the control of the metal by speeding up or slowing down the rate of cooling.

CLASSES OF ALLOYS

To the steelmaker and foundryman, an alloy is a meld of two or more metals, or one or more metals and one or more nonmetallic elements that are soluble in each other in the molten state.

There are five classes of alloys, of which two, Class IV and Class V, are better known as *emulsions* and *compounds*, respectively. Emulsions (Class IV), such as lead and zinc or lead and copper, are not soluble in either the liquid state or the solid state. Compounds (Class V) are of importance to the welder as hardening agents in alloys. These alloys are usually found as constituents, such as *cementites* (Fe_3C), used as hardeners in steel; $CuAl_2$, used as a hardener in aluminum; and the carbides of chromium, vanadium, and tungsten found in hardfacing alloys. The remaining three classes of alloys are:

Class I: The constituents are completely soluble in both the solid state and the liquid state.

Class II: The constituents are completely insoluble in the solid state but are soluble in the liquid state.

Class III: The constituents are partially soluble in the solid state but completely soluble in the liquid state.

Class I Alloys

We realize that steel is of greater concern to the welder but in order to work somewhat gradually into the enormous complexities of the iron–carbon alloys and their further alloying with other metallic and nonmetallic elements, we shall begin with one of the simpler alloys. The constitution diagram for one of the common nickel–copper alloys is shown in Figure 5.5.* It is a 60% copper and 40% nickel alloy.

Separately, the two metals behave thus:

Point B: The right boundary of the diagram represents the cooling of pure nickel. Nickel begins to solidify at 1452°C (2646°F), its melting point.

---

*The straight-line diagrams in Figures 5.5, 5.7, 5.9, 5.10, and 5.12 are for illustrative purposes only and should not be used where absolute accuracy is required.

Classes of Alloys

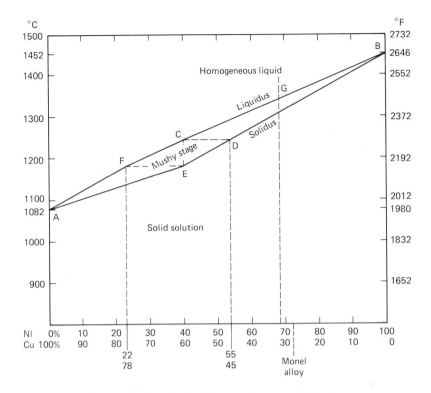

Figure 5.5 Constitution diagram of copper–nickel alloys.

The temperature of the melt must then remain at that temperature until solidification is complete.

Point A: The left boundary represents the cooling of copper. Copper begins to solidify at 1082 °C (1980 °F), its melting point, and the temperature of the melt remains constant until all the copper has solidified. (Recall that pure metals solidify at a constant temperature.)

The alloy of nickel and copper behaves differently. Figure 5.5 shows the behavior of a 40% Ni–60% Cu composition from the molten state through the solidification range.

Point C: Solidification of the combinations starts. Small crystals appear in the liquid.

Point D: The crystals are not 40% Ni–60% Cu but have the composition 55% Ni–45% Cu. (See vertical dashed line *D*.)

Line DE: This descending line shows the changing composition of the solid portion as the meld cools from *C* to *E*.

Line CF: This descending line shows the composition of the liquid portion at corresponding temperatures of line *DE*.

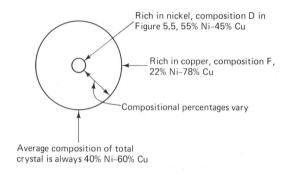

Figure 5.6 Copper-nickel crystal composition.

Thus the liquid and the solid become richer or poorer in nickel or copper as more liquid solidifies. As the crystals grow, they form a center that is rich in nickel, with their composition shading gradually to a copper-rich boundary (see Figure 5.6). The centers of the crystals will be composition D (55% Ni–45% Cu). The outer boundary of the crystal will be composition F (22% Ni–78% Cu). The composition of that portion of the crystal intervening the center and the boundary will be richer or poorer in nickel or copper depending on its proximity to either the center or the outer periphery.

Compare Figure 5.6 with Figure 5.2. Crystals first form on the primary axis. These crystals will be composition D formed at C (Figure 5.5). As the crystals grow, they are progressively richer in copper along their secondary and ternary axes. At the point of complete solidification of the meld, the two metals will begin to homogenize into a uniform structure of 40% Ni–60% Cu throughout (completely soluble in the solid state, as is characteristic of Class I alloys). This homogeneity, however, requires slow cooling, or may be accomplished by annealing. Therefore, if we cool the metal too rapidly, a complete diffusion will not have time to take place; and if complete homogeneity is the desired structure, we will have to reheat the metal up to the top of the transformation range (just below the *solidus* in Figure 5.5) and control the cool-down. As welders, we must begin to see that when we remelt a metal, we are going well above its transformation range, and its cool-down becomes important.

Monel alloy, is another common Ni–Cu alloy, consisting of 67% nickel and 28% copper, with iron, cobalt, and manganese existing as impurities. Monel is sometimes called a natural alloy, because it exists in a natural state as an ore. As can other Ni–Cu alloys, the Monel cool-down may be read from the diagram, but because of the percentage of impurities, some adjustment may be necessary. Perhaps using 70% Ni–30% Cu on the diagram will give us sufficient accuracy (see dashed line G).

Class II Alloys

For our Class II alloy, we have selected antimony and lead as the alloying constituents. Figure 5.7 diagrams the cool-down and separation of the constitu-

Classes of Alloys

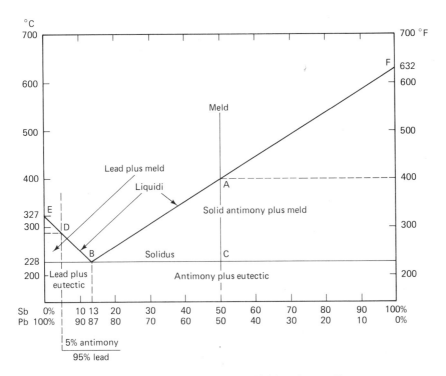

Figure 5.7 Constitution diagram of lead–antimony alloys.

ents in the solid state. We have included two alloys (combinations of antimony and lead) in order to illustrate the lever principle in mathematics relative to the ratio of eutectic to the preponderant constituent. It is essential at this point (and in the study of steel and cast iron) that we understand the eutectic.

Class II alloys are sometimes called *eutectic alloys*, the word "eutectic" being coined from Greek, meaning *easily melted*. To welders, however, the meaning is much more complex. Such definitions as "the intersection of two descending liquidi;" are quick and to the point but assume a great deal of prior knowledge. A eutectic needs to be explained rather than defined.

A eutectic can exist only in an alloy of two or more metals wherein two of the alloying constituents are insoluble in each other in the solid state. A eutectic is that combination of two metals (or elements) which has the lowest melting point of any possible combination of those two metals. A eutectic is an alloy but freezes at a constant temperature, as does a pure metal. The eutectic alloy does not exist as a homogeneous alloy such as our Ni–Cu alloy in Figure 5.5 but as separate and distinct crystals of each of the alloying constituents. A eutectic has a lower melting point than the pure metals that formed the eutectic. A eutectic may exist as an alloy by itself (as *pure eutectic*, such as *pure pearlite* in steel) or may exist as a matrix for the constituents over and above the percentage required to form the eutectic, or may exist as part of a more complex alloy. A eutectic has the highest and best physical properties of any combination of the two constituents. We have attempted to incorporate all of the functions of a eutectic in Figures 5.7 and 5.8.

Another metallurgical phenomenon we encounter in this alloy is the fact that any metal added to another metal reduces that metal's melting point. Referring to Figure 5.7, we can easily understand that by adding lead, with a melting point of 327 °C (620 °F) to antimony, with a melting point of 632 °C (1170 °F) we can reduce the melting point of antimony. It is a little difficult to see that antimony will reduce the melting point of lead. We shall see later that carbon, melting at 3648 °C (6600 °F), reduces the melting point of iron, whose melting point is 1527 °C (2780 °F).

Bearing in mind that the two constituents are *in solution* in the liquid state, their spontaneous separation during the cool-down of our 50-50 alloy proceeds as follows:

Point A: Crystals of pure antimony start to form. As the temperature drops, the crystals grow larger—the solidifying metal always being pure antimony. As the antimony solidifies from the meld, the meld becomes richer in lead.

Line AB: The composition of the meld varies as the material cools from A to C.

Point B: At temperature C, the meld has composition B (13% antimony–87% lead). This alloy is called the eutectic. The temperature remains at 228 °C (443 °F) until all of the eutectic has solidified. The antimony crystals formed here are much finer than those formed along line AC.

In Figure 5.8, composition X represents the larger antimony crystals (A) formed along line AC in Figure 5.7. Composition Y represents the eutectic formed at B (Figure 5.7), with a representing the smaller antimony crystals and l representing the lead crystals. This is our *matrix*. Composition Z represents the final structure, as if we placed composition X on top of composition Y and pressed the two together to form composition Z. Composition Z may be compared to a concrete aggregate with a being the sand, l being cement, and A being the gravel.

Referring to our 95-5 alloy in Figure 5.7, pure lead starts to separate at point D [293 °C (560 °F)], thereby enriching the meld in antimony. Our eutectic still forms at B, the composition of the eutectic always being 13% Sb and 87% Pb. To analogize, substitute L (lead) for A (antimony). Composition Y, of course, remains the same.

In the 50-50 alloy, the ratio of eutectic to antimony can be determined by using the eutectic as a fulcrum:

$$\frac{87 - 50}{87} \quad (\times\ 100\ =\ 42.5\%\ \text{antimony}\ (57.5\%\ \text{eutectic})$$

Classes of Alloys

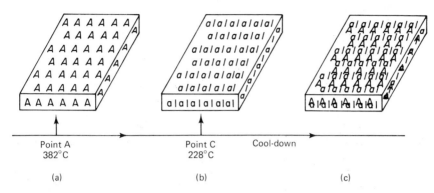

Figure 5.8 Analogy for Figure 5.7: (a) composition X; (b) composition Y; (c) composition Z.

For the 95-5 alloy:

$$\frac{13-5}{13} \quad (\times 100 = 61.5\% \text{ lead (38.5\% eutectic)}$$

Class III Alloys

We have chosen three lead-tin alloys for our Class III example. We have chosen them because of similarities with the Class I and II alloys, and for the important phenomenon that structural changes do occur while in the solid condition (below the solidus line). Coupled with our study of Class I and II alloys, this Class III alloy will ease our way into an understanding of the iron-carbon alloys that follow.

A few salient points should be understood before tracing alloy in Figure 5.9.

1. Solid tin will not dissolve any lead.
2. Lead will hold tin in solution: 15% at 184 °C (365 °F) but only 2% at room temperature.
3. The eutectic of lead-tin occurs at 184 °C (365 °F) with a composition of 66% tin and 34% lead.
4. Where the eutectic in Class II was a mixture of pure metals, the eutectic of lead-tin is composed of fine crystals of pure tin and fine crystals of lead that contain (hold in solution) the 15% tin of item 2.
5. Recall that no structural changes occurred in the solid condition in Class I and II alloys. (The homogeneity transformation shown in Figure 5.6 is not a structural change.)
6. Particularly note that in this Class III alloy a structural change takes place along the *JLM* curve, wherein the solid solution of *J* precipitates its tin until at *M*, it holds only 2% tin in solution.

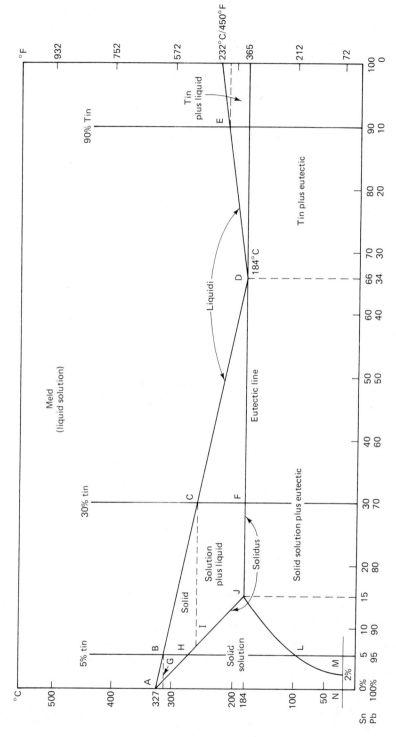

Figure 5.9 Constitution diagram of Class III lead-tin alloys.

7. The precipitation above takes a relatively long time; hence we can speed up the cool-down and arrest or retard the precipitation.

Ninety percent tin alloy. Tracing this alloy in Figure 5.9, we see:

Point E: Pure tin starts to solidify. The meld is therefore richer in lead and reaches the composition at D (66% Sn-34% Pb). This composition has a higher percentage of lead than we started with because of the loss of the tin that separated by solidifying.

Point D: This composition is our eutectic, which will freeze at the constant temperature 184 °C (365 °F). Unlike our previous eutectic (Class II), this eutectic is not composed of the two pure metals in combination but is composed of pure tin crystals and lead-tin crystals of composition J (15% tin).

Points J-L-M: The lead-tin crystals formed at D will precipitate their tin as the solid cools to room temperature. Thus the percentage of tin held in solution can be traced on the *JLM* solubility curve from a maximum of 15% at 184 °C to a maximum of 2% at room temperature.

The final structure of the 90% tin alloy, if allowed to cool naturally, will be composed of the large crystals that separated out of the meld from E to F, the fine crystals that formed at D (in the eutectic) and the even finer crystals that formed along the *JLM* curve. In our final structure, our tin (except 2%) takes three forms: crystals formed along EF, crystals formed at D, and crystals precipitated along *JLM*. Our lead is in the crystals of the eutectic at D but now contains only 2% tin in solution.

Thirty percent tin alloy. Solidification begins at point C. The first crystals to form will be the lead-tin composition at point I. As the solid cools from C to K, the crystals will grow, and their average composition will vary from I to J. They are nonhomogeneous, similarly to the Ni-Cu crystal of Figure 5.6. Their centers are composition I and their outer boundaries are composition D. At D, the remaining meld will solidify as the eutectic. As the solid cools from K to room temperature, the lead-tin crystals will precipitate their tin along the solubility curve *JLM*.

Five percent tin alloy. Solidification begins at point B. The composition of the first solid will be G, and solidification is complete at point H. The solid will cool without change to point L, where the solid solution will precipitate its tin along the curve *LM*.

IRON-CARBON ALLOYS

Although iron has come down to us as the symbol of strength, as in the expression "a man of iron," it is a relatively soft, weak metal very similar to copper.

By the addition of a little carbon, however, we can change this soft, weak metal to any hardness we wish up to about 9.5 on Mohs' scale of hardness (between sapphire and diamond) and increase the strength to 105 kg/mm^2 (150,000 psi) without adding an element other than carbon. This extreme hardness is obtained by saturating the iron with 6.67% carbon to obtain pure iron carbide (Fe$_3$C). The effect of very small amounts of carbon can be appreciated by the fact that a rolled steel of up to 0.15% carbon has a tensile strength of about 60,000 psi and a Brinell hardness of 110, but a rolled steel of 1.05% carbon has a tensile strength of 150,000 psi and a Brinell hardness of 300 (again without adding other elements). By special heat treatments made possible by the allotropic natures of iron and carbon, we can harden and toughen steels containing relatively small amounts of carbon, yet soften and weaken cast irons containing larger amounts of carbon.

Before studying the complex iron-carbon equilibrium diagram of Figure 5.10, we should take note of several salient characteristics of our iron-carbon alloy.

1. Iron is allotropic; that is, it exists in more than one form, alpha iron (bcc) and gamma iron (fcc) being our prime concerns.
2. Carbon is allotropic, existing as carbon, diamond, and graphite.
3. As with our Class II and III alloys, the iron-carbon alloy has a eutectic.
4. Our iron-carbon alloy has a *eutectoid*, so called because it appears to be a eutectic and behaves as a eutectic.
5. Iron cannot hold carbon in solution in the solid state (less than 0.015%, that is).

As we study the diagram, we shall need to know or be acquainted with the following:

1. The eutectic occurring at 4.3% carbon at 1130°C (2066°F).
2. The saturation point of gamma iron at 1.7% carbon at 1130°C.
3. That austenite is gamma iron (fcc) and dissolved carbon.
4. The eutectoid formed at 700°C (1292°F) is composed of ferrite and cementite which separated from the austenite. This alloy is called *pearlite*, due to its appearance when ruptured.
5. Cementites (also called iron carbides, or Fe$_3$C) are unstable and will decompose to graphite under controlled cooling.

Reading the equilibrium diagram (Figure 5.10), we find:

1. Alloys from 0.0 to 4.3% carbon solidify along the line *AB* by separation of solid-solution (austenite) crystals from the liquid.
2. Alloys with more than 4.3% carbon begin to solidify with the separation of cementites from the liquid along the *BD* line.

Iron-Carbon Alloys

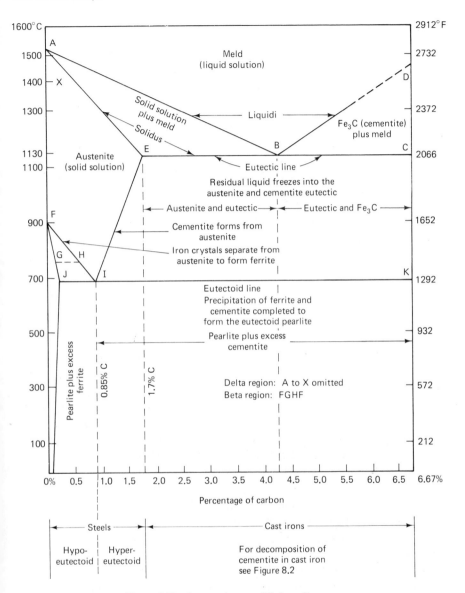

Figure 5.10 Iron–carbon equilibrium diagram.

3. Alloys with 1.7% carbon freeze out austenite from *AB* to *AE*.
4. Area *AEIF* is an alloy in a homogeneous state, austenite.
5. Alloys with 1.7 to 4.3% carbon freeze-out austenite beginning at *AB* and continuing to *EB*. The remaining liquid then solidifies as the *eutectic*. This eutectic is composed of alternating layers of cementite and austenite in a finely divided state.
6. Therefore, alloys containing 1.7 to 4.3% carbon just below the *EB* line

are mixed crystals of austenite and eutectic. The composition of this mix is from 0.0% eutectic and 100% austenite in the case of a 1.7% carbon steel to 100% eutectic and 0.0% austenite in the case of a 4.3% alloy.

7. In alloys with more than 4.3% carbon, cementite freezes out on *BD* to *BC*. Residual liquid at *BC* is eutectic and solidifies as eutectic. Thus alloys just below *BC* are composed of mixed crystals of cementite and eutectic.

8. Alloys with 0.85% carbon are pure eutectic (pearlite). *Note:* In alloys discussed previously, no appreciable changes occurred below the solidus line except diffusion in a heterogeneous solid-solution phase. In iron–carbon alloys, however, changes occur below the *EBC* line due to the allotropic phenomena of iron and carbon: gamma iron can change to alpha iron, and carbon can change to graphite.

9. In area *AEIHFA* (except the delta region) and area *EBCKIE*, iron is gamma iron. Gamma iron dissolves carbon. Its grain size depends on temperature, time, and working. It is nonmagnetic and is denser than alpha iron. Alpha iron has a body-centered cubic lattice, is magnetic, dissolves little carbon, and its grains do not change under normal conditions.

10. On the ferrite curve, pure gamma iron exists from *X* to *F*. At *F*, gamma iron changes to beta iron (bcc). Beta iron can exist only to *G*, where it changes to alpha iron. This alpha iron can hold in solution considerable amounts of elements such as nickel, silicon, phosphorus, and so on, but very little carbon (less than 0.015%). *Ferrite* is not pure iron; it is a solid solution in which alpha iron is the solvent. Along the *FHI* line, alpha iron (now ferrite) separates from gamma iron at 900 °C (1652 °F), but when carbon is in solid solution with gamma iron (austenite), the temperature of separation is 700 °C (1292 °F) for an alloy containing 0.85% carbon. Thus the ferrite separates from austenite along line *FHI*.

11. The cementite curve shows that at the eutectic temperature [1130 °C (2066 °F)], austenite holds 1.7% carbon in solid solution. Cooling below 1130 °C, cementite precipitates from the austenite along line *EI*. The maximum solubility at 1130 °C is 1.7% carbon and at 700 °C (1292 °F) is 0.85% carbon (see point *I*).

12. Thus, as noted above, at the eutectoid line (regardless of the percentages of carbon started with) austenite at 700 °C (1292 °F) can contain only 0.85% carbon (on slow cooling, of course). But cooling below 700 °C, all of the solid solution (austenite) separates into crystals of ferrite and cementite.

Note that steels containing less than 0.85% carbon are structured as a pearlite matrix in which free ferrites are embedded and that alloys with more than 0.85% carbon are composed of pearlite and free cementites (except, of course, where cementites are decomposed into graphite). We can determine the percentages of each constituent by using the eutec-

Heat Treatment of Steel

toid as a mathematical fulcrum—for example, for structural steel having 99.75% iron and 0.25% carbon:

$$\frac{0.85 - 0.25}{0.85} \times 100 = 71\% \text{ free ferrite, } 29\% \text{ pearlite}$$

Salient points to remember regarding iron–carbon alloys are:

1. Ferrite is a solid solution with alpha iron as the solvent.
2. Cementite is a compound, Fe_3C, between sapphire and diamond on Mohs' scale of hardness.
3. Pearlite is a combination of ferrites and cementites similar to the eutectic in Figure 5.8b.
4. Steels of 0.85% carbon are pure pearlite (the toughest).
5. Steels below 0.85% carbon are composed of pearlite in which free ferrites are embedded (see Figure 5.8c).
6. Steels with more than 0.85% carbon are composed of pearlite in which free cementites are embedded (see Figure 5.8c).
7. Ferrites make steel softer and more ductile, but weaker.
8. Cementites make steel harder and stronger, but more brittle.

HEAT TREATMENT OF STEEL

Understanding Austenite

The understanding of austenite is the focal point from which all understanding of the heat treatment of steels must follow. Austenite is gamma iron (fcc) and carbon in solid solution.

We are all familiar with salt and water in solution; we call it *brine*. We may also call it *salt water* or *saline solution*. We know that when the water freezes, the salt is forced out of solution and the water freezes as salt-free ice. By analogy, we can then visualize that the water is iron and the salt is carbon, and the two are in solution in the liquid state. The carbon, however, is not frozen out of solution upon solidification (freezing) of the steel. The carbon is retained in solution with the iron in a solid state. We call this a *solid solution* (sometimes redundantly referred to as *solid-solution austenite*).

Figures 5.10 and 5.12 show that a 1.2% carbon steel begins to freeze at 1430°C (2605°F) and is solid at 1300°C (2372°F). The structure of the iron is face-centered cubic (gamma iron), which allows room for the carbon atoms to dwell. The diagrams also show that at 840°C (1544°F) the iron again changes to a bcc structure (alpha iron) and the carbon is forced out of solution. There simply is not room for the carbon atoms to rove about among the iron atoms.

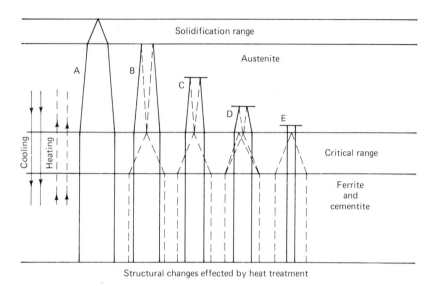

Figure 5.11 Structural changes effected by heat treatment.

They must migrate to the crystal boundaries, and there each carbon atom combines with three iron atoms to form the compound Fe_3C (also called cementite or iron carbide).

At the top of the austenitic range (Figures 5.11 and 5.12), the crystals (grains) grow rapidly, then progressively more slowly down to the top of the transformation range (also called the *critical range*). This growth, however, can be controlled in several ways:

1. We can stop the cooling at some point in the austenitic range, and the grain growth will stop. As cooling is resumed, grain growth will resume but at a slower rate.
2. We can *hot-roll* the metal as it cools through the austenitic range to whatever point we wish.
3. We can *forge* the metal as in point 2.
4. We can allow the metal to cool normally to its condition in Figure 5.11 and then reheat the metal to a desired point and control the cooling.

Hot finishing to the top of the critical range results in maximum *refining*. (*Finishing*, to the steelman, means *making fine*.)

Figure 5.13 summarizes the consitution and heat treatment of iron–carbon alloys.

Control of Steel Properties

The control of the physical properties of steel by heat treatment is based on changing the conditions under which the transformations take place (refer to Figures 5.11 and 5.12).

Heat Treatment of Steel

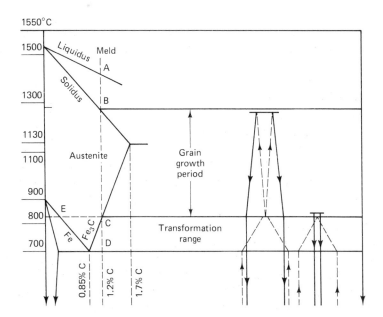

Figure 5.12 Ordinary structural composition at selected temperature of heat treatment.

1. Heating to the top of the critical range destroys the grain structure, as indicated by the convergence of dashed lines at the top of the critical range.
2. Heating above the critical range, the grains begin to grow as indicated by the diverging dashed lines.
3. Grains do not grow at temperatures below the top of the critical range (see the solid lines).
4. If we heat as in Figure 5.11E, grain structure is destroyed and re-forms as a fine grain.
5. If we heat as in Figure 5.11B, the grain structure is destroyed at the top of the critical range; then grains grow as heating continues into the austenitic range, and grow even more as the metal cools.
6. By comparing graphs A through E in Figure 5.11, we can see that grain structures may be controlled as desired by varying the reheat within the austenitic range.

Note: The dashed lines denote heating; the solid lines denote cooling. The space between the dashed lines and the solid lines below the critical range show the degree of change in size of grains.

From Austenite to Pearlite

In passing from austenite to combinations of ferrite–pearlite and pearlite–cementite, there are four stages.

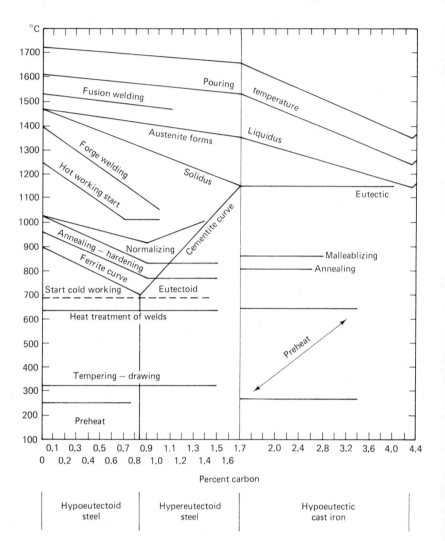

Figure 5.13 Constitution and heat treatment of iron–carbon alloys.

1. *Austenite:* Soft, ductile, tough, wear-resisting. A true solid solution of carbon in gamma iron.
2. *Martensite:* The first stage of the decomposition of austenite, characterized chiefly by its physical properties. Under the microscope it looks like austenite. Martensite is the hardest constituent in steel except cementite; it is very brittle. Its best use is in such items as razor blades, paper cutters, and so on.
3. *Troostite:* The decomposition of austenite into cementite and ferrite is complete, but these constituents are in a very fine state of division. Troostite

is not as hard as martensite but is tougher. A typical use is in files.
4. *Sorbite:* Cementite and ferrite are small particles and can be distinguished under the microscope. Sorbite is not as hard as troostite but is tougher. At this stage steel has the highest yield point. A typical use is in spring steel. It is at this stage that the excess cementite or ferrite above the eutectoid begin to segregate.

Note: The terms *troostite* and *sorbite* have given way to the terms *bainitic structure* and *bainite* (see the Glossary), but many old-timers are certain to carry them along for a few more years.

5. *Pearlite:* The eutectoid structure (see point *I*, Figure 5.10). This structure is normal in slowly cooled steels. It is the toughest and has the highest breaking strength of any of the steel structures. Excess ferrite or cementite segregates into individual grains apart from the pearlite grains.

Since these changes occur in the solid solution, the transformations require time. Therefore, if we can lower the temperature rapidly enough, we can bring the steel to ordinary (room) temperature before the transformations can go to completion and get a steel in one of the intermediate stages above pearlite. There are two ways in which these unstable structures above pearlite may be obtained.

1. The rate of cooling may be so adjusted that the changes can progress only to the state desired. This method is, in many cases, impossible.
2. We can cool the steel as rapidly as possible and then reheat it to the temperature that will allow only the desired change to take place. The result would be austenitic, martensitic, or troostitic, and the tempering would then allow the steel to assume the troostitic, sorbitic, or pearlitic state.

Heat-Treatment Processes

Stress relieving. Steel undergoes a uniform heating below the critical range followed by uniform cooling. This treatment is the most common treatment used by welders. The range of temperatures is from 150 °C (300 °F) to as high as 650 °C (1200 °F) for cast steel, cast iron, and welded structures. Some residual stresses are even relieved at room temperature, but this relief may take months. If the 650 °C (1200 °F) temperature is used, we must cool slowly in a furnace down to 425 °C (800 °F), then cool in air. Temperature and time are the factors to be considered. We must never reach the critical temperature. (*Note:* Stress relief is all too often erroneously referred to as *annealing, tempering,* or *normalizing.* We should be more careful with our choice of words, as all of these treatments have their differences. *Tempering* and *drawing* do have the same meaning.)

Annealing. Heat to 40 °C (104 °F) above the transformation range and hold there (called *soaking*) long enough for the carbon to distribute itself uniformly throughout the austenite. Soaking should be 1 hour for each 2.5 cm of thickness. Cool in furnace or cover with some insulating material. Annealing is used to refine the grain (see Figure 5.11) to remove stresses, to soften steel for easier machining, and to prepare for cold working.

Normalizing. Similar to annealing, but faster. Heat to 40 °C (104 °F) above the critical range, hold there very shortly, and cool in air. Annealing sometimes makes low-carbon steels too soft and gummy for machining. Normalizing leaves steel just about right for free machining.

Spheroidizing. Heat just below the transformation range for a number of hours and cool slowly throughout the upper part of the cooling range. The cementite collects into tiny globular or spherical particles, leaving a matrix of ferrite. Reheat again to put the carbides into solution in the austenite. If desired, the steel may then be hardened by quenching, since spheroidizing makes high-carbon steels softer and more easily machinable than does annealing.

Quenching. This is the rapid cooling from an elevated temperature by immersion in or blanketing with a liquid, gas, or solid medium that absorbs the heat or conducts it away. The speed with which the metal is transferred from its heat source to the quenching medium and the efficiency of the quenching medium both affect the cooling rate. The cooling rate and time control the depth of hardness. (See the discussions of water, oil, and air hardening of tool-and-die steels in Chapter 7.)

Quench hardening. Heat to 40 °C (104 °F) above transformation range and hold there long enough for the center to become austenitic (overheating and prolonged heating may cause grain growth; see Figure 5.11). Quickly cool in oil, water, or other medium (see the successful quench shown in Figure 7.2). Full hardening can convert the austenite to martensite all the way to the center. Retempering will reduce the possibility of stress cracks. Remove immediately from quench to furnace.

Tempering. Heat to some point below the transformation range and cool in air or water. This process relieves internal stress and toughens the metal, but it also reduces the hardness. The degree of hardness governs the temperature to which steel is reheated. Refer to Chapter 7 for typical tempering of air-, oil-, and water-hardening steels. Tempering and drawing are the same, but *tempering* is the preferred term.

Martempering. See Chapter 7.

Austempering. See Chapter 7.

Case hardening. *Casing* is any hardening process that hardens the surface layer of metal to a greater degree than the *core*. The best known product of case hardening is the *case knife*, wherein the cutting edge and exterior are hard and somewhat brittle but the softer core allows flexibility. Case hardening may be done by carburizing and quenching, cyaniding, nitriding, and induction or flame hardening. (Case hardening is a form of differential hardening).

Carburizing. This process is the addition of carbon to a solid ferrous alloy. The metal is heated in contact with a carbon material—solid, liquid, or gas—to a temperature above the transformation range and held at that temperature. The metal is then quenched. Soaking in the carbon material under heat allows the carbon atoms to migrate to the center of the metal.

Cyaniding. Case hardening is accomplished by heating in molten cyanide. The metal absorbs carbon and nitrogen simultaneously. Quenching produces a harder, deeper case.

Nitriding. Case hardening is accomplished by heating in an atmosphere of ammonia, or in contact with other material containing nitrogen.

Flame Hardening. This process is a local or full hardening by heating with an oxyacetylene or other flame and quenching in a jet of water.

SUMMARY

As we mentioned in our introduction, this chapter should be studied in conjunction with Chapter 7, and, perhaps, with each chapter on a specific group of alloys.

At this point, if you have browsed through this chapter and have found something of interest, that something should be pursued for a better understanding. Browse on through other chapters and then return to this one for another casual study.

If you have studied this chapter diligently and are thoroughly confused, congratulations. You must know a great deal about such a complex subject to be confused, and to be confused is to be on the brink of a breakthrough.

How to study the diagrams. Do as we have done with Figures 5.11 and 5.12. Note that the *effect-of-heat* diagram (Figure 5.11B) has been superimposed on Figure 5.12. As our dashed lines heat up our 1.2% carbon steel, the grain structure is destroyed at 800°C (as indicated by the convergence of dashed lines) and the grains begin to grow as we heat up into the austenitic range (diverging dashed lines). At the same time, everything that happened during cool-down on the left half of Figure 5.12 has been undone. The eutectoid of ferrite and cementite and the excess cementites have lost their carbon and are back to austenite (gamma iron and carbon in solution). When our heat is removed, cool-down begins, and the grains begin to grow again (diverging solid lines) and the steel again pursues its course as before. Ferrites form, cementites precipitate out of austenite, the eutectoid is re-formed at 700°C, and the excess cementites are again sifted throughout the eutectoid matrix.

Note that the space between 0.85% carbon and our 1.2% carbon shows a difference of 0.35% C. This 0.35% represents the carbon that is combined (Fe_3C) in those cementites that roam free in the eutectoid matrix (pearlite). Note that the space between a 0.7% carbon steel and the eutectoid shows a difference of 0.85% minus 0.7%, or 0.15% carbon. This difference means that we did not have enough carbon to form a pure pearlite (eutectoid) and some of our alpha iron must exist as free ferrites within the pearlite matrix just as did the free cementites in the 1.2% carbon steel.

Referring to Figure 5.10, we see that a 0.3% carbon steel at room temperature is a hypoeutectoid structure consisting of free ferrite and pearlite. No structural changes occur as the steel is heated up to the *JI* line. At the *JI* line, the pearlite changes to a fine-grained austenite, the alpha iron changing to gamma iron. This change affects only the original pearlite (about 35% of the structure). Above the *JI* line, the free ferrite also recrystallizes from alpha iron to gamma iron, and this gamma iron is absorbed by the austenite. The recrystallization remains fine-grained up to the *FHI* line. Above the *FHI* line, grain growth sets in and continues up to the *AE* line; the slower the heating, the greater the grain growth.

During a normal slow cool-down, no change occurs until cooling to the *FHI* line. At the *FHI* line, the austenite begins to precipitate fine ferrite crystals to the boundaries of the austenite crystals. This precipitation continues to the *JI* line with a gradual growth in the size of the ferrite grain. The austenite grain size is thereby reduced. At the *JI* line, the remaining austenite recrystallizes to form pearlite (a fine lamellar mixture of ferrite and cementite crystals). At slightly below the *JI* line, no further changes occur; but if the structure is cooled very slowly, the ferrite grains may enlarge and cementite layers in the pearlite may spheroidize.

Figure 5.10 also shows us that a 1.5% carbon steel consists of pearlite and free cementites. When heated, no change occurs below the *IK* line. At the *IK* line, the pearlite changes to austenite, with its alpha iron converting to gamma iron and its cementites dissolving into the fine-grained austenite. The excess cementite is unaffected at this temperature. Further heating up to the *IE* line, however, dissolves the excess cementite into the austenite. Very little grain growth occurs during heat-up from the *IK* line to the *IE* line. At the *IE* temperature, all of the cementite is dissolved and further heating above the *IE* line results in rapid grain growth.

In order to read that portion of Figure 5.10 to the right of the 1.7% carbon point, study Figures 8.1 and 8.2 and their accompanying explanations.

6

IRON-CARBON STEELS

According to *Webster's Intercollegiate Dictionary*, steel is a "commercial iron that contains carbon in any amount up to about 1.7% as an essential alloying ingredient." The *Welding Encyclopedia* does better: "an iron-carbon alloy which may or may not contain other alloying ingredients aside from those which appear as impurities." The carbon percentage of 1.7% as the cutoff point for steels, which appears in most definitions and on most constitutional diagrams, has been surpassed by a later breed called *graphitic steels*, steels which, but for the highly specialized production techniques, would be cast irons. There are literally hundreds of steels, and for a thorough understanding of steel, the welder should thoroughly study metallurgy for welders, welding tool-and-die steels, and welding the stainless steels, which are the subjects of Chapters 5, 7, and 9.

For the purpose of this chapter, steel has been divided into three main categories: low-, medium-, and high-carbon steel.

LOW-CARBON STEELS

Low-carbon steel is our most useful and abusable metal. With it, we have built ships for peace and war, bridges for our widest rivers, skyscrapers for our offices and dwellings, machinery for our factories, and millions of appliances, gadgets, and toys—some useful and some nonsensical. We have bent it, twisted it, drawn it, hammered it, punched holes in it, sheared it, heated it white hot and plunged it in cold water, and subjected it to uncountable other tortures, all without corrupting its docile nature.

Certainly, then, the welder would expect to have little or no trouble with this obedient servant. Unfortunately, however, this docile human creation must also obey certain rules laid down by nature, and if the welder wishes to understand the sometimes contrary behavior of the many varieties of steels, he would do well to begin his understanding of metallurgy with this least recalcitrant of all steels.

Steel is an alloy of iron and carbon, and may be additionally alloyed with other elements, such as nickel, chromium, manganese, molybdenum, and others. We classify low-carbon steels as those having less than 0.30% carbon and only the minutest amounts of other alloying ingredients. These steels contain minute amounts of "impurities," such as oxygen, nitrogen, hydrogen, sulfur, phosphorus, alumina, and others—some unavoidably, some purposefully.

1. *Oxygen:* It is paradoxical that this essential element is both friend and enemy. Under normal conditions, steel is one of our metals least able to defend itself against this destroyer of all things. Other metals may have a greater affinity for oxygen, but their oxides form a protective coating over the metal that resists further oxidation. Iron oxide (rust) does not seem to protect iron from further oxidation. The iron oxides in steel subject the steel to shrinkage-cracking, causes porosity in the weld, and lowers the ductility.

2. *Nitrogen:* The amount of this element unavoidably contained in cast or rolled steel is less than 0.30% and has little or no effect on weld metal. Nitrogen is an abundant constituent of air and should not be introduced either purposely or accidentally.

3. *Hydrogen:* This element constitutes 11% of water by weight. The amount normally found in steel does not affect welding. Excess amounts cause cracking during cool-down from the molten state.

4. *Phosphorus:* Steel usually contains less than 0.30% and presents no problem for the welder. Most low-carbon steels are well below the high of 0.30% (0.01 to 0.02%). Phosphorus is increased in some low-carbon low-alloy steels to increase tensile strength. If welding high-phosphorus low-carbon steel, use low welding current and fast travel.

5. *Sulfur:* Although this element promotes cracking and porosity, amounts less than 0.035% do not affect weldability. Excess amounts of sulfur are often introduced in steels to increase machinability. These free-machining steels should be welded with low-hydrogen-type electrodes.

6. *Selenium:* This element has the same effects on steel as those of sulfur with respect to welding and machinability.

7. *Silicon:* This element is both friend and enemy. The amounts usually found in steel (0.15 to 0.35%) actually aid in making sound welds; but on the high side (toward 0.35%), surface holes are likely to occur when using E-6010 and E-6011 electrodes.

8. *Alumina* (actually aluminum oxides): The small amounts usually present

show no appreciable effect when welding. Over 1%, however, causes an annoying surface film while gas welding.

One of the laws of nature that metals must obey is that matter must expand when heating up and contract when cooling down. This expansion-contraction phenomenon is probably the single greatest concern for the welder. A simple experiment may be used to demonstrate why the welder who does not understand these opposing forces does not always get what he wants in the completed weldment.

1. Fix a piece of flat bar (about 1 × 12 × 1/4 in.) in a vise and heat one side rapidly in one confined area only, and observe that the bar will bend toward the cold side.
2. When the heat is removed, the bar will bend back in the opposite direction and will bend in the reverse direction beyond its original alignment.

This reverse bend occurs because the expansion on the one side compresses the metal on the opposite side, and the heated metal has expanded beyond its elastic limit in compression, and is "set" (cannot return to its original dimension). In cooling, the shrinkage must equal the expansion, and the result is a bent bar. If the heating is continued until it is equally distributed in the bar, the stress will be equalized and the bar will not be bent. (For more about control of distortion, see Chapter 13.)

MEDIUM-CARBON STEELS

At this point, the maintenance welder may say: "How am I to know how much carbon is in the steel? I don't know who made it or where it came from."

There are several means of giving an educated guess: the spark test, the file test, the sound test, the hammer test, and the torch test, to name a few. These tests can only be reasonably accurate—based on the welder's experience and basic knowledge of metals. No attempt will be made to paint pictures with words. The welder must observe the differences in various steels and carry that picture in his mind, and occasionally refresh his memory as he works in the trade.

1. *Spark test:* Every welding book written probably has a picture showing the different spark patterns given off by different steels under the grinder. These pictures are factual and, perhaps, useful, but no picture can equal the actual comparison of steels by the welder on his shop's grinding wheel. If at all possible, he should experiment on the wheel with known types of steel. In fact, the harder the steel, the higher the carbon; in theory, the softer the steel, the deeper the bite of the wheel and the longer the spark.
2. *File test:* The softer the steel, the more the file bites the steel; the harder the steel, the less the cutting effect. On very high carbon steels, especially water-hardened steels, the metal may wear the file.

3. *Sound test:* Strike the steel with a hammer, lightly, as the clapper of a bell. The duller the sound, the softer the steel; the harder the steel, the more ring. This applies to iron–carbon steels; other alloying elements such as nickel and manganese profoundly affect the vibration frequency.
4. *Hammer test:* Using the sharp point of the slag pick, strike the various specimens with equal blows and compare the depth of the compression (similar to the Rockwell test). This test can only be reasonably accurate and then only if the welder has a very flexible wrist and a delicate sense of power. Place the two metals side by side, establish a pendulumlike rhythm by several blows on a third piece, then, using the same power, strike the two metals to be compared.
5. *Torch test:* The accuracy of this test depends on the welder's ability to see the scintillation of the carbon particles when heated. Using the oxyacetylene torch with a neutral flame, heat a spot to a dull, then a bright cherry red. The carbon particles can be observed as very minute scintilla. The higher the carbon, the more the scintillation. The difference between very low carbon steels and structural steels is very pronounced. Experienced welders can also see the differences in hardness as they flame-cut various steels.

Medium-carbon steels range from a low 0.30% carbon to a high of about 0.5% carbon. It is with these steels that the welder encounters his first real challenges to the making of a successful weld. He should realize now that the more the carbon, the more likely the failures. It is with these steels that the welder's knowledge of metallurgy will be a definite advantage over the less knowledgeable welder.

If we take a 0.5% carbon steel for our example, we can see from Figure 5.13 that our welding puddle will reach a temperature of 1535°C (see fusion-welding range). The center of the puddle will be much higher.

Figure 5.10 shows that, under normal cool-down, our 0.5% carbon steel begins to solidify at about 1510°C and remains as austenite as the cool-down progresses to 756°C. During this period, the grains grow in size and provide room for the carbon to migrate. At 756°C free ferrite (alpha iron) begins to separate out of solution and will join with the pearlite formed at 700°C, giving us a relatively tough "structural" steel with a composition of pearlite (65%) and ferrite (35%).

But can the welder control the cool-down to approximate the normal cool-down? The example above is for a mass of steel that is uniformly heated and cools slowly in the ambient air. The welder's molten steel, however, is not only cooled by the ambient air but is also cooled by the parent metal of the weldment, which extracts heat from the weld metal much more rapidly than air. Thus the time for grains to grow, carbon to migrate, pearlite to form, and cementites to compound from the displaced iron atoms and the precipitated carbon is greatly reduced.

Medium-Carbon Steels

As the welder contemplates this problem, he can quickly deduce that it is not likely that the metal he has deposited and the parent metal he has remelted will be of the same pearlite-ferrite structure. If his weld cools too quickly, some of the carbon, if not all, will be trapped in the austenite; and he may well wind up with martensite—certainly, an undesirable composition for structural steel.

All 0.5% carbon steels are not "structural" steels. Some are hardened by heat treatment to produce steels for such uses as tools and dies. If the welder encounters a hardened 0.5% carbon steel, he must not only be concerned about the cool-down but also with heat-up.

These hardened steels are not the normal pearlite-ferrite structure but are the martensite that we have suggested the welder has created in his weld in the paragraph above. We say that these metals are in an "unstable" condition (strained) because they are anxious to return to their natural composition of pearlite-ferrite. In this unstable condition, the metal is likely to crack under sudden heat from the electrode or torch. The metal cracks because it does not have the ductility necessary to withstand the stresses caused by unequal heating. Some structures are in such an unstable state that when the welding engineer cautions a group of welders: "Don't even let your electrode or holder make a mark on the walls," they must not believe that he has lost his senses—he means it!

As a general rule, there will be no problem with the medium-carbon structural steels if ordinary welding precautions are observed. If the steel being welded is thought to be on the higher end of the carbon range, the better practice is to go to the low-hydrogen type of electrode. Specifications, however, sometimes call for an E-6010 root pass. In this event, the second pass should burn through the root pass.

Figure 6.1 shows the probable result of the welder's simultaneous deposit of filler metal and the meltdown of the parent metal using a low-carbon-type electrode. The hardened part will be just below the fusion line. This area is subjected to two hardening processes:

1. The metal here passes through the pearlite formation range (700 °C) (1390 °F) more quickly, and the softer pearlite has less time to form.
2. This area has been at a high heat longer, thus causing the grains to grow. (A coarse-grained structure hardens more readily than does a fine-grained structure.)

This hardening is not always considered a bad effect; it may even be desirable and may be called the "strengthened zone." The ill effects, however, may outweigh the good.

1. The machinist does not like hardened zones because they may damage the cutting tool.

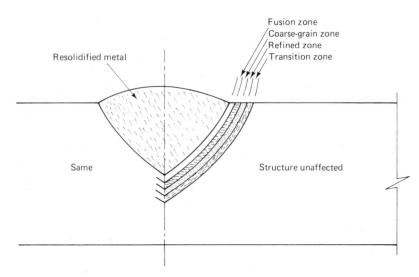

Figure 6.1 Deposit of filler metal and meltdown of the parent metal.

2. Hardening tends to produce stresses that may cause cracking in service.
3. If cooling is too rapid, the weld may crack several hours after welding.

CONTROLLING HARDNESS

Since hardness results from the quenching effect that the parent metal has on the weld metal, we can readily see that the hardening can be reduced by slowing up the cool-down. Since the quenching effect occurs because the parent metal is "cold," the obvious solution is to heat up the parent metal. The more of the parent metal that is heated, the slower will be the cool-down.

There are three ways we can heat up the weldment.

1. Use the higher weld-current range for the size of electrode used and slow down the travel so that the parent metal has time to absorb the heat.
2. Preheat the parent metal. (Weigh the cost of preheat against faster weld time.)
3. Use multiple-pass welding. Each pass preheats for the subsequent pass and tends to anneal or temper the preceding pass (not necessarily true on relatively long welds).

Note: Again we have dealt in generalities. The welder must see at this point that the thicker the section to be welded, the greater will be the quenching effect. Preheating should take place under the following conditions.

1. *With steels under 0.2% carbon:* never on plate less than 1 in. thick and rarely on plate more than 1 in. thick

2. *With steels from 0.2 to 0.4% carbon:* never on plate less than $\frac{1}{2}$ in. thick and sometimes on plate over $\frac{1}{2}$ in. thick
3. *With steels of more than 0.4% carbon:* always, on all thicknesses

HIGH-CARBON STEELS

These steels (those with more than 0.45% carbon) cannot be welded by the same procedures used in welding mild steel but are not considered "difficult to weld" by welders with a basic understanding of metallurgy.

Some characteristics of high-carbon steels are:

1. They do not stretch as readily as mild steel.
2. They harden under a relatively slow cooling rate.
3. They have a lower transformation range (thus they have a larger heat-affected zone).
4. They tend to become porous.

To make a successful weld:

1. Deep penetration is undesirable. Proper joint preparation reduces the need to melt down the parent metal. Low-hydrogen electrodes (with iron-powder coatings) have less digging action. Some penetration is, of course, necessary and if kept low will give no trouble on steels below 1.0% carbon. Above 1.0%, however, enough carbon may be picked up from the parent metal to dangerously reduce the ductility of the weld metal. With these 1.0 + % steels the better practice is to go to the high-alloy electrodes, such as E-309 or E-310. On larger welds where electrode cost is a factor, the joint may be buttered with the high-alloy electrode and filled with a lower alloy type. The E-309 and E-310 types have a minimal penetration characteristic, show very little carbon pickup, remain austenitic regardless of rate of cool-down, and stretch considerably before exerting a stress on the parent metal.
2. Deposit relatively large beads, avoid concave beads, and fold back craters.
3. Preheat and/or postheat as necessary (see Chapter 7).

Some individual problems may be encountered and are listed here together with their solutions.

1. *Crater cracks:* Do not leave craters at ends of beads, fold back on bead; use side swing for joining restarts.
2. *Root crack:* Cracking of the first (root) bead will carry up through subsequent layers at least through the second layer; and entire weld may "tear" later. Avoid root cracks by (a) preheating, (b) changing joint design, or (c) leaving members free to move (see Figures 13.15, 13.16, and 13.20).

3. *Underbead crack:* Weld "peels" away from parent metal. Preheat, or weld slowly so that the weld cools slowly, or butter with high-alloy E-309 or E-310.
4. *Radial crack:* Cracks occurring at right angles to the fusion zone and extending down into the parent metal. The solution is the same as for item 3.
5. *Porosity:* Melted high-carbon steel readily absorbs hydrogen, carbon monoxide, nitrogen, and so on. If using E-6010 or E-6011 electrodes, weld slowly to give gases time to escape from the molten puddle. Low-hydrogen types practically eliminate porosity.
6. *Hardening of parent metal:* Some hardening is inevitable unless weldment is preheated and kept at uniform heat during welding. The alternative is to heat-treat after welding.

Note: The welding of high-carbon steels is covered more comprehensively in Chapter 7.

STICK-WELDING CARBON STEELS

If the welder is not relatively expert in stick welding, a study of shielded metal arc welding is recommended, with an emphasis on welding techniques and arc action. Except for some of the higher-carbon steels, the following selection of electrodes should cover most of the ordinary carbon steel applications. The specialized electrodes will be covered under specific metals.

E-6010. This is an all-purpose, all-position electrode for the lower-carbon and low-alloy steels, featuring a relatively thin coating, spray-type transfer across the arc, an excellent digging and purging action, and a fast-freeze solidification. These characteristics result in a weld bead that is somewhat irregular in appearance and flat in contour, whose overall appearance and soundness depend on the skill of the welder in controlling the penetration, travel lines, and weld metal distribution. The thin coating, proven to provide adequate shielding of the weld area, permits good visibility of the boiling effect at the center of the puddle and does not demand a slow, stay-in-the-puddle travel or weave. Hence, one of the attributes of the electrode most liked by experienced welders is the fact that it can be whipped in and out of the puddle on troublesome vertical-up welds and whipped ahead of the bead to clear away a galvanized or painted surface. Since no slag blanket forms on the tip of the electrode, it is excellent for tacking. Its ductility makes for easy hinging-off of tacked-down *dogs, hangars,* and other fitting devices. E-6010 is used where welds of high quality are needed, such as shipbuilding, construction of bridges and buildings; and on tank, pipe, and pressure-vessel work. Its general utility in maintenance work is unsurpassed. The E-6010 functions properly only on dc reverse polarity. Its straight-polarity performance is so poor that it can be used for cutting all metals by setting the current 25 to 50% higher.

Stick-Welding Carbon Steels

E-6011. This is an all-purpose, all-position electrode having similar characteristics to the E-6010, and may be considered an E-6010 adapted to ac application. Its rate of deposition is slightly higher than that of E-6010 in proportion to power consumption. Although developed for alternating current, it may also be used on dc reverse polarity. It is often preferred to E-6010 for thin-gauge metals and for some thin-to-thick applications due to its typical ac arc action. The E-6011 of some electrode manufacturers so closely resembles the pipe rod, E-6010P, in arc action, that some pipe welders will choose it over the E-6010 when E-6010P is not available.

E-6012. The E-6012 is a somewhat specialized electrode for the lower-carbon steels, featuring a relatively rapid deposition rate and an almost complete absence of spatter. Although the numbering would indicate it to be an all-position electrode, welders generally find it troublesome on overhead butt welds and vertical-up fillets. It is designed for ac and dc straight-polarity application but may be used on reverse polarity. It is excellent for machinable buildups of worn areas. It is less ductile than E-6010 and E-6011, and has a higher yield strength. It is excellent, also, for most steel fabrication, where the weldment may be positioned for all flat or horizontal welding. Because of its shallow penetration, joints must be more open for good weld metal distribution. The larger diameters are susceptible to arc blow when used with dc. Although the E-6012 is sometimes called an *idiot rod* by highly skilled welders, its ease of application can prove deceptive to the not-so-skilled welder. Because of the heavy slag blanket that freezes before the weld metal completely solidifies, the welder must be able to visualize the quality of weld bead underneath the cover. As in all heavily coated electrodes, he must be able to discern between the slag and the metal while both are in the molten state.

E-6013. This ac DCSP electrode is quite similar to the E-6012. The coating is heavier, deposition faster, bead contour flatter, and the ripple more closely packed. Although it is designated as all-position, it has the same problems with overhead butts and vertical-up fillets. It has replaced E-6012 in most high-speed production work but has not supplanted it in buildup applications. Easy arc starting, easy slag removal, and good arc stability make the smaller diameters excellent for thin-gauge steels. Its lack of digging, purging action makes it unsuitable for welding through galvanized and other coatings, and unsuitable for general maintenance work. It may be used on dc reverse polarity.

E-7014. The E-7014 is an ac-DCSP/DCEN electrode of the E-6012 and E-6013 class, but it is more fluid and is designated as a fast-fill electrode as compared to the fill-freeze types E-6012 and E-6013. The coating of the E-7014 is much thicker, due to its iron-powder content. This iron powder becomes a part of the weld deposit, hence the deposition rate of E-7014 is much higher. It is excellent for the lower-carbon steels in production welding on plate of medium thickness.

E-7018, E-7015, and E-7016. We have lumped these three low-hydrogen electrodes together because the E-7018 has all of the characteristics of the others and has virtually eliminated them from the welding scene. The E-7018 coating contains a high percentage of iron powder, which becomes part of the weld deposit. Since it is a low-hydrogen electrode for high-quality welds, a very short arc is essential to its purpose. In controlling the arc, we should never whip the arc and should keep it in the puddle at all times. Limited weaving is permitted, but the weave must be slow and steady. The arc should be so tight that we can feel the coating lightly contacting the weld pool. In vertical-up welding, the arc should never be pulled away from the weld pool but, instead, should be pushed into the notch of the fillet. To be successful with the E-7018, we must be able to discern the weld metal from the slag while both are in a molten state (the slag is more fluid, brighter, and remains liquid longer). E-7018 should be used on medium- to high-carbon steels (*caution:* see Chapter 7) and on low- to medium-alloy steels. They are also best for the free-machining steels. By adding alloying constituents to the coating, these low-hydrogen electrodes have attained greater strengths, such as the E-8018 to E-12018.

E-7028. Characteristically, a low-hydrogen fast-fill electrode featuring a higher iron-powder content in the coating (50%) and a faster deposition rate than the E-7018 group. It should be used only in the flat and horizontal positions. The penetration is shallower, and the weld bead is flat to concave.

GAS-WELDING CARBON STEELS

Oxyacetylene welding of steel is usually confined to thin sections, small-diameter pipe, and brazing. Braze welding is often advantageous on certain medium- to high-carbon steels that do not take kindly to a remelt with the torch, such as office furniture, bicycle and tricycle fenders, auto bodies, and others. A general-purpose, low-melting, nickel-bearing filler rod is excellent for braze-welding these steels, especially those steels which except for their thinness, the maintenance welder would arc-weld with E-309 or E-310. Their tensile strengths range as high as 100,000 psi. The oxyacetylene torch is also used for many wearfacing applications on the higher-carbon steels, especially the thin layers of many of the carbides (see "Sweating," and the discussion of powder brazing, Chapter 2). Except for the sweating technique ($1\frac{1}{2}$ to $2\times$ flame) and contrary recommendations by the rod manufacturer, use a neutral flame, bearing in mind that, except for the brasses and bronzes, an oxidizing flame stirs up more troubles than does the slightly carburizing flame.

TIG-WELDING CARBON STEELS

All of the weldable carbon steels may be welded by the TIG process. TIG finds applications similar to those in gas welding and is used extensively for high-quality welds on pipe and pressure vessels. Its deep penetration ideally suits it to the root pass on pipe. With or without a consumable backup ring, joints can be designed that permit a skilled welder to approach the quality of a machine weld in the smoothness of the *burn-through* of a *root opening*. On many highquality welds, the root pass is made with TIG, and the joint then filled and covered with stick or MIG.

MIG-WELDING CARBON STEELS

The MIG process produces sound, clean welds on carbon steels and is by far the most economical where large deposits are made (see Chapter 4). Selection of shielding gas and arc action are discretionary. Table 6.1 is offered as an introductory guide. Table 6.2 may be used discretionally as a suggestive guide in electrode selection and special treatment in the welding of carbon and low-alloy steels.

TABLE 6.1 Gas and Arc Selection

Thickness of Metal	Shielding Gas	Arc Action[a]
3 mm ($\frac{1}{8}$ in.) and up	Inert	Spray
3 mm ($\frac{1}{8}$ in.) and up	CO_2	Buried arc, flux core
Light gauge	Inert	Pulsed arc
20 gauge to 6 mm ($\frac{1}{4}$ in.)	CO_2	Short arc

[a]Buried arc and flux core are recommended for flat and horizontal welding only; all others are all-position.

If a complete analysis of the steel is available, bear in mind the following.

1. As the carbon content increases above 0.25%, the tensile strength rating of the electrode increases and the more necessary becomes the preheat and postheat treatments.
2. As the percentage of the other alloying ingredients are increased, the more consideration must be given to their constituency, such as Class 3115 (1.10 to 1.40% Ni) and Class 3316 (3.25 to 3.75% Ni). These constituents increase the carbon equivalent.
3. The suffixes of low-hydrogen electrodes indicate the electrode's approximation to the actual composition of the metal being welded. Since exact matches are not always available, use the suffix that gives the closest approximation.
4. Because of the interplay of alloying ingredients, the necessity for special treatment of welds can be adjudicated more precisely than in note 1 above, by obtaining the carbon equivalent from the formula

TABLE 6.2 Electrode Selection

Series	Class	Electrode	Treatment for Best Results
1008 to 1025	Nonsulfurized carbon steel	E-60xx to 70xx	None required
1027 to 1050	Nonsulfurized carbon steel	E-7015, 16, and 18	Preheat 150–260 °C
1050 to 1095	Nonsulfurized carbon steel	Same as for 1027 to 1050, or stronger	Preheat 210–325 °C Posttreat: anneal (maximal); stress relieve (minimal)
11xx	Sulfurized carbon steel (free machining)	E-7015, 16, and 18, or special	Exercise care to minimize porosity and cracking
13xx	Low alloy (manganese)	Low-hydrogen 70xx or stronger with suffix A-1, D-1, or D-2	Preheat 125–150 °C Postweld: stress relieve
23xx	Low-nickel steels	Low-hydrogen 80xx to 100xx, preferably with suffix C-1 or C-2	Preheat if carbon content is over 0.15% Postweld: stress relieve
25xx	Low-nickel steels	Same as for 23xx, with suffix C-2	Same as 23xx
3100 to 3300	Low-alloy Ni-Cr steels	Low-hydrogen 80xx or 90xx with prefix C-1 or C-2	Thin sections: no preheat Heavier sections: Preheat 0.20% carbon 95–150 °C and higher carbon 315 °C Posttreat: anneal (maximal); stress relieve (minimal)
40xx	Low-alloy Mn–Mo steels	Low-hydrogen E-80xx or stronger	Preheat heavier sections if carbon exceeds 0.25% and stress relieve
41xx	Low-alloy Cr–Mo steels	Low-hydrogen E-80xx or stronger	Preheat 315 °C Posttreat: anneal (maximal); stress relieve (minimal)

Series	Class	Electrode	Treatment for Best Results
43xx	Low-alloy Ni–Cr–Mo steels	Low-hydrogen E-80xx plus; match suffix to Cr content	If carbon exceeds 0.25%, preheat to 400 °C Posttreat: anneal (maximal); stress relieve (minimal)
46xx	Low-alloy Ni–Mo steels	Low-hydrogen E-80xx plus; match suffix to Ni content	Same as for 31xx
48xx	Low-nickel steels	Same as for 46xx	Same as for 23xx
5xxx to 5xxxx	Low-alloy Cr steels	Low-hydrogen E-80xx plus; suffix B	Same as for 43xx
61xx	Low-alloy Cr–V steels	Low-hydrogen E-80xx plus	Same as for 43xx
8xxx	Low-alloy Cr–Mo	Same as for 61xx	Same as for 41xx
92xx	Low-alloy Si steels	Same as for 61xx	Preheat to 400 °C Posttreat: anneal (maximal); stress relieve (minimal)
93xx	Low-alloy Ni–Cr steels	Low-hydrogen 80xx plus; match to Ni content	None, generally
94xx	Low-alloy Ni–Cr steels	Low-hydrogen 80xx plus	Same as for 43xx
97xx to 98xx	Low-alloy Ni–Cr steels	Same as for 94xx	Preheat 400 °C Posttreat: anneal (maximal); stress relieve (minimal)

$$CE = \%C + \frac{\%Ni}{15} + \frac{\%Cu}{15} + \frac{\%Mn}{6} + \frac{\%Mo}{4} + \frac{\%Cr}{5}$$

As the carbon equivalent (CE) exceeds 0.40%, the higher the CE, the more necessary is special treatment.

If the maintenance welder does not have the proper selection of electrodes, or cannot give his weldment special treatment, or finds that his low-hydrogen electrodes are not doing the job, most of the so called "problem" steels can be adequately welded with E-309 or E-310 electrodes, which have tensile strengths up to 67 kg/mm^2 (95,000 psi) and elongations of 35% and 45%, respectively. It should be borne in mind, however, that the necessity for heat treatment becomes increasingly important as the carbon content exceeds 0.25% or the CE exceeds 0.40%. Other high-nickel-chrome electrodes of a balanced austenitic-ferritic structure, such as Marco 200, Eutectic 680, and others, give tensile strengths up to 84 kg/mm^2 (120,000 psi) and elongations in excess of 35%.

With the exception of the free machining (high-sulfur) steels, all of the steels can be welded by oxyacetylene, TIG, or MIG, and all can be brazed or braze-welded. Exceptionally strong braze welds are possible with the nickel-bearing brazing alloys.

WROUGHT IRON

Although wrought iron cannot be called steel and should not be classified with iron-carbon alloys, it is included in this chapter because the welding techniques involved do not differ greatly from the welding of low-carbon steels.

True wrought iron is becoming increasingly rare. It is a relatively pure iron containing up to 3% ferrosilicate, sometimes called "slag," which increases rust resistance and strength. This slag appears in the iron as a stringy thread-like fiber threaded through the iron in the direction of the rolling compression used in working it. Wrought iron is readily weldable, but in fusion welding, some care should be exercised to ensure a thorough fusion.

1. The fusion temperature will be in the range 1480 to 1540°C (2700 to 2800°F), and the welder should not be taken in by the fluxy, greasy appearance that first appears in the range 1150 to 1200°C (2100 to 2200°F) and occurs due to the fluxing action of the slag content.
2. A molten puddle should be maintained constantly, and the deposition of the electrode or rod should be *in the puddle.* In SMA welding, this means keeping the arc in the puddle at all times; and in torch welding, it means that the filler rod is fed into the center of the molten puddle.
3. The molten puddle should be disturbed as little as possible. Avoid stirring and rubbing of the puddle with the electrode or filler rod.
4. The higher heat required for fusion welding wrought iron may be obtained

Wrought Iron

by reducing the amperage and travel speed. Simply stated, stay in the puddle longer and boil out some slag.

5. Avoid introduction of carbon into the weld area from the electrode composition and/or a carburizing torch flame (bearing in mind, of course, that an oxidizing flame is even more detrimental).

Since much of the silicate slag is necessarily boiled out in order to get a sound weld, the weld area will not be as oxidation resistant as is characteristic of wrought iron.

Wrought iron is most positively identified by the nature of its rupture, which will occur similarly to our toughest woods, such as hickory. No heat treatment is necessary. Use the lowest-carbon steel electrodes and filler metals that are available. E-6010 or E-6011 should suffice.

7
TOOL-AND-DIE STEELS

Although a basic understanding of metallurgy is not necessarily a prerequisite to the successful welding of tool-and-die steels, such an understanding not only will reduce the chances for failure but also will make the welder's approach to the problem more interesting than mysterious. Therefore, a study or a review of Chapter 5 is recommended, with an emphasis on the iron–carbide equilibrium diagram (Figure 5.10) and the heat-treatment diagram (Figure 5.13). The tool-and-die steels are those ranging from 0.5 to 2.5% carbon. (Some tools are made of the lower-carbon steels, but these do not present a particular problem for the welder.) Additionally, the tool-and-die steels are often alloyed with other elements, principally chromium, molybdenum, tungsten, vanadium, cobalt, and manganese. These alloying elements reach as high as 30% of the steel's constitution.

SLOW (NATURAL) COOL-DOWN OF HIGH-CARBON STEELS

A review of the natural cool-down of the iron–carbon alloys will show that:

1. Upon completion of solidification at 1130 °C (2066 °F), the austenite will hold 1.7% carbon in solution, and cooling down from this temperature, Fe_3C compounds (also called cementites or iron carbides) precipitate out of solution.
2. AT 700 °C the maximum solubility is 0.85% carbon.
3. If an alloy contains more than 1.7% carbon, the cementite decomposes into iron and graphite, both of which soften and weaken the steel.

4. At 900 °C iron crystals precipitate out of solution and join with the cementites to form the eutectoid pearlite at 700 °C.
5. At 700 °C alloys containing more than 0.85% carbon have a structure of pearlite and cementite (up to 1.7% C) or a structure of pearlite–cementite–graphite (beyond 1.7% C).

From the review above, the welder should retain the following:

1. Pearlite is tough. It is a "eutectoid" of ferrite and cementite.
2. Ferrite is soft, ductile, and lower in tensile strength.
3. Cementite is hard, brittle, and raises tensile strength but reduces ductility. It is a compound of iron and carbon.
4. Graphite is soft. It softens and weakens the structure. It is an allotrope of carbon and is formed by precipitation from a supersaturated austenite. It is an excellent antigalling agent.

If tool-and-die steels range from 0.5 to 2.5% carbon, how do we account for the high tensile and compressive strengths of a tool-and-die steel of 2.5% carbon when the cementite of alloys between 1.7 and 2.5% carbon decomposes into iron and graphite, both of which weaken and soften the structure? It would appear that with all the free graphite available in a 2.5% carbon alloy, we should have a gray cast iron.

The answer is, of course, that we cannot allow the natural changes to take place. If these changes require time (and they do), we can speed up the cool-down and interfere with the natural development of the structure. We also have the option of introducing other alloying elements that inhibit those natural transformations.

The transformation range is divided into three intermediate structures between austenite and pearlite.

1. *Martensite:* In this first stage of the decomposition of austenite the transition occurs suddenly and is not discernible under the microscope. It is the hardest constituent of steel except cementite. It is the most important single constituent in our study of welding tool-and-die steels.
2. *Troostite:* The decomposition of austenite into ferrite and cementite is completed but in a finely divided state. It is not quite as hard as martensite. This transition is also very sudden. As a comparison, we may point out that troostite is common to the manufacture of files and martensite to razor blades.
3. *Sorbite:* Cementite can be distinguished under the microscope. This structure is typical of spring steel.

Where the transformations from austenite to martensite and martensite to troostite are quite sudden, the changes from troostite to sorbite to pearlite are not clearly defined—they merge gradually. We need mention or treat with the structures other than martensite in order to point out that they are the natural course of a cool-down and they are structures that we wish to avoid.

Referring to Figure 7.1, we see that by heating to the top of the transformation range, the grain structure is destroyed (converging dashed lines) and a cool-down from there results in a fine-grained steel because the grains cannot grow at temperatures below the top of the transformation range. Figure 7.1A–C shows that heating well into the austenitic range causes the grains to grow (expanding dashed lines) and that the grains again grow during the cool-down to the top of the transformation range (expanding solid lines). Thus the higher we heat above the transformation, the coarser will be the grain structure.

The American Society for Testing Materials classifies steels as 1 through 8 on the basis of number of grains per square inch at 100 × magnifications:

1: Up to $1\frac{1}{2}$
2: $1\frac{1}{2}$ to 3
3: 3 to 6
4: 6 to 12
5: 12 to 24
6: 24 to 48
7: 48 to 96
8: 96 and more

If martensite is the structure we want for our tool-and-die steel, and if we can obtain this structure by a rapid cool-down, it would seem that we could pour our steel, allow it to solidify as austenite, hot-work it to the top of the transformation range, and then cool it so quickly that only the martensitic transition could take place. This method, however, presents several problems.

1. The massive bulk of the ingot, slab, billet, or plate would make the necessary 1-second cool-down impossible except on the outer surface of the mass.
2. The hard, brittle structure of martensite would make the shaping of the steel into tools and dies very difficult because of its resistance to deformation and machining.
3. The majority of our tools and dies are composites; that is, only part of the tool or die need be hard, and in many cases we want a tool or die with a hard, wear-resistant edge and a tough shock-resistant backup (austenite).
4. Many composites are made of two or more steel structures welded together.

The simplest procedure becomes one of selecting the alloy we want, allowing it to cool down naturally, forming the tool or die while our alloy is in

Slow (Natural) Cool-Down of High-Carbon Steels

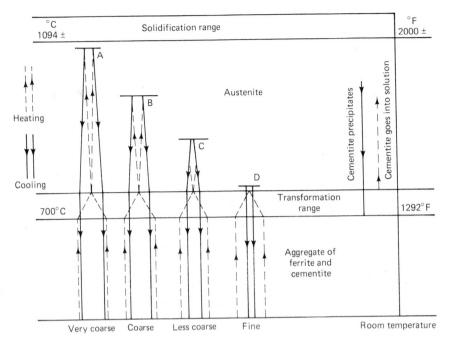

Figure 7.1 Steel transformation range.

this natural state, then heat-treating the tool or die to obtain the desired structure and hardness, and the desired depth of that hardness.

This hardening procedure is accomplished in two operations.

1. Heat the steel above the critical range as shown in Figure 7.1.
2. Quench the steel back to room temperature too quickly for the softening transformations to take place (see Figure 7.2).

In the heating process, the alpha iron again becomes gamma iron which will hold carbon in solution. We call this solution "austenite" (see Figure 7.1). It is the same austenite that we started with in the initial cool-down. Steels of less than 1% carbon become pure austenite, steels of more than 1% carbon become a structure of austenite and free cementites—the percentage of free cementites depending on the amount of carbon in the steel and the temperature to which it is heated. Heating the steel has accomplished two things: We have dissolved the cementite and refined the grain structure (see Figure 7.1).

Hardening the steel is accomplished by cooling it so rapidly that we reach the M-point without any change in the austenite. If our cooling is successful, at the M-point the austenite transforms to martensite—a hard, brittle structure in an unstable condition. The martensite retains the carbon or cementite in a dissolved or nearly dissolved state. Although martensite is said to be the first stage of decomposition of austenite, the microscope cannot discern the finely divided state.

The true structure of martensite and the reasons for such hardness are something of a mystery, but from our study of metallurgy, we know that metals are soft and weak because of the ease with which the crystals deform by the slipping or sliding of the atomic layers within the grains. Therefore, we can deduce (or surmise, or theorize) that the greater resistance along the grain boundaries is enhanced by refining the grain structure. The more grains we have per square inch, the more times the slipping will have to change direction per unit of its linear travel. We know also that the precipitated cementites in such a finely divided state disperse themselves along the slip lines and interfere with the slipping in much the same way that a grain of sugar or salt will interfere with the sliding of a deck of cards that is fanned out on the table by a card-sharp. This slip resistance is similar to that of the precipitation hardening of aluminum–copper alloys ($CuAl_3$). The unstable condition of the structure also contributes to its hardness.

Although the cooling rate is critical, the transformation to martensite does not occur abruptly at 94 °C, as shown in Figure 7.2, but over a range of temperatures dependent on the composition of the steel. Some austenite (1 to 30%) may be retained at room temperature. This residual austenite may then be transformed by tempering or by cold-treating (chilling to -20 to -40 °C).

Since all steels do not transform at the same temperature, the point at which a steel begins to transform is usually shown on diagrams as the steel's

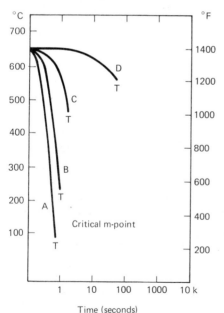

Figure 7.2 Expected result in a 0.85% carbon steel when quenched to point T and allowed to proceed at room temperature. A, martempering (60+ Rockwell C); B, austempering (56 Rockwell C); C, fine pearlite (about 45 Rockwell C); D, medium pearlite; T, transformation begins.

M-point. For example, a steel of 0.80% carbon finds its M-point at 260 °C (500 °F) and the transformation is 80% complete at room temperature, resulting in a structure of 80% martensite and 20% austenite. We must transform the remainder by tempering or by cold treatment. This retention of austenite is especially true of low-alloy steels and special high-alloy steels.

All steels do not require the same speed of cooling to reach their maximum hardness. We classify them as to their quenching medium.

1. Carbon steels generally require a fast rate of cooling, such as a water quench. We call them "water-hardening steels."
2. Many of the low-alloy steels may be hardened at a slower rate and are quenched in oil. We call these "oil-hardening steels."
3. Some high-alloy steels have a very slow rate of transformation and are cooled in the ambient air. We call them "air-hardening steels."

Liquid quenching (oil or water) happens in three stages.

1. Hot steel is immersed in the bath and the liquid produces a vapor that envelops the steel. This envelope of vapor keeps the cooling liquid from contact with the hot metal and slows up the cool-down. This vapor must be broken up.
2. The vaporous envelope collapses and exposes the hot metal to the liquid, and boiling action begins. This agitation speeds up the cooling.
3. As the temperature of the hot metal approaches the boiling point of the bath, the cooling action is slowed up again. This stage is of little consequence since the maximum transformation takes place in the second stage.

The first stage is the cause of most failures, such as soft steel, soft spots, warping, and cracking. That is why the heat-treater will be seen to move the steel to and fro in the bath if the steel is hand-held by tongs. Whatever the method used, the liquid should be agitated during the cooling.

If water-quenched, the bath becomes more efficient by the addition of about 5% sodium chloride. Sodium hydroxide (3 to 5%) is even faster.

The oil quench is best for thin sections (knives or razor blades) and is also best for those steels having a slower transformation rate and deep hardening (many of the alloy steels). Oil cools steel much more slowly in the third stage, reducing the danger of internal stresses, warping, and cracking. Since the third stage is below the M-point for most steels, there are no undesirable effects. Oils are different and should be selected carefully as to flash point, boiling point, density, and specific heat. The oil-bath volume should be computed as 8 liters of oil for 1 kg of steel for 1 hour.

Hot-quenching treatment is usually done if hardening and tempering are to be accomplished in one operation, as in martempering and austempering.

HARDNESS OR TOUGHNESS?

Hardening, if successful, gives the following:

1. Smallest grain size
2. Maximum hardness
3. Minimum ductility
4. Internal stresses and strain (unstable structure)

Tempering gives the following results:

1. Increased toughness
2. Decreased hardness
3. Relief of stresses
4. Stability of structure
5. Change of volume

Tempering

Quench-hardened steel is in an unstable condition. Normally, austenite undergoes its first stage of decomposition into martensite at or near the top of the transformation range. Successful quenching prevents this transformation until the steel has reached almost room temperature. This martensite is also known as alpha martensite and has a tetragonal atomic structure (thus strained). In obedience to the laws of nature, such martensite is impatient to change to its more natural pearlitic structure and will do so if given the opportunity.

Heating this alpha martensite above 95 °C (205 °F) changes the structure back to the more normal body-centered cubic lattice of ordinary cool-downs (as in alpha ferrite). This structure is called beta martensite. Heating beyond 95 °C (205 °F), the beta martensite begins to precipitate carbon in cementite form noticeably at just above 205 °C (400 °F). Such precipitation of the in-solution carbon decreases the steel's hardness: for example, decreasing a 1.10% carbon steel from a 60 Rockwell C at 150 °C (300 °F) to a 56 Rockwell C at 315 °C (600 °F) and a 35 Rockwell C at 595 °C (1100 °F).

Generally, the longer the time of soaking at the given tempering temperature, the better the results. Soaking seems to release the locked-in stresses and increase the plasticity and toughness without reducing the hardness appreciably.

Tempering is best carried out by liquid baths, such as oil, salt, or lead, or in an air-tempering furnace. The more precisely controlled is temperature and time, the more successful the tempering.

Table 7.1 lists various temperatures for tempering together with the oxide color associated with each temperature. These colors *are not* the color of the heated steel. To use these color guides, we must first polish a portion of

Hardness or Toughness?

TABLE 7.1 Typical Temperatures for Tempering

°F	°C	Oxide Color	Some General Uses of Steels
430	220	Faint straw	Taps, lathe tools, twist drills
460	240	Dark straw	Punches and dies, flat drills
500	260	Bronze	Hammer faces, rock drills, shear blades
540	280	Purple	Axes, wood-carving tools
570	300	Dark blue	Chisels, knives
610	320	Light blue	Springs, screwdrivers, wood saws
630	335	Steel gray	Springs; cannot be used for cutting tools

the steel. The changing colors are caused by the change in the thickness of the oxide film.

It must be remembered that the purpose of tempering is not to reduce hardness but to increase toughness. The ideal would be to increase toughness without decreasing the hardness. When tempering the welder should consider the following:

1. Tempering above 200 °C (390 °F) reduces hardness.
2. Tempering near room temperature increases the hardness, usually.
3. Hardness increases with time (aging) when tempered at the low temperatures. This is due, probably, to some retained austenite transforming to martensite.
4. Although tempering at temperatures above 200 °C (390 °F) tends to decrease hardness, toughness does not increase progressively over the entire tempering range. The range from 260 to 315 °C (500 to 600 °F) may show a loss of toughness (called the "brittle temperature range"). Chrome-nickel steels with relatively high carbon content become brittle if tempered at or near 620 °C (1150 °F).
5. The longer the time the steel is kept at the temper, the better are the results. Increasing the time at lower temperatures is better than increasing the temperature and shortening the time.
6. Conversion to martensite results in an increase in volume. Tempering tends to bring the volume back to the original size. (Change in volume is one of the stress raisers.)
7. The maximum hardness of steel is pure martensite at a 67 Rockwell C. But this hardness is too brittle for practical use.
8. A 56 Rockwell C is a good combination of hardness and toughness, with some practical applications of a 60 Rockwell C where toughness is less critical.

Martempering

Heat the steel in the usual way to the austenitic range, but instead of water- or oil-quenching, quench in a hot-salt bath. Keep the steel in the bath long enough for the temperature to equalize throughout the steel. Remove the steel from the bath and cool it to room temperature in still air. Temper by conventional methods.

Martempering differs from austempering (discussed in the next section) in that no changes occur in the austenite until the steel is lower in temperature than the quenching bath. The austenite is quenched at a rate equal to or faster than the critical quench rate (see M-point, Figure 7.2) to a temperature above 205 °C (400 °F). The time in the bath should be long enough for the steel to approximate the temperature of the bath. During the cool-down from near 260 °C (500 °F) to room temperature, the austenite will transform to martensite.

The temperature of the quenching bath should be kept slightly above the M-point of the steel being tempered. The M-points for different steels may vary but satisfactory martempering results from keeping the bath at 205 °C (400 °F).

Martempering is preferred over austempering when maximum hardness is desirable, and is much more easily accomplished with heavier sections. Both result in more stable structure (less strained) than that attainable by the more conventional method.

Austempering

By the usual methods of quench hardening, we can obtain only two structures of uniform hardness in carbon steels: a 40 to 42 Rockwell C or a 64 to 65 Rockwell C. In order to get a 56 Rockwell C (tougher, more plastic) we must first harden our steel to a 65 Rockwell C, then temper it to the 56 Rockwell C.

If we can successfully undercool austenite to 288 °C (550 °F), the austenite will transform to a very fine pearlite with a hardness of 56 Rockwell C. This direct tempering of austenite requires that it be held at 288 °C (550 °F) for about 1 hour. This process is called "austempering." It is much tougher than steel tempered to 56 Rockwell C by the usual method described above. Additionally, by never going to the martensitic stage, the structure is more stable.

Austempering, however, has the problem of how to prevent the transformation at 540 °C (1000 °F). Austempering is done with a hot quench (molten salt) and it is a problem to quick-quench through the 540 °C (1000 °F) range rapidly enough when using a hot quench. The heavier the section of the steel, the more difficult is the successful quench. (Knives, pins, razor blades, and needles are no problem.)

TYPES OF TOOL-AND-DIE STEELS

As a best rule, the maintenance welder working with tool-and-die steels should follow the recommendations of the manufacturer of the product. In the ab-

Types of Tool-and-Die Steels

sence of these recommendations, he must rely on his knowledge of steels and how to weld them, and how to treat them after welding. It is hoped that by his study of the heat treatment of carbon steels to this point, he has gained a relatively comprehensive understanding of tool-and-die steel that will reduce his chances for failure when welding them and, even more important, will enable him to understand why he failed and to correct his fault on his next try.

Tool-and-die steels are divided into groups corresponding to their method of hardening.

Water-Hardening Steels

The W-group steels vary in carbon content from 0.50 to 1.5%. Some are straight carbon, some are alloyed with chromium (Cr) or vanadium (V): chromium for heat and corrosion resistance, vanadium for grain refinement, resistance to grain growth during heating, and fatigue resistance. These steels are shallow hardening (differential hardening) or case hardening. The hard outer case makes these steels wear resistant on the outside and shock resistant at the core, and especially adapts them to use in percussion and impact tools such as cold header dies and shear blades. They are suitable only for cold work because of their low red hardness. These steels are often classified as to "temper," which reflects their carbon content: that is, 5 (0.5% carbon) to 15 (1.5% carbon). Best known or most used are the tempers 9 through 11, used for blanking dies, trimming dies, and so on. On the other end, 5 to 8, they are most used where toughness and shock resistance are more important than abrasion resistance. On the high end, 10 through 15, they are used where hardness, abrasion resistance, or edge strength are necessary, such as in form cutters and engraving tools. With a successful quench the austenite should be almost totally untransformed at 232 °C (450 °F) and convert to martensite at that temperature. The quickly formed case (shell) of martensite insulates the core and slows the cooling of the core so that it becomes a ferrite–pearlite–bainite structure. The lower the carbon, the deeper will be the case hardness. These steels are subject to a change of volume since martensite occupies more space than austenite.

Oil-Hardening Steels

The O-group steels are alloyed with 0.9 to 1.5% carbon, 1.0 to 1.6% manganese, and a small percentage of chromium or molybdenum or tungsten in some grades. They are medium-deep hardening when oil-quenched and full-hardened by heating above the critical range and immersing in oil. They are subject to internal stresses, but these stresses are relieved by tempering after quenching. They are suitable for dies and other tools having sharp changes in mass. Other characteristics are lower machinability, low red hardness, lower impact resistance, and are limited to cold work. Their structure is ferrite plus carbides, with virtually no carbon in solution. Transformation to austenite begins at 743 °C (1370 °F). Above 743 °C (1370 °F) more carbon goes into solution and the grains coarsen.

The upper limit of heating should be 788°C (1450°F). Although slower, oil quenching through the range 732 to 482°C (1350 to 900°F) traps the austenite without precipitation of the carbon. Untransformed austenite reaching 260°C will begin to transform to martensite. Since martensite occupies more space than austenite, internal stresses are locked in and must be relieved by tempering. Additionally, if the steel is removed from quench to tempering furnace, cooling after tempering will cause residual austenite to transform to untempered austenite, creating the danger of cracking.

Air-Hardening Steels

A group. These steels are uniformly deep-hardening with only air quenching. They do not distort during heat treatment. In the annealed condition, their machinability is 65% of that of water-hardening tool steels. They are used for blanking and forming punches and dies, extrusion and drawing dies, and so on. This group has 1 to 1.5% chromium and 1% molybdenum. The addition of 3% manganese increases impact strength measurably, and raising the chromium content to 5% gives medium red hardness.

D group. Some of these steels are really oil-hardening and, perhaps, we should classify them as high-carbon high-chromium die steels. A typical analysis would be 1 to 2.25% C, 12% Cr, and 0.8 to 1% Mo (or 0.9 to 1% V). These steels harden from 982 to 1010°C (1800 to 1850°F) and require long soaking. They harden slowly and do not distort during heat treatment, but their martensitic structure is highly stressed, resulting in low impact strength and susceptibility to thermal cracking. They are unmachinable when hardened. Grinding must be done carefully, since the high abrasion resistance causes heat during grinding that may cause thermal cracking. These steels are subject to carbide precipitation during heating.

High-Speed Steels

Steels of the M and T groups have high red hardness and are especially adaptive to metal-cutting tools that become hot in service, in such uses as drills, taps, milling cutters, and reamers. As their name indicates, they keep a sharp edge in spite of the frictional heat generated by high-speed metal removal. They have a balance of red hardness, abrasion resistance, and a high resistance to shock and vibration. Basically, their structure is a composite of wear-resistant carbides suspended in a tough heat-resistant matrix. They are deep-hardening and have little or no distortion during heat treatment. Most of them are oil-quenched, but some may be air-quenched (only the M group has this option).

Typical analysis of the tungsten–carbide group (T group) shows 0.70 to 1.50% C, 12 to 18% W, 4.0 to 4.75% Cr, and 1 to 5% V. Some may have 5 to 8% cobalt to give a super red hardness at a sacrifice of toughness. Some may have 0.75 to 1.0% molybdenum.

Typical analysis of the molybdenum type (M group) shows 0.80 to 1.15% C, 4.5 to 8.5% Mo, 4.0 to 4.25% Cr, and 1.0 to 5.0% V. Tungsten may be present up to 6.5% and cobalt may be present in some grades from 5.0 to 12%. The M group is more sensitive to heat treatment and overheats easily. The non-cobalt M grades are tougher than those containing cobalt.

All tool steels are subject to decarburization at elevated temperatures. Significant amounts of chromium, cobalt, or silicon contribute most seriously to this loss of carbon at the surface. The welder should be aware that this decarburization is a function of time and is significant only when the steel is soaked for a long time above 760°C (1400°F).

Hot-Working Die Steels

The H-group steels are characterized by medium to high impact strength, medium-high to high red hardness, and other special properties.

1. *Chrome types:* 0.35 to 0.45% C; 5.0 to 7.0% Cr; 1.5 to 5% Mo or 1.5 to 7% W or 0.4 to 1.0% V. Nickel may be used to reduce chromium content. These steels are air-hardening. They are used for high-impact work such as in mandrels, hot shears, and forging dies. They are easily welded.
2. *Tungsten types:* 0.25 to 0.5% C; 9.0 to 18.0% W; 2.0 to 4.0% Cr. Chromium may go as high as 12%. They have high resistance to softening and deformation at high temperatures but have lower impact strength than the chrome types. They are used generally in punches, shear blades, and extrusion and casting dies. They deep-harden in air or oil, but air hardening promotes scale. They tend to crack from thermal shock. They cannot be water-cooled in service and must be preheated before use at higher temperatures.
3. *Molybdenum types:* 0.55% C; 5.0 to 8.0% Mo; 4.0% Cr; up to 6.0% W; 1.0 to 2.0% V. Their tendency to decarburize makes tight control of hardening temperature critical; otherwise, their character is that of the tungsten grades.

Low-Alloy Tool Steels

The L-group steels need no particular caution for the welder except to consider their carbon content of 0.5 to 1.2%.

Mavericks and Renegades

Some tool-and-die steels are called mavericks and renegades only because they do not fall under the standardization systems. Their performance may go as high as excellent. One such tool steel is the graphite type, so-called because of the free-graphite constituent. The controlled free-graphite content gives good machinability in the annealed condition, high abrasion resistance after hardening, high resistance to frictional wear, an excellent antigalling property, good resistance to stress cracking, and good weldability. The group includes air-, water-, and oil-hardening types.

MAINTENANCE WELDER'S VIEW OF TOOL-AND-DIE STEELS

Since the exact composition and the characteristics of tool-and-die steels often is not available to the welder, he must rely on his general knowledge of them gained by study and experience in welding them. His approach to welding them becomes an educated guess, and the more comprehensive his education, the fewer failures he will have.

When the composition and character of the steel is not known, the welder should first take into account the particular service the steel is performing: That is, is the steel resisting abrasion, or impact, or both? Is it becoming hot in service? Must it maintain a sharp edge? Are there abrupt changes in mass? From these questions he may deduce whether the steel is air-hardening, oil-hardening, hot-working, or other, and carry out welding and heat treatment accordingly.

Some characteristics are common to all tool steels.

1. All have a relatively high carbon content.
2. All have a good hardenability and low ductility.
3. All require considerable time at high temperatures for carbides to go into solution.
4. They are notch-sensitive.
5. They are susceptible to weld cracking, underbead cracking, decarburization, and contamination.

Preventing Cracked Welds

Cracks result from too-rapid cooling. If a weld cracks longitudinally ("down the middle"), the weld metal has shrunk more rapidly than the base metal. Sometimes only the first or root bead will crack, but this crack will continue up through the weld after the tool has been returned to service. If the weld shrinks faster than the base metal, the obvious preventive measure is to preheat the base metal. An alternative measure, if preheating is impractical, would be to use a more ductile filler metal, provided, of course, that the weld is not across or not part of a wearing or impacting surface. Carbon pickup by the weld deposit contributes to weld cracking—nonferrous metals do not pick up carbon; therefore, use a higher-alloy electrode. To help prevent cracking in service, stress-relieve the joint.

Underbead cracking may result from two possible causes:

1. If the base metal is in the annealed condition, it has a quenching effect on the overheated base metal near the weld, and this metal transforms to martensite, which occupies more space, while the untransformed metal shrinks to its original volume. These opposing forces produce locked-in tensile and compressive stresses that tend to crack the brittle martensitic crystals in the hardening area.
2. If welded in the hardened condition, the welding anneals the base metal outside the transition area and results in an area of untempered martensite parallel to the fusion line. This area is prone to microcracking and is susceptible subsequently to fracture in service.

Underbead cracking can be avoided or lessened by preheating, by keeping the welding heat as low as possible, cooling slowly, and stress relieving. The safer way is to preheat whether the steel is in the annealed or hardened condition. Other recommended procedures may increase the chances for successful welds without preheating but should not be depended upon fully—the omission of preheating more often results in-service fracture.

Preheating Temperature

The proper preheating temperature for welding is not to be confused with the preheat temperatures usually given in manufacturers' heat-treatment recommendations. These temperatures usually show the temperatures at the top of the hardening process. Useful guidelines for preheating temperatures are:

1. *Preheating for hardened tool steel:* Do not exceed the heat of the last tempering or drawing. If unknown, heat into the lower half of the steel supplier's tempering range for the steel, or preheat to the lower limit of the tempering range typical of that class of steel.
2. *Preheating for annealed tool steel:* Same as above, except heat to the upper limit of the tempering range.

Preheating must be uniform. Best is a bath of brine or oil brought to the temperature slowly or use of a temperature-controlled furnace. If a makeshift furnace is built of bricks and asbestos cover, make sure that the flame is soft and flows around the weldment uniformly. The harsh flame of the oxyacetylene torch should be avoided. Metal outside the weld area should be covered to retain the heat during welding. Welding-heat input should be kept as low as possible by very shallow penetration, the smallest-diameter electrode that will do the job properly, intermittent welding, and short stringer beads. If weldment loses heat during welding, reheat to the proper temperature before continuing welding.

WELDING TOOL-AND-DIE STEELS

Shielded metal-arc welding is the most widely used process in welding tool steels. Except for some silver brazing, the oxyacetylene process should be avoided.

The inventory of electrodes needed may be kept to a minimum by selecting electrodes for an entire class of tool-and-die steels. As a general rule, as stated previously, choose an electrode whose weld deposit responds to the same heat treatment as the base metal, that is, an air-hardening electrode for air-hardening steels. This selection will reduce the inventory to a maximum of 6 to 8: one each for oil-hardening steel, water-hardening steel, air-hardening steel, high-speed steel, hot-working steel, and E-310, and a balanced austenite-ferrite electrode of E-310 alloy content.

Oil-hardening tool-and-die steels. The electrode should have a 56 to 60 Rockwell C hardness. Anneal at 775 °C (1452 °F), harden at 775 to 800 °C (1425 to 1475 °F), and draw at 120 to 205 °C (250 to 400 °F). Preheat carefully. Do not exceed draw-range temperatures during welding. Use stringer beads. Peen while hot and soft. When replacing cutting edges, the weld should proceed in one direction to within a short distance of the other end, then proceed in the opposite direction and overlap the other bead. If facing the entire forming or cutting edges of draw rings, extrusion dies, or circular dies, use the skip-weld technique. If the type of tool steel is unknown, the welder may make an educated guess based on the work the steel performs. These steels are used in cold blanking, forming, and trimming dies; cold cutting and shearing; and wherever shrinking and deforming during treatment must be held to a minimum.

Hot-working dies. The as-welded hardness should be 52 to 56 Rockwell C. They should anneal at 815 to 870 °C (1500 to 1600 °F), harden at 955 to 1010 °C (1750 to 1850 °F), and quench in air. They should draw at 485 to 650 °C (900 to 1200 °F) and cool in still air. Welding preheat is approximately 260 °C (500 °F). When repairing existing die units, preheat to 485 °C (900 °F) and do not drop below 370 °C (700 °F) during welding. These steels are used for hot-working punches, headers, trimmers, and so on.

Air-hardening tool steels. The as-welded hardness should be 58 to 60 Rockwell C. Anneal at 815 to 870 °C (1500 to 1600 °F), harden at 955 to 1010 °C (1750 to 1850 °F), and quench in air. Draw at 150 to 540 °C (300 to 1000 °F). If welding 5% chromium steel, preheat to 205 to 260 °C (400 to 500 °F). If repairing high-chrome steel, preheat to 370 to 480 °C (700 to 900 °F) and keep the heat during welding. For maximum toughness, temper at 480 °C (700 °F). These steels are used in cold extrusion dies, blanking dies, drawing dies and rings, hot-trimming dies, and burnishing rolls.

Water-hardening tool steels. These steels usually take an electrode with an as-welded hardness of 58 to 60 Rockwell C. Anneal at 775 to 788 °C (1425 to 1450 °F), harden at 775 to 788 °C (1425 to 1450 °F), and water quench. Draw at 135 to 230 °C (275 to 450 °F). Preheat within the drawing range. Do not exceed the drawing range during welding. Temper after welding. These steels are used where abrasion resistance is desirable on the surface and impact resistance is desirable in the core. They are shallow-hardening. (See differential hardening and "Case Hardening," Chapter 5, pages 74-75.)

High-speed steel. Due to the high heat required for hardening, choose an electrode that will repair or rebuild the HSS without rehardening—60 to 66 Rockwell C. Anneal at 870 °C (1600 °F), harden at 1205 to 1218 °C (2200 to 2225 °F), and quench in oil at 120 °C (250 °F). Draw at 540 °C (1000 °F). Follow recommended welding procedures. For composite fabrication of cutting tools, machine or grind surface, allowing for 5 mm of finished weld metal. Preheat to 205 °C (400 °F). Butter the surface with one layer of E-310 alloy or a balanced austenite-ferrite structure of E-310 Cr-Ni content. The buttering prevents pickup of impurities from the parent metal. Temper at 540 °C (1000 °F).

SILVER-BRAZING TOOL-AND-DIE STEELS

Many shapes of tool steel fractures lend themselves to successful silver brazing. Use only the high-grade silvers. The better practice is to tin both fractured surfaces heavily, fit the pieces back together as broken, heat until tinning remelts, and press together to squeeze out residual flux. Although silver brazing means heating to 650 °C (1200 °F), generally this heat is not maintained long enough to be harmful.

Some hazards that may be encountered in silver brazing are:

1. Copper-based alloys form chromic oxides on the surface of the chrome-type steels and prevent good wetting out on the surfaces to be joined.
2. Silver brazing cannot be used where in-service temperatures at the weld area reach 260 °C (500 °F).
3. Silver brazing cannot be used where high stresses occur at frequencies that promote fatigue.

On many occasions, the welder will not have facilities for the heat treatment required, or the broken tool does not need an identical structure at the fracture site. In such cases, the welder, by following proper stainless steel welding procedures, may weld the tool steel successfully with a reasonable preheat, an E-310 electrode, and slow cooling. A balanced austenite-ferrite electrode of E-310 alloy content may pick up where E-310 fails; or, perhaps, should be used in all cases where the outcome may be in doubt. The weld deposit of such electrodes, however, cannot form a part of the cutting or working edge.

8

CAST IRON

Although cast iron is a more complex alloy than steel, in its simplest form it is an alloy of iron, carbon, and silicon. The carbon content, however, is much higher, running from about 2.5 to 4%. The silicon content will usually run from slightly more than 3% to less than 0.5% in inverse ratio to the carbon content. The ratio of carbon to silicon is the most important factor in determining the final structure of a particular cast iron, with the allotropic nature of carbon being the factor that makes the structural variations possible.

There is some manganese, sulfur, and phosphorus in all cast irons, and special alloying elements, such as copper, molybdenum, nickel, and chromium, are sometimes added in order to give the iron special chemical and mechanical properties, but it is the control of the carbon that determines whether the cast iron will be white, gray, or malleable iron.

The principal difference between steel and cast iron is plasticity. Where steel is plastic and forgeable, cast iron is rigid and is unforgeable at any temperature. It is cast iron's refusal to bend or stretch that is of the greatest importance to the welder. Where steel will distort and accommodate many welded-in stresses, cast iron will fracture. Where a welder can repair a cast steel or forged steel motor housing and apply great pressures as he bolts it down to its foundation, the cast-iron housing must be perfectly aligned and mate with its base perfectly. The welder who has welded intake or exhaust manifolds and tried to force-fit them to a truck engine knows that the result is a second fracture and another repair weld.

Gray Cast Iron

UNDERSTANDING CAST IRON

To understand the control of structure in cast iron, it is necessary to understand the constituents *ferrite, pearlite,* and *cementite* of Figure 5.10 and add *graphite* to that prerequisite understanding. Graphite is an allotrope of carbon and appears in cast iron as flakes or nodules formed from carbon that precipitates out of solution during the decomposition of austenite and cementite.

Regardless of their complexities, all cast irons may be classified as *gray cast iron, white cast iron,* and *malleable cast iron*. They are often referred to simply as gray, white, or malleable iron. Gray iron and white iron get their names from the appearance of the face of the metal when ruptured, and malleable iron from the fact that it will bend very slightly before rupturing.

GRAY CAST IRON

As in steel, the carbon in gray cast iron plays the most important role in the control of its structure. Some of the carbon is combined, but the greater amount is distributed in the iron as flakes of graphite that appear as discontinuities in the structure. Control of the structure is accomplished by controlling the shape, size, amount, and distribution of the combined and graphitic forms of carbon.

The constituents that determine the character of the iron are graphite, ferrite, pearlite, and cementite. The constituents pearlite, ferrite, and cementite are the same as those found in steel. Graphite is carbon that has separated out of solution to roam free in the form of flakes (large or small) or, by special treatment, as nodules or spheroids.

The constituents graphite and pearlite are found in varying amounts in all gray irons. Free ferrites or free cementites in gray iron depend on the amount of combined carbon. If there is less combined carbon than that needed to form an all-pearlite structure, free ferrites will appear and if the combined carbon exceeds the eutectoid amount, cementites will appear. This condition may be compared with the pearlite plus excess ferrite and pearlite plus excess cementite of Figure 5.10.

Graphite is far and away the most important constituent in gray cast iron. The size, shape, amount, and distribution of the graphite will determine the final character of the iron. The density of graphite is about one-third that of carbon; therefore, it will occupy about three times the amount of space occupied by the same percentage of carbon. It is this lack of density that causes graphite to weaken the iron. The size and shape of the graphite also determine the strength of the iron. The larger the flakes, the weaker the iron; and by means of a spheroidizing treatment, the graphite will take on a nodular shape which will tend to strengthen the iron. It naturally follows that the smaller the spheroids, the stronger the iron. This nodular graphite is an important characteristic of malleable irons.

The amount of graphitic carbon in gray iron is influenced by:

1. The silicon-to-carbon ratio, the silicon promoting the graphitization of carbon
2. Adding steel to the meld, which will reduce the total amount of carbon
3. Very high melting temperatures, which will dissolve any graphite residual in the meld
4. Rate of cool-down; the faster the cool-down, the less time for graphitization to take place

The distribution of combined and graphitic carbon can be critically influenced by the shape and section of the casting because the rate of cool-down may not be uniform throughout the casting, the thin sections cooling more rapidly than the heavier sections. The addition of nickel as a constituent tends to speed up graphitization in thin sections and chromium tends to retard graphitization in the large sections.

Pearlitic Gray Iron

The pearlite in gray iron is the same pearlite we find in steel: alternate ferrite and cementite. A casting of pearlite and graphite is called pearlitic gray iron. The pearlitic type is the strongest of the gray irons, yet it is easily machinable. The structure may be compared with steel in that the pearlite (eutectoid) is sifted with graphite just as the hypoeutectoid steel is pearlite sifted with ferrite. As in steel, finer pearlite and finer grains increase the strength and hardness of the casting.

As in steel, free ferrite in gray iron will soften and weaken the casting but, unlike steel, will not increase ductility because the graphite breaks up the continuity of the ferrite matrix.

Free cementite in gray iron reduces machinability and makes the iron more brittle. If present by design, it is there for the purpose of increasing hardness and wear resistance. Free cementite is characteristic of white and chilled irons.

The importance of cool-down may be illustrated by the cooling of a wedge-shaped casting in a sand mold (Figure 8.1). The thin part of the wedge will cool more rapidly than the heaviest portion. Thus if a foundryman wishes to produce a pearlitic gray iron, he must slow down the cooling of the thin section and speed up the cooling of the heavy section, or accomplish his goal by alloying with other elements (such as nickel and chromium).

Addition of Various Metals to Gray Iron

Nickel in gray iron. Nickel promotes graphitization of the carbon. It is only one-third as effective as silicon, but where silicon is ineffective in percentages above 3%, nickel's effectiveness extends well beyond 3%. Nickel in

Gray Cast Iron

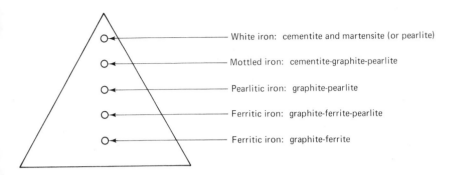

Figure 8.1 Cooling of a wedge-shaped casting in sand.

small amounts will reduce the effects of chilling and will soften thin cast sections. It also promotes density and freedom from porosity by permitting lower silicon content without causing hard spots in light sections. Unlike silicon, nickel uniformly hardens the matrix of the iron by changing a coarse pearlite to a finer and harder pearlite, and finally to martensite. Nickel helps refine the grain and disperses the graphite in a finely divided state.

Chromium in gray iron. As we know from our study of tool-and-die steels and wearfacing, chromium is a carbide-forming element. Chromium carbides are much more stable than iron carbides (cementite). By stabilizing the cementite, chromium acts opposite to nickel and silicon, and inhibits graphitization, thus promoting hardness and strength.

Molybdenum in gray iron. Molybdenum has very little effect on graphitization but increases the strength of gray iron by inhibiting the decomposition of austenite to pearlite. Molybdenum decreases the rate of transformation during the cool-down in cast iron in the same way as it does in steel. The result is a much finer pearlite—almost martensite. Molybdenum increases strength, hardness, impact strength, fatigue strength, and strength at elevated temperatures.

Copper in gray iron. Copper functions primarily as a graphitizer. Its effect is about one-tenth that of silicon. A large percentage of the copper forms a solid solution with the ferrite and the remainder is dispersed as microscopic or submicroscopic particles in the iron. Copper refines pearlite, increases tensile strength and hardness, improves wear resistance, acid resistance, and, when combined with chromium, increases strength and corrosion resistance at elevated temperatures.

Pearlite in gray iron. As noted above, the pearlite in gray iron is the same pearlite as that found in steel. It is possible to have all of the carbon in gray iron as graphite, with no pearlite. Such an iron will be soft and weak—a ferrite-graphite structure. As the combined carbon increases, the iron changes from ferrite and graphite to pearlite and graphite. The finer the ferrite and cementite layers, the stronger the iron.

Although the advance of welding techniques has replaced many castings with welded structures, especially the more complex structures, gray cast iron has retained a great respectability because of its general utility. Some of its desirable attributes are:

1. It can be easily cast into complex shapes.
2. Its tensile strength can be varied from 15 kg/mm^2 (20,000 psi) to well beyond 50 kg/mm^2 (70,000 psi).
3. Its compressive strengths vary from three to four times the tensile strengths.
4. It can be varied in stiffness and hardness.
5. It has an excellent damping capacity for vibration, making it especially suitable for use in machine tools where vibrations may cause inaccuracies in precision machining or may cause a rough finish.
6. It has comparatively good corrosion resistance.
7. It is easily machinable.
8. It has excellent wear resistance and reduces galling in metal-on-metal wear.

WHITE CAST IRON

White or chilled cast irons have almost all of their carbon in the combined state. White iron generally runs from 2 to 2.5% carbon. The silicon content is relatively low in order to minimize graphitization. These irons are cast in sand molds and slow-cooled to room temperature without graphitizing any of their carbon. Their structures will be pearlite and cementite, similar to hypereutectoid steels, but they will contain much more free cementite. Most iron of this type of structure will be used to produce malleable iron by a special heat treatment called malleabilizing. (See "Malleable Cast Iron," this chapter.) If alloys such as nickel are used, the structure will be martensite and free cementite since the alloys will inhibit the change from austenite to pearlite. These irons will be used where high hardness and excellent wear resistance are desirable, such as in crusher jaws and hammers.

White iron is also made by rapid cooling (chilling) of iron that would otherwise be graphitic iron (gray iron). Localized hardening can be effected by a localized chill in the mold. Thus a casting can have a hard white iron surface and a softer impact-resistant base, such as that found in railroad wheels. The depth of the chill is determined by the carbon-to-silicon ratio, the thickness of the casting, the thickness of the metal acting as a chill, the time the casting is in contact with the chiller, and the use of alloying elements. Other uses for these irons are in ore crushers, cement grinders, steel-rolling mills, and the like.

MALLEABLE CAST IRON

Malleable iron, in spite of its higher cost, is used in plows, harrows, rakes, automobile parts, small tools, pipe fittings, and wherever greater toughness and shock resistance are desirable.

Malleable iron is made of the lower carbon–lower silicon irons, mentioned in our discussion of white cast iron, by an extensive heat treatment that may last up to 7 days.

The malleabilizing treatment changes the combined carbon (cementite, Fe_3C) to a graphitic carbon called *temper carbon*. Cementite is an unstable condition and at red heat [845 to 870 °C (1545 to 1600 °F)] will decompose to graphite and ferrite under slow cooling. In the malleabilizing treatment, the white iron is heated very slowly to reach the annealing temperature, soaked at that heat for 48 to 60 hours, cooled at a rate not exceeding 5.5 °C (8 to 10 °F) per hour down to 700 °C (1292 °F), held at 700 °C (1292 °F) for about 24 hours, then somewhat slowly air-cooled to room temperature. The malleabilizing is successful if all of the combined carbon has decomposed to graphite. The final structure should be ferrite and graphite. Specialized furnaces can reduce the time required for malleabilizing.

Fracture of this iron will display a very dark core framed as a picture with a very light halo. For this reason, it is called *black heart malleable*. The light surface is caused by a burnout of surface carbon by the annealing heat. By using a carbon-type packing material or a well-controlled furnace, surface burnout can be avoided.

A *white heart malleable* iron can be produced by using iron oxide as a packing material, but because of its difficult machinability and inferior properties, it is seldom manufactured.

The usual tensile strength of malleable irons is 38 kg/mm^2 (54,000 psi). These irons can be strengthened by the addition of such alloying ingredients as copper and molybdenum without decreasing their machinability. By retarding graphitization (control of silicon-to-carbon ratio and the addition of alloying elements), pearlitic malleable irons can also be produced.

The welder can readily see that if he fusion-welds a malleable casting, neither the weld nor the fusion zone will be malleable iron. Since a remalleabilizing treatment is generally impractical, malleable iron is braze-welded [below 815 °C (1500 °F)].

NODULAR IRON

Nodular iron gets its name from the nodular or spheroidic shape of its graphitic carbon. It is sometimes called "ductile iron." Its structure is largely pearlitic but may have some free ferrite and free cementite. These irons are higher in strength and greater in ductility than gray iron. They have excellent machinability, greater shock resistance, greater thermal shock resistance, and excellent rigidity. They are also much higher in price. The spheroidizing of the graphitic carbon is produced by the addition of magnesium and/or cerium.

WELDING CAST IRON

Since the welder will be reheating a casting regardless of the welding process he chooses, he should understand that a reheat is a reversal of a cool-down and, as in the case of steel, heat-up into and beyond the transformation range can change the structure of the iron considerably unless the cool-down regime closely approximates that used in the foundry.

The typical cool-downs of two cast irons through the transformation range are shown in Figure 8.2.* The complexity of our treatment of the subject is not intended as a scare tactic. It is intended, instead, to give the welder as much knowledge of his adversary as he may wish to acquire. The welder should be, however, sufficiently knowledgeable to approach his adversary with a degree of confidence; with a healthy respect for cast iron's seeming unpredictability, perhaps, but not as an arrant coward.

All of the cast irons can be successfully welded, with the possible exception of the white irons, and most of them can be welded by more than one of the welding processes common to the maintenance welder.

When welding cast iron, the welder should remember four things:

1. Cast iron cannot deform or stretch like steel—it breaks.
2. Heat-up and cool-down greatly affect the structure.
3. Deep penetration is undesirable.
4. Preheats must be uniform and never exceed 785 °C (1450 °F).

The old-timer's motto, "Heat it all or not at all," should be understood and carefully observed.

Transformation of 3% and 5% Cast Irons

In order to understand the transformation, Figure 8.2 should be studied in conjunction with the equilibrium diagram, Figure 5.10. The reader should strive for a prior or simultaneous understanding of the term "solid solution" and the difference between *decomposition* of cementite and *precipitation* of cementite.

The 3% alloy begins to freeze at point A and is solid at point C. The first crystal to form is composition B. Cooling from A to C, the solid portion changes to composition D. This solid solution is the same austenite as that seen on the equilibrium diagram at 1.7% carbon. The remaining liquid is composition E

---

*This straight-line diagram is for illustrative purposes and should not be used where absolute accuracy is required.

Welding Cast Iron

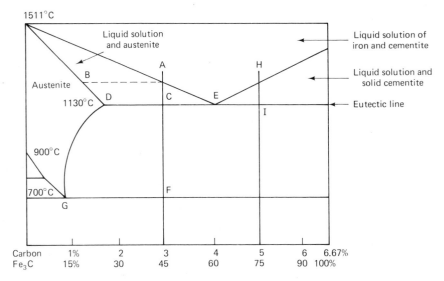

Figure 8.2 Transformation of 3% and 5% cast irons.

and solidifies as the eutectic of cementite and austenite (composition D). As the mass cools from C to F, the austenite formed from B to D and which formed in the eutectic at E, precipitates its cementite along D-G. At G, the austenite precipitates the balance of its cementite and forms the ferrite-cementite eutectoid (pearlite). The cementite may remain as cementite and give us white cast iron or it may decompose, wholly or partially, and produce a pearlite-cementite-graphite or pearlite-ferrite-graphite or ferrite-graphite structures, depending on the completeness of the decomposition, which, in turn, depends on the rate of cooling.

The 5% alloy begins to freeze as cementite at point H and continues to point I, and concentrates the austenite to the eutectic point E. At E, the material finishes solidification as the austenite-cementite eutectic. (Recall here that a eutectic freezes at a constant temperature the same as a pure metal.) The austenite has composition D. As the mass cools below the eutectic line, the austenite precipitates cementite along D-G. The cementite will appear in three forms: (1) that formed from H to I (2) that formed in the eutectic at E, and (3) that precipitated along D-G. The decomposition of cementite will follow the same pattern as the 3% alloy.

As shown in the equilibrium diagram (Figure 5.10), pearlite is a eutectoid of cementite and ferrite at 0.85% carbon. The hypoeutectoids are alloys in which the pearlite is surrounded by free ferrites. The hypereutectoids are alloys in which the pearlite is surrounded by free cementites and exist up to 1.7% carbon.

In alloys from 1.7 to 4.3% carbon, austenite freezes out of the melt beginning at A-B and continuing to E-B. At E-B the remaining liquid, which is the eutectic composition (austenite-cementite) freezes at a constant temperature, forming the eutectic mixture. This eutectic consists of alternating layers of cementite (Fe_3C) and saturated solid solution (austenite). The austenite is saturated gamma iron, meaning that it holds in solution the maximum 1.7% carbon. Hence alloys containing 1.7% carbon to 4.3%, just below the E-B line, consist of mixed crystals of austenite and eutectic. The structure may vary from 100% austenite and 0.0% eutectic at 1.7% carbon to 100% eutectic and 0.0% austenite at 4.3% carbon.

In alloys containing more than 4.3% carbon, cementite freezes out upon cooling from B-D to B-C. The residual liquid at B-C then freezes as the eutectic. Thus alloys just below the B-C line consist of mixed crystals of eutectic and cementite.

Joint Preparation

Weld-groove preparation is important in welding cast iron regardless of the process used. The groove should accommodate distribution of the filler metal with a minimum of meltdown of the parent metal.

Single or double Vee. The comparative strengths of the iron and the filler metal should be taken into account. For example, if the filler metal has twice the strength of the parent metal, as is the case of a steel, nickel, or bronze filler applied to ordinary gray iron, the depth of the weld need not be more than half of the bevel dimensions. (We know from our study of the wedge-shaped casting that the stronger part of the casting is nearest the surface.) When welding steel to cast iron, the leg on the cast iron should cover a greater area than the leg on the steel.

The method of preparation should not generate enough heat to transform the structure, not to any great depth at least. Fortunately, oxyacetylene cutting is so difficult that very few welders attempt it; it should not be used to prepare welding surfaces. Preparation by machining, sawing, or chipping are preferred. Grinding may be used, but it tends to smear the graphite, which then interferes with gas welding, especially braze welding with the torch.

The openness of the cast iron structure permits deep impregnation by oily substances, moisture, and other contaminants; hence the weld area should be thoroughly cleaned, preferably with a solvent. A good preheat into the higher end of the preheat range will drive out all contaminants, including residual gases.

Heat Treatment

With very few exceptions, hot welding of cast iron must be accompanied by heat treatment. The following procedure may not accommodate all castings, but is a typical procedure that is adaptable to most situations.

1. Preheat the casting between 250 and 650 °C (480 and 1200 °F). The preheating must be uniform throughout the entire casting.
2. Hold the heat during welding.
3. Re-cover the casting after welding and cool slowly.

The general practice of maintenance welders is to build a brick furnace around the weldment and cover the top with a sheet or two of asbestos. Bricks may then be removed at strategic points to provide apertures for the heating flame or flames. The number of apertures will vary with the size and shape of the casting. For large castings, several flames should be used simultaneously. With small castings, the flame may be moved from one aperture to another rotationally. The flame should be soft and should flow gently around the casting. Harsh, blasting-type flames should be avoided. Natural gas and ambient air pressure will provide a soft, widely dispersing heat.

The casting should be so placed that the weld area is accessible when the asbestos cover is removed. Whether part of the asbestos cover is laid back or bricks are removed, only the immediate weld area should be exposed. The heating flames may be removed or reduced discretionally, but the weldment must remain at or near the preheat temperature throughout the welding cycle. If the weldment cools too low, the casting should again be covered and the heat should be brought back up to the proper temperature.

Upon completion of the weld, cool slowly as follows:

1. Re-cover the furnace opening and reduce the heating flames.
2. Remove the heating flames.
3. Remove the asbestos cover for air cooling.

Welding Gray Cast Iron

Fusion welding with steel electrodes. Stick welding of cast iron gives strong but unmachinable welds using steel electrodes. Preheating as described above is necessary, and electrodes high in carbon and silicon are recommended: high in carbon to compensate for the carbon lost in the superheating action of the electric arc, and high in silicon to help promote graphitization.

The welding joint should be designed so that very little meltdown of the cast iron is required. Deep penetration is undesirable. The digging action of the arc should be kept in the molten puddle at all times, and the puddle permitted to flow out as closely approximating the sweating technique as possible. The arc length should be such as to reduce the digging action through the puddle. A slight downhill weld will help keep a puddle between the arc and the weldment. (If the weld joint cannot be inclined, use a 20° pushing angle with the electrode.) Restrikes should be made in the weld crater, not on the weldment.

Steel electrodes are recommended for inferior or low-grade castings, where economy is a major factor, where color match is desired, and especially where overhead welding is necessary.

Fusion welding with oxyacetylene gas. Using cast-iron filler rods, the oxyacetylene process will provide a weld approximating the structure of the weldment itself. This process is essentially a recasting process. The filler rod should be high in silicon to promote graphitization. Follow the heat treatment and joint design as recommended above. Position the weldment so as to weld horizontally or slightly uphill wherever possible. Use only cast-iron fluxes; do not use brazing fluxes.

Although cast iron is very fluid [melting at 1260 °C (2300 °F) and lower] the melting point is not as discernible as the melting point of steel as it is approached under the torch, and thin weldments may collapse suddenly. Bring weldment to orange-yellow and begin testing with a wiping action of the filler rod. Continue heating, and when the filler rod displaces some of the parent metal, begin feeding the filler into the puddle as in gas welding of steel. (*Note:* The filler rod should be heated to a red heat before probing in the puddle. Fluxing is required throughout the weld.)

Skip welding on long welds spreads the heat and aids in keeping the weldment uniformly heated. When approaching the edge of the weld joint, pause about 1 to 2 in. from the edge, and weld from the edge inward. On very short welds, a good practice is to begin at the center and backhand-weld toward the ends of the joint. Both of these techniques help ensure a square edge.

Blowholes are a hazard of oxyacetylene welding. Those caused by gas pockets inherent in the casting are easily worked out by manipulation of the filler rod. Those caused by faulty welding technique are very difficult or impossible to remove. They should not be permitted to happen. They are caused by overheating the iron or by the torch blowing into the puddle too long. They can be avoided by keeping the torch moving, intercepting the blow with the filler rod.

Hard spots are caused by dipping a cold filler rod into the molten puddle. The cold metal chills the weld puddle and produces spots of white or chilled iron. Hard spots will not win you the admiration of your machinist! Allow the torch to redden the end of the filler before sticking it into the puddle. Other causes of hard spots are improper fluxes and/or too much flux, and filler rods too low in silicon. Fluxcore filler rods assure a more consistent metal-to-flux ratio and eliminate the sometimes annoying dipping of a filler rod into a flux container.

Fusion welding with cast-iron electrodes. Although fusion welding with cast-iron electrodes is feasible, the process is rarely used. If attempted, all of the preheat and postheat recommendations should be observed, and the procedure should then approximate the procedure previously covered in "Fusion Welding with Steel Electrodes." Some specialized wearfacing is done with cast-iron electrodes.

Braze welding with oxyacetylene gas. Oxyacetylene braze welding of cast iron is far and away the preferred method of experienced welders. It is accomplished at a heat below the critical range and provides a weld that is stronger than gray iron in all cases and, by proper choice of filler bronzes, can equal or exceed even the strength of many highly specialized irons. Unless remalleabilizing is possible, braze welding (arc or gas) is the only method recommended for the malleable irons. (If unfamiliar with the braze-welding process, read "Braze Welding" in Chapter 2.)

Observe recommendations for joint design and heat treatment outlined earlier. Note that grinding of cast iron will smear the graphite, which will interfere with good tinning. Use a neutral flame unless the rod manufacturer has specified oxidizing or carburizing flames. If possible, align the weldment so that the joint can be welded uphill; the degree of incline to be that which best accommodates the welder and the shaping of the deposit. In the inclined position, the filler metal tends to stack up instead of running ahead of the tinning. When the filler runs ahead of the tinning, the welder may tend to flow it out with the torch flame and thus overheat the deposit to the point of volatizing the zinc; thus producing a porous weld. This result can be discerned by the whitish smoke that is given off and the loss of the natural shine of the weld deposit. The tinning should precede the weld bead by 1 in. or more. Good tinning is apparent when the filler metal wets out on the surfaces of the joint by natural capillary action.

In depositing subsequent layers of filler metal (if necessary) care should be taken not to burn up the first layer (volatizing of the zinc). The welder should note that on a second layer, he is welding brass to brass (or bronze to bronze) on part of the joint and is braze welding on the uncovered cast iron.

The proper bonding temperature of bronze to cast iron is said by experienced welders to be "black heat"; that is, the iron is a dull red heat under the torch and when the torch is lifted, the iron is "black." Alternate raising and lowering of the torch should produce the alternate black and red condition in the iron.

Ordinary commercial "cast-iron brazing flux" is usually sufficient to break down the surface tension and produce a sound weld, but some welders will use a tinning flux ahead of the bead, dipping their filler rod alternately in each of the flux containers. Flux-coated filler rods are available at a slightly higher price. The flux-coated rod not only eliminates the sometimes annoying need to remove the rod from the puddle but also assures a consistent ratio of flux to metal (see *hard spots* page 118.)

Braze welding with bronze electrodes. Stick welding of cast iron with bronze electrodes is not only feasible, but is widely used. It is fast, and the heat-up of the weldment does not get into the critical range, as does welding with nickel and/or steel electrodes. Some of these electrodes are relatively high in price but are lower in price than nickel electrodes, and their rapid rate of deposition may well offset the cost by a savings in labor costs.

Horizontal welding (down-hand) is no problem. Vertical welds are difficult, and overhead welding should be avoided. Although electrodes are available for ac application, most electrodes are designed for dc reverse polarity (dc +).

The general procedure is as follows:

1. Follow the heat-treatment recommendations. Design the joint to accommodate a fast-fill, low-purging action. Remember that only the electrode melts (the cast iron does not) and that this is a braze weld, not a fusion weld.
2. Tilt electrode 10 degrees* in the line of travel and maintain a close arc. The electrode must be fed into the puddle faster than the E-6010 types. The heat (amperage) should be set somewhat higher than for E-6010, size for size (from 100 to 140 A for the $\frac{1}{8}$ D, for example).
3. On heavy sections requiring more than one pass, lay in the first pass with the electrode alone, but on subsequent passes, feed a silicon-bronze brazing rod into the puddle with your free hand. The welding current may be increased to accommodate the additional mass of the melt.
4. Since the purging action of bronze electrodes is minimal, clean the weld area thoroughly between passes.

Welding White Iron

Although we have recommended that white iron be avoided like the plague, it can be welded if absolutely necessary. If so, it should be welded with a white-iron filler rod. Mechanical strengthening will greatly enhance chances for success. With a little ingenuity, the maintenance welder may devise a joint that is largely mechanical—as the overlay on a white-iron roll, wherein the buildup is not really a weld but a shrunk-on sleeve, the first layer being hard and brittle but by the third layer becoming machinable. If the weld joint is more mechanical than fused, the electrode may be chosen discretionally. Thermit welding is also possible.

Welding Malleable Iron

As mentioned in our discussion of braze welding of gray iron, braze welding (arc or gas) is the only method recommended for malleable iron. Malleable iron can be fusion-welded with a cast-iron filler but must then be subjected to the long and expensive remalleabilizing procedure that made it malleable in the first place. Since malleable iron began as white iron and was made malleable by special heat treatment, fusion welding or any heating that exceeds the transformation range will reconvert the malleable iron to white iron, and a normal cool-down will leave it in a white-iron structure.

Beware of the careful, painstaking, "cold" method of welding with a nickel electrode; it will avail the welder nothing except, perhaps, a thinner section of white iron between the weld deposit and the unaffected section of iron. There-

---

*Many welders will tilt the electrode as much as 45° in the line of travel. Their argument that this technique will preheat the iron ahead of the bead may have merit and is worthy of a trial.

fore, if you cannot remalleabilize the iron after welding, do not weld it with nickel or steel electrodes, and do not fusion-weld it by any process. When braze-welding malleable iron, whether using arc or gas, work the bronze filler up the sides of the joint without melting the surface of the iron.

Observe the usual heat-treatment procedures outlined earlier. The preheat should be 480 °C (900 °F).

Welding Nodular Iron

As mentioned earlier, nodular iron is a high-grade cast iron with a specialized structure, and since heating above the transformation range always reconverts or drastically changes a structure, fusion welding cannot be used without losing some of the special properties of the iron unless the filler is of an identical alloy and the identical heat treatment is followed after welding.

Fortunately for the maintenance welder, it is not always essential that the iron remain in its exact original state. Such is the strength and ductility of nodular iron that some loss of the special properties can be tolerated. Strong, ductile welds can be made that do not seriously injure the structure if care is exercised in welding.

As usual, the normal heat treatment of cast iron should be observed. Preheat should be about 315 °C (600 °F) when arc-welding and about 540 °C (1000 °F) when gas-welding. This treatment limits the formation of martensite and reduces the incidence of underbead cracking. To transform any residual martensite and reduce the carbides, postweld-anneal at about 650 °C (1200 °F) (more than 540 °C, less than 650 °C, never above 650 °C). The best welding procedure is the shielded metal-arc process, using nickel electrodes (at least 60% nickel). Observe also the recommendations for joint preparation and use the minimum current consistent with good bead shaping. Buttering and skip-welding the iron surfaces before filling is a good practice to follow.

Figure 8.3 shows a typical double-V butt weld for cast iron. The depth of the weld (B) need not be as great as dimension A because of the fact that weld metal is much stronger than the iron and the structure of C is much weaker.

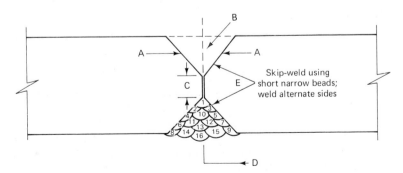

Figure 8.3 Typical double-V butt weld for cast iron.

(Gray iron, especially, contains more and larger graphite flakes at the center, due to slower cooling.) Plan beads (D) so that transverse shrinkage is one bead wide. Butter the cast-iron surface first on alternate sides as shown. If "cold" welded, shrinkage should be beads 1, 10, 13, and 16. (Numbering is illustrative only; the procedure will vary from job to job.) Weld alternate sides (E) if double-V, to distribute shrinkage around the neutral axis.

Using Nickel Electrodes

Fully machinable welds may be made on cast iron using the special nickel electrodes for cast iron. The welding procedure should be the same as that described under "Fusion Welding with Steel Electrodes" except that it is even more important that no meltdown of the walls of the joint occur. The heat treatment may be modified to a preheat of only 260 to 315 °C (500 to 600 °F).

Electrodes are available containing from 60 to 99% nickel. It would seem that, since nickel will not pick up carbon, the 99% nickel electrode would be the better choice. Many shops, when cost is a factor, will butter the cast-iron surface with a "100%" nickel electrode and fill or finish the weld with the lower-alloy electrode.

If the weld metal is applied properly, the fusion area should be fully machinable. Machinability may be improved by tempering the hardened zone at the fusion (or diffusion) line. This can be done by allowing the weldment to cool below 95 °C (200 °F) (to give the martensite time to transform) and then laying a cover pass without touching any part of the cast iron.

"Cold"-Welding Cast Iron

The welding of cast iron without preheat is possible, but due to the very slow welding procedure, it should be used only where preheat and postheat are impractical. Excluding the heat treatment, all procedures outlined previously should be followed. Additionally, observe the following:

1. Use the skip-weld procedure, using short, narrow beads.
2. Do not deposit additional weld until the palm of a bare hand will endure momentary touching to the proposed weld site.
3. Peen each deposit *lightly*, while hot, with the sharp point of a chipping hammer.

Nickel electrodes are the most desirable due to the stretching ability of the deposit, but the steel electrode has been successful for nonmachinable welds.

Studding the joint. Studding is a method of welding cast iron cold. Large sections of heavy machinery, such as bending brakes and roller mounts, may require studding in order to make a repair of sufficient strength. Studding is done by drilling and tapping holes in the metal and inserting studs, as shown in Figures 8.4 and 8.5.

Welding Cast Iron

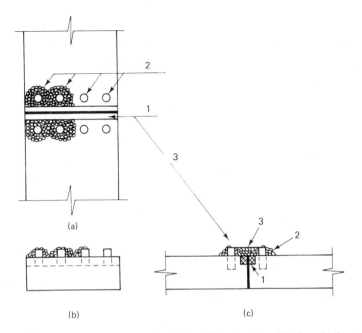

Figure 8.4 Square-grooved method of studding the joint: (a) plan; (b) front view; (c) side view.

The studding procedure as illustrated in Figure 8.4 is as follows:

1. Weld the groove flush with the top surface.
2. Weld around the studs. Build the weld around the stud in a conical shape, as shown in (b) and (c).
3. Fill the grooves as in (c).

The cross-sectional area of the studs should be 25 to 30% of the weld surface. The depth of the stud into the weldment should equal the diameter of the stud. The stick-out of the stud is determined by the type of weld joint. In the double-V joint (Figure 8.5), the stick-out need be only 5 to 6 mm ($\frac{3}{16}$ to $\frac{1}{4}$in.), whereas in the square-grooved example (Figure 8.4) it must be considerably more.

Studding should be used where preheat is impractical and sections are rigidly restrained. Follow procedures outlined for welding of cast iron cold with steel electrodes.

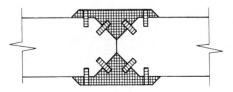

Figure 8.5 Double-V method of studding the joint.

BRAZING CAST IRON

The welder is often tempted to try brazing small cast-iron parts. Although feasible, it presents problems and should be avoided if possible. However small the part, the better practice is to vee-out and braze-weld. By far the greatest deterrent to brazing is the fact it is almost impossible to heat the joint uniformly and hot enough for the brazing filler to flow through the joint without seriously overheating the iron.

If brazing must be done, there are two procedures that may alleviate the overheating or eliminate it.

1. Heat the weldment with a very gentle heat, giving the center of the casting time to absorb heat from the surface without getting the surface too hot.
2. Butter the fractured ends individually, place them in close proximity, heat the area until the filler metal remelts, press the parts together, continue the heating and the pressure until some of the filler squeezes out, withdraw the heat, and maintain the pressure until the filler has cooled below its plastic range.

In all cases, the filler metal must have a melting point below the critical range [below 785 °C (1450 °F)]. Brazing of malleable iron should not be done unless done before or during the malleabilizing process.

SOLDERING CAST IRON

As with many metals, the difficulty with soldering cast iron is one of reluctance to accept tinning. The problem may be solved in several ways.

1. Plating of the surface with copper is the preferred solution.
2. If plating facilities are not available, mix 2 ounces of copper sulfate, 1 ounce of sulfuric acid, and 1 pint of water. Brush the mixture on the surface to be soldered, rinse with water, and allow to dry.
3. Several manufacturers of specialized maintenance alloys have developed solders (and their pertinent fluxes) for cast iron. They especially recommend them for filling defects on the surfaces of castings.

As in all soldering, the surface should be cleaned thoroughly, including the removal of oxides.

THERMIT-WELDING CAST IRON

Thermit welding is one of the few welding procedures that may warrant the description of "spectacular," and it is unfortunate that so few of our readers will have the opportunity to experience such a weld because the process has been supplanted almost entirely by less costly methods. We say that the weld is spectacular because the scheduling of the weld, if a relatively large one, takes on a social significance. Without exception, these welds are attended by the foremen and superintendents of other shops, the shop welders, and any others who are able to leave their jobs for the very few minutes it requires for the weld. The thermit welder can be excused if he "hams it up" a little and takes a curtain call after his performance.

The thermit weld of a casting (iron or steel) can closely approximate the structure of the parent metal itself by control of the alloying ingredients of the weld metal and control of the cool-down after welding. The weld is essentially a recasting process wherein the weldment is brought to a red heat, and superheated metal is poured around and through or on the weld area. The number of simultaneous welds that can be made or the complexity of the weld are limited only by the ingenuity of the welder in devising heating gates, pouring ducts, risers, and vents within the mold. Steel mill welders make eight simultaneous welds when refacing the wobbles of spindles and mill rolls.

The "disappearing wax" method is generally used in maintenance work. The wax is applied to the weld area and shaped into the desired weld contour. Refractory material (a special sand and clay mixture) is packed around the weld area to a thickness of several inches; and the rest of the mold is packed with inferior material, including the material salvaged from previous welds. The heating cycle melts the wax and disperses it into the sand, which the heating flames have hardened into a refractory state. The amount of wax used to shape the mold is used as a guide to determine the amount of thermit mixture needed. The weld area is brought to its red heat by way of heating gates placed at strategic points around the mold. The welder also uses these gates to observe the progress of the heating process. At red heat, the heating flames are removed, and the heating gates are plugged with balls of the refractory sand and clay mixture. A disaster results for the careless welder who may push the plugs so far into the heating duct that the plug interferes with the distribution of the weld metal around the weld area.

The thermit mixture (for cast iron) is composed of ferrosilicon, a compound of aluminum in a finely divided state, iron oxide, and mild steel in small, easily melted pieces. This mixture is placed in a crucible which is suspended over the pouring spout of the mold. On a relatively large job, the crucible is suspended by an overhead crane.

This is our welder's finest hour. The spectators have gathered. The weld is imminent. It is, perhaps, a travesty that our welder protagonist's hour upon the stage, after so much preparation, is reduced to a scant 90 seconds or so. With as much éclat as he can muster, our welder takes a steel rod in one hand and an oxyacetylene torch in the other, and heats the end of the rod white hot and dripping. He plunges the molten end of the rod deep into the mixture in the crucible and looks at his watch. The crucible shakes and rumbles, and emits a shower of sparks as the violent chemical reaction boils inside it. The boiling

TABLE 8.1 Cast Iron Summary

Type	Method	Treatment	Result
Gray	Fusion-weld, cast-iron filler	Preheat, cool slowly	A recasting, same as original
Gray	Braze-weld, bronze filler	Preheat, cool slowly	Weld stronger than cast iron, heated zone as good as original
Gray	Braze-weld	No preheat	Weld stronger, base hardened
Gray	Steel electrode	Preheat	Weld stronger, base hardened
Gray	Steel electrode	No preheat "cold" weld, skipweld	Weld stronger, base hardened
Gray	Steel electrode, studded	No preheat, weld studs first	Strong as original, base hardened
Gray	Nickel electrodes	Preheat	Strong joint, machinable
Gray	Nickel electrodes	No preheat, butter, skipweld	Strong joint, machinable
Malleable	Cast-iron filler	Preheat, remalleabilize	Good weld, malleability restored, costly
Malleable	Braze-weld, bronze filler	Preheat	Strong weld, heated area not as malleable
White	Fusion welding, not usually recommended; see "Thermit-Welding	Strengthen mechanically, use steel electrodes	Dependent on mechanical strength
Nodular	Nickel electrode	Preheat, postheat	Strong and ductile, loss of some special properties
Nodular	Nickel electrode	No preheat, observe cold-weld techniques	Strong and ductile, all special qualities lower in quality
Most	Thermit	Foundry	Strong weld, structured as designed

continues for approximately 45 seconds, then subsides much as does the popping of corn in a saucepan. But the welder cannot trust his ears as casually as does the popper of popcorn. He times the procedure to the last second, and at the appointed time he "taps" the crucible, and the molten mixture empties into the pouring spout of the mold. It is over. The welder is grateful if just a few of his spectators acknowledge his act with a nod, a gesture, or a word of approval as he assures himself that all went well within the mold, and contemplates the tedious job of dismantling his creation on the morrow.

The thermit process is a fascinating one for the welder who takes an interest in the metallurgy of his trade. The aluminum in the mixture, upon ignition by the molten steel rod, literally steals the oxygen from the iron oxide so rapidly that tremendous heat is generated; and the ferrosilicon, residual iron from the iron oxide, and the steel pieces are not only melted but are superheated to above 2700 °C (5000 °F). The oxygen of the iron oxide is now in the aluminum. The iron in the iron oxide is pure iron—the same as the pure pig iron that is poured directly into an open-hearth or electric furnace. So we now have a mixture of iron, aluminum oxide, and a little silicon and carbon. The pouring spout always induces the molten mixture to the lowest point of the weld area. The aluminum oxide, being lighter than the iron, is floated up the riser into a slag basin at the top of the mold. The superheated iron easily melts the red-hot weld area and blends with it into one amalgated mass.

SUMMARY

Table 8.1 summarizes the expected results for various types, methods, and treatments of cast iron.

9

STAINLESS STEEL

UNDERSTANDING STAINLESS STEEL

The designation "stainless steel" is given to a wide variety of corrosion-resistant steels. The first of these steels was a chromium–iron–carbon alloy for the making of high-grade cutlery and tools. It began in Sheffield, England, and the name Sheffield steel won as much acclaim as its predecessor Damascus steel. The advantages of Sheffield steel's corrosion resistance were quickly recognized by industry as a whole, and a great variety of corrosion-resistant steels evolved.

The principal alloying constituents of stainless steels are chromium and nickel. The carbon content may range from 0.03% to as high as 0.40%. Other elements may be added for specific properties, such as molybdenum for its anti-pitting-type corrosion resistance.

The use of nickel in carbon steels in the United States was introduced in 1891 by the U.S. Navy to produce armor plate. Its superiority was soon recognized worldwide. When nickel is added to steel, it dissolves into solid solution with iron and lowers the critical range, thus making a steel that can be made pearlitic, martensitic, or austenitic by simply varying the percentage of nickel. Nickel slows down the rate of grain growth at elevated temperatures, improves resistance to fatigue failure, increases corrosion resistance, and improves toughness and impact strength. Of greatest benefit to the welder is the fact that nickel greatly improves the weldability of the harder, tougher steels.

Chromium dissolves in both gamma and alpha irons, but also combines with carbon to form chrome carbides. Chromium gives steel hardness, corrosion resistance, wear resistance, useful magnetic properties, and better corrosion resistance at higher temperatures. As with the nickel content, the effects vary with the percentage of chromium. Chromium raises the critical temperature of the steel and slows up the rate of transformation from austenite to pearlite, thus permitting deeper hardening. To obtain maximum corrosion resistance, the chrome carbides must be kept in solution or the carbon content must be kept low. (See "carbide precipitation" in the Glossary.)

AISI CLASSIFICATION

Metallurgically, stainless steels fall into three major groups: martensitic, ferritic, and austenitic. The American Iron and Steel Institute has assigned a numerical identity to individual alloys within a series identification (200, 300, 400, 500) such as 304 and 304Se, the Se telling us that selenium has been used instead of phosphorus to obtain a free-machining quality.

2xx Series—Austenitic

These steels were developed during a critical nickel shortage, wherein manganese was substituted for a great percentage of the nickel. As we know from our experience with Hadfield steel (austenitic manganese steel), manganese inhibits the transformation of austenite as does nickel. When encountered, the welder may treat them in the same manner as their counterpart in the 300 series; that is, weld 201 the same as 301, and 202 as 302. These steels are sometimes called "high-manganese steels."

3xx Series—Austenitic

These steels are far and away the better known types by the welder, the number "308" and the word "stainless" being almost synonymous. (It is hoped, however, that by the reading of this treatment of stainless steel, the welder who says, "If I don't know what it is, I weld it with 308," will reconsider his simplistic philosophy and try a few others.) These steels are austenitic in structure, hence nonmagnetic. They contain from 16 to 26% chromium and 6 to 22% nickel. The carbon content varies from 0.03 to 0.25%. The most prevalent of this group are 302, 304, and 308, their chrome-nickel content varying only slightly. These stainless steels are often identified by welders as to their chrome-nickel content: 18-8, 25-12, or 25-20.

The 3xx series present no problem in welding except, perhaps, 303 and 303Se, due to their high sulfur or selenium content. They require no preheat and no annealing after welding. They do not harden by heat treatment, but then can be hardened considerably by cold working. They should be annealed after rolling, drawing, and other manufacturing processes. To assume their maximum corrosion resistance, they should be heated to 1040 to 1120° C (1900 to 2050° F) and quick-quenched. Because of their low thermal conductivity and high thermal expansion, distortion may become a problem in welding.

When heated to 425 to 815° C (800 to 1500° F) the corrosion resistance of these steels is lowered due to the loss of chromium at the grain boundaries, which is caused by the formation of chrome carbides containing up to 90% chromium. This compounding is called "carbide precipitation" and the loss of corrosion resistance is called "intergranular corrosion." (When the carbides form, they take their chromium from the metal adjacent to the grain boundaries and thereby reduce the corrosion resistance along the boundaries.) When welding these steels, an area near the fusion line must necessarily reach the critical range of 415 to 815° C (800 to 1500° F), the temperature range of carbide precipitation. There are several ways in which this undesirable phenomenon can be prevented.

1. Lower the amount of carbon as in the extra-low-carbon steels such as AISI 304L.
2. Add an element, such as titanium, columbium, or tantalum, that stabilizes the metal. (These elements have a greater affinity for carbon than chromium has.) Columbium or tantalum may be added to the welding electrode to stabilize the weld deposit, but titanium does not transfer across the arc.
3. Reduce the depth to which the parent metal is heated to the critical range. The metal should be cleaned and the joint prepared so as to eliminate the deep melting down of the walls of the joint, and the puddling should be controlled so as to reduce the digging action of the arc. Again, as in cast-iron welding, deep penetration is undesirable.
4. Welding current should be 10 to 20% lower than E-6010 size for size of electrode.
5. Faster welding speeds reduce heat input and thereby reduce the precipitation.
6. Lower amperage and shorter electrodes keep the electrode from getting too hot. (Some welders use half an electrode and lay it aside to cool. A rotation sequence of four or five electrodes will suffice.)

Exceptions and variations of the general character are:

1. AISI 301 and 302 have higher carbon and are more subject to carbide precipitation.
2. AISI 305 tends to crack when cooled.
3. AISI 302B (high silicon) needs 309Cb electrode.
4. AISI 303 and 303Se (free-machining) require low-carbon electrodes (ELC types).
5. AISI 316L (high antipitting-type corrosion resistance) needs molybdenum and columbium in the electrode.

4xx Series—Martensitic

The martensitic types are high in chromium and contain little or no nickel. Their chromium content ranges from 11.5 to 18%; their carbon from 0.15% to as high as 1.20%. They are also called "straight chromium" types and "chrome irons." They are air-hardenable: heated to 815° C (1500° F), they form austenite, which, when rapidly cooled in air, transforms to martensite. This martensite may be toughened by tempering. Local hardening along a weld may be toughened by stress relief.

The welding characteristics for this series are:

1. The more carbon, the harder
2. The harder, the more likely to crack as the weld cools
3. The harder, the more you need heat treatment

As a general rule, if a martensitic stainless steel is to be postheated or used under extreme temperature conditions, it should be welded with an electrode of the same structure (a 400 series electrode). If postheating is not to be used, weld with a series 300 electrode. (Since their thermal expansion rates are radically different, welding 400 series with 300 series electrodes should be limited to steels used at room temperature or within a narrow constant-temperature range.)

4xx Series—Ferritic

Like the martensitic types, these steels have little or no nickel. Their ferritic structure is due to their low carbon-to-chromium ratio.

The AISI 430 is borderline to martensite. A 430 electrode may be used with a local preheat of 150° C (300° F). If the weldment is to be used as-welded, use a 300 series electrode.

The general rule for these steels is that if they are not to be heat-treated, and used as-welded, use a 300 series electrode that will give a weld deposit with a chromium content similar to that of the parent metal.

WELDING STAINLESS STEEL

Stick-Welding Stainless Steel

Shielded metal arc welding of stainless steel is still the most used method. Its preponderance over MIG is due, in part at least, to its versatility and ease of getting the equipment to the job. The welder who is skilled with the E-7018 electrode will have no problem controlling the puddling of the stainless steel electrode. The fluidity of the puddle and the arc penetration are somewhat less than the E-7018, but the electrode manipulation is quite similar.

For best results, proceed as follows:

1. The joint should be prepared so as to minimize the meltdown of the parent metal. (Vee-out heavy sections.)
2. Deep penetration is undesirable.
3. Use small-diameter electrodes and short beads.
4. The metal should be as clean as possible.
5. Set the weld current at about 15% less than for a comparably sized E-6010 electrode.
6. Although the electrode may be weaved slightly, the weave must be steady. Do not whip the electrode.
7. Do not make restarts with a short, hot electrode. Lay it aside to cool for use later.
8. Clean each pass thoroughly. The purging action of the arc is not as lively as the E-7018.
9. If the electrode "sticks," do not try to shake it loose by using the electrode holder as a lever. Release the holder immediately because the high resistance—or low electrical conductivity—of the electrode causes rapid overheating.
10. Because of its low thermal conductivity and greater thermal expansion, stainless steel is more subject to distortion (refer to Chapter 13).

Gas-Welding Stainless Steel

Sound welds can be produced using the oxyacetylene torch by observing the same precautions as those that should be used in the gas welding of ordinary steels. Good welding technique is more important in the welding of stainless. Where faulty manipulation of the torch and filler rod in welding ordinary steels (low-carbon) may produce an unsound weld, faulty technique when welding stainless steel produces a disaster.

It must be remembered, too, that the same distortion encountered in stick welding of stainless steel is even greater in gas welding due to the slower welding speed and the difficulty in concentration of the heat. Distortion can be mitigated by the following:

1. Keep the weld bead as small as possible.
2. Observe the applicable procedures outlined in Chapter 13.
3. Concentrate the heat as narrowly as possible.
4. Use chill bars to absorb the heat.

The procedure for best results is as follows:

1. Adjust the flame to slightly carburizing—not more than $1\frac{1}{4} \times$. The prin-

cipal reason for the carburizing flame is to make sure that the flame does not inadvertently creep up to an oxidizing flame. Perhaps the neutral flame may be ideal, but a little excess acetylene causes much less trouble than does an excess of oxygen. The feather should be no longer than 2 mm ($\frac{5}{64}$ in.) beyond the tip of the cone. *Do not exceed this feather.* Keep in mind that chromium will pick up carbon from the excess acetylene and form chromium carbides which embrittle the weld. On the other hand, keep in mind that an oxidizing flame will form chromium oxides which are not easily soluble. You are choosing the lesser of two evils. Use the proper tip size, the tip that will give a soft flame. Avoid the harsh flame of a tip that is too small. The author has found that the excess acetylene feather is generally enough shielding so that no flux is required for the weld puddle. However, the underside should be coated with flux in heavy paste form.

2. Forehand welding seems to give the best results with the least trouble. Proceed the same as in welding ordinary (low-carbon) steel but be sure to do the following:
   (a) Keep the flame in the puddle and avoid any sweeping or grandiose torch movements.
   (b) Advance the torch steadily.
   (c) Keep the filler rod nearly perpendicular and keep the tip of it in the shielding effect of the excess acetylene at all times.
3. Keep in mind at all times that if the metal is clean, all troubles in the weld puddle will derive from oxidation. Oxidation will make the weld deposit appear burnt and porous. A good puddle will appear as a shiny, wet pool of metal that flattens out slightly; and close observation will disclose a thin film of dross being floated out and away from the puddle by the force of the flame. This film is formed from the surface oxides remaining after cleaning, the oxides formed during welding, and some impurities in the metal itself. The finished weld will not have the brilliantly clean look of that obtainable by the best stick electrodes in the hands of a skilled welder, but the weld can be polished to a bright luster.

Fusion welding of the 4xx and 5xx series with the OAW process should be avoided if possible.

TIG-Welding Stainless Steel

Gas-tungsten arc welding produces excellent welds on stainless steels. As in the case of aluminum, the transition from oxyacetylene to TIG will be gratifying to the welder who takes pride in sound, clean welds. A thoriated-tungsten electrode is recommended and may be used on ac or dc straight polarity. DCSP is generally preferred because of its deep penetration. Alternating current is preferred for thin gauges. In both currents, a high-frequency unit should be used:

with ac because of arc stability and dc because touching is not required to start the arc. Argon or helium, or a mixture of the two, may be used for shielding. Argon is generally preferred. (*Note:* Many welders complain of an upset on the underside of welds on thin sections due to deep penetration of the arc. Too often they do not think of lengthening the arc, changing to ac, or increasing the size of the electrode and switching to DCRP, all of which changes will reduce or eliminate the upset.) *Do not use carbon electrodes.* Carbon electrodes will introduce carbon in the weld in the form of chromium carbides. (For detailed information on gases, stick-out, and CFH, see "TIG-Welding Aluminum," Chapter 11.)

MIG-Welding Stainless Steel

The MIG process produces excellent welds on stainless steel, and the welder who is experienced with MIG on ordinary carbon steels will find the transition a painless one. MIG features ease of application, high speeds, and the absence of welding fumes. The deep penetration (actually a submerged arc, in that the arc is deep in the puddle) and absence of slag make a relatively wide weaved vertical-down weld surprisingly easy. As one welder put it when describing the act of lining a reactor, "You can throw it on like brushing paint." (See "Short Arc," Chapter 4.) Argon is generally used for shielding and is more economical than helium. Mixtures with other gases usually find the argon upward of 90%.

BRAZING STAINLESS STEEL

Excellent results can be had in brazing stainless steel if proper brazing techniques are followed, but due to the scaling tendency when heated, brazing must be done somewhat more carefully than with many other metals. As the old-timer says, "You gotta braze that stuff cold." He means, of course, that the welder must approach the brazing temperature carefully lest he overheat the metal.

Brazing may be done at bonding temperatures ranging from 1175° C (2150° F) down to 635° C (1175° F). The upper range are the nickel–chromium group of alloys, which have that unusual ability to surface-alloy with both sections being joined, the filler metal in the joint becoming a new alloy. These brazing alloys should be used where the weldment encounters high temperatures in service, such as in jet aircraft.

The silver-bearing brazing alloys, often called "silver solders," range in melting points from 815° C down to 635° C (1500 to 1175° F). These lower temperatures are preferred because the higher heats increase the scaling tendency. Lower temperatures also lessen the distortion. The standard grades 1 through 8, ranging from 80% silver down to 10% and bonding from 870 to 760° C (1600 to 1400° F) are often used, but perhaps, should not be used on stainless steel

SMAW Electrodes for Stainless Steel

due to their high copper content and the consequent electrolytic action. In these grades the lower the silver content, the higher the copper content. The higher the silver, the better the wetting characteristic. If using copper-type alloys, extra care should be taken not to overheat. The copper will penetrate the parent metal by following the grain boundaries and thus cause checks and/or cracks upon cool-down.

The silver–zinc–cadmium group melt in the range 760 to 705° C (1600 to 1300° F) and should be used where copper is objectionable. (*Caution:* Zinc and cadmium fumes should be avoided. Work in a well-ventilated area if brazing time is extensive.)

For joint-preparation and heating techniques, see the discussion of oxyacetylene brazing in Chapter 2.

SOLDERING STAINLESS STEEL

The soldering of stainless steel may require a little extra cleanliness, a more corrosive flux, and a little more attention to overheating but otherwise should present no problem for the welder. Solders are available with melting points from 92 to 370° C (200 to 700° F) and should be chosen with regard to in-service conditions to which the joint will be subjected, with special consideration given to melting points, plastic range, and constituents that may be objectionable (such as lead not being a desirable metal around foodstuffs). Melting points are important in that the joint is weakened as the in-service temperature approaches the plastic range of the alloy used in making the joint.

SMAW ELECTRODES FOR STAINLESS STEEL

The American Welding Society (AWS) numerical system for stainless steel electrodes is different from the system for ordinary steels, such as E-6010. The first three numbers are, of course, the type of stainless steel core, whose composition compares with the table showing the analyses of corrosion-resistant steels. The hyphenated numbers refer to welding positions and type of coating. Thus a 309-15 is a "25-12" in welding vernacular (25% Cr–12% Ni) all-position (the number "1") with a lime-type coating (5), and a 309-16 is the same in all respects except that the coating is of the titanium dioxide (titania) type (6).

The type 5 coating (lime) is for dc reverse polarity only and is a low-hydrogen type, with relatively shallow penetration and somewhat sluggish arc action. Thus it cannot be the first choice of the welder but should be used when weld specifications call for it and when underbead cracking is a problem. Lime is used to reduce hydrogen because the same ingredients used to make the E-7018 low-hydrogen cannot be used with stainless steel due to causing carbide precipitation. Some welders will use the lime-coated electrode for the root pass and fill-and-cover with the titania type: lime for the root pass to prevent underbead cracking and titania to get the better-appearing concave cover bead.

The type 6 coating (titania) produces an arc that is more penetrating, livelier, and more stable. It is preferred by all welders for vertical and overhead welding, for work in difficult positions, and certainly where a little unremovable surface contamination calls for a livelier purging action. The weld bead will be slightly concave and better appearing than the lime type. The titania type may be used on ac or dc, although it is available for dc reverse only.

Lime-titania electrodes are also available. They are used in some instances for the straight-chromium and chrome-molybdenum types of stainless steel.

Since the maintenance welder addressed in this book will generally use a weldment "as-welded," Table 9.1 is offered as a guide to successful welding without special treatment.

Because of the many similarities between stainless steel and the nickel-base alloys, the welder should give some attention to "Welding Techniques" in Chapter 10.

TABLE 9.1 Welding Guide[a,b]

Type	Electrode[c]	Recommended Heat Treatment
301	308	
302	308	
302B	309Cb	
303	309	
303Se	309	
304	308	
304L	308ELC	
305	308	
308	308	
308L	308ELC	
309	309	
309S	309	
310	310	
310S	310	
314	310	
316	316	
316L	316	
317	316	
321	308ELC	
347	308ELC	
403	309/310	
410	309	400° C (750° F) pre- and post-
414	309/310	
416	309/310	
416L	309/310	
416Se	309/310	
420	309/310	
431	309	
440A	316	
440B	316	

(Continued)

SMAW Electrodes for Stainless Steel

TABLE 9.1 *(Continued)*

Type	Electrode[c]	Recommended Heat Treatment
440C	316	
501	316	
502	310	
405	309	400° C (750° F) pre- and post-
430Ti	308ELC	
430F	308/309	
430FSe	308/309	
430	309	400° C (750° F) pre- and post-
442	309	400° C (750° F) pre- and post-
446	309/312	400° C (750° F) pre- and post-

[a]See Appendix D, Table D.6, for constituency table of stainless steels.
[b]Weld 200 series the same as 300 series counterpart.
[c]Inventory can be minimized by going to the higher alloy for general use. For example, 309-16 and 310-16 are superior in most cases and adequate in all others except:

1. Stay with 316-16 on stainless steel where product contamination must be kept to the absolute minimum.
2. In chemical plants and oil refineries, use 347-16 on AISI 321 and 347.
3. If carbide precipitation problem is critical in the 304L and 308L, use 308ELC16.
4. The 308ELC16 may also do double duty on 321 and 347 if 347-16 is not on hand. The 309 and 310 types are best for welding stainless steels to carbon steels.

10

OTHER METALS

NICKEL-BASED ALLOYS

We use the term "nickel-based alloys" for those alloys wherein, with very few exceptions, nickel is the principal element. These nickel alloys are used where corrosion resistance is of the greatest importance, and the metal is exposed to extremely unfriendly media, such as heat, cold, and acid; and may be subjected to electrolysis and severely cyclical heating conditions. These alloys are most likely to be found in the chemical, petroleum, food, space, and aircraft industries.

We have made a partial list of these alloys that should prove sufficient for the maintenance welder to interpolate or extrapolate the data in order to adapt his welding procedures. These alloys are generally identified by trade names and suffixes, such as MONEL alloy 400, INCONEL alloy 600, IN-COLOY alloy 800, HASTELLOY alloy C-276, and such names as Permanickel, Duranickel, and others.

In many instances, the maintenance welder will be required to weld according to absolute specifications, in which case, the filler metal and welding procedures are spelled out for him. Where he must make a choice, Tables 10.1 to 10.3 and Appendix D can be used.

If the welder can identify the parent metal, he may match the filler metal to the parent metal as closely as possible. If he cannot identify the metal, other than that it is a high-nickel alloy, the better practice is to stay with the higher nickel content and choose the other alloying elements according to his knowledge of what those elements might contribute to the weld deposit under the weldment's in-service environment. A good substitute for a highly educated guess

Nickel-Based Alloys

would be the use of highly alloyed filler metal such as ERNiCrMo-3 or ERNiCrMo-4.

Since the welding of the nickel-based alloys somewhat approximates the welding of the stainless steels, a review of Chapter 9 is recommended.

Welding Techniques

For successful welding of nickel-based alloys, observe the following.

1. Shielded metal-arc welding (stick) is the most common process in use. Use dc reverse-polarity (DCRP)/(DCEP) stringer beads at the lowest heat consistent with good arc action. Shape the beads slightly convex, and grind out crater cracks before restarts.

2. Crater cracks may generally be avoided by the foldback, side-swing, or forward-swing techniques; or, in extreme cases, a restart can be made about 1 cm ($\frac{1}{2}$ in.) behind the foldback and the weld bead then cascaded over the weld deposit down into the joint. The latter technique will require grinding off the high part of the restart.

3. Since the nickel welding puddle does not spread out as readily as do steels or stainless steels, joint design is important. The joint must be more open to permit distribution of the filler metal without excessive meltdown of the parent metal.

4. Surface preparation is very important. Remove all oxides, lubricants, and other adhered material (including ink or marking crayons) from an area at least 5 cm (2 in.) on each side of the joint. Surface oxides may have a melting point about 540° C (1004° F) higher than the melting point of the parent metal.

5. Although preheating is seldom required, the weldment should be brought up to about 21° C (70° F) in an area 25 cm (about 10 in.) on each side of the joint.

6. Remove the flux residue completely before restarting the arc in a crater and before laying additional beads. Give particular attention to the edges of the bead. Scratch the edges with a sharp instrument—old saw blades are good—and brush the bead briskly. Grind if necessary.

7. Choose the electrode size for good weld quality instead of welding speed.

8. Choose SMAW electrodes according to base metal requirements and good welding characteristics instead of cost per pound.

9. Arc should be kept tight, but do not touch the puddle. Some electrodes may be held so tight as to feel the crumbling of the coating into the puddle. Try not to tilt the electrode more than 20° out of the perpendicular.

10. Vertical-up and overhead welding should not be a problem for the experienced welder.

11. Spray transfer in the MIG process may be difficult except in the flat position, but pulsed arc and short arc permit good welding in any position.

12. When TIG welding, avoid contact between the electrode and the filler rod and between the electrode and the weldment. Use dc straight polarity (DCSP/DCEN). Use a pointed electrode, 2% thoriated tungsten. (See the discussion of TIG welding of stainless steel, Chapter 9.) Use 4.7 to 7.0 liters per minute (10 to 15 cfh) or higher gas flow and use a larger cup.
13. Oxyacetylene welding of some nickel alloys is feasible but is confined to small parts and thin gauges. (See the discussion of oxyacetylene welding of stainless steel, Chapter 9.) Gas welding is not recommended for HASTELLOY alloys.
14. When welding nickel alloys to stainless or other metals, the better practice is to choose the filler metal of the higher alloy content.
15. Nickel overlays on cheaper metals can extend the life of those metals, save maintenance costs, and adapt the cheaper metal to an unfriendly environment. Millions of dollars are saved in nuclear and oil-refinery reactors, for example, by cladding or lining with a relatively thin layer of stainless steel.
16. Nickel alloys cannot be cut with oxy-fuel equipment; use powder, air arc, TIG, plasma arc, or cutting-routing electrodes. Grind away all dross and dress up the heat-affected surfaces.
17. Most nickel-based alloys may be brazed. Clean thoroughly and flux the entire weld area. Silver brazing presents no problem but weldment should not be kept long in the critical range. Consult the manufacturer for recommendations regarding the brazing of materials at higher temperatures, such as those brazing alloys in the range 1130° C (2100° F). (See "Brazing Stainless Steel," Chapter 9.)
18. In the submerged arc process (SAW), the high heat input to the parent metal and the slow cooling can lower the ductility of many of the alloys and promote cracking.
19. The very nature of the use of the high-nickel alloys almost ensures that some contaminants harmful to welds are impregnated in the prospective weldment. The type of cleaning away of these contaminants will necessarily depend on the nature of the contaminant. Whatever the type of cleaning used, the residue of the cleaner may also need removing; that is, if caustic cleaners are used, their residues should be purged with hot water.
20. The surface oxides of these alloys have a considerably higher melting point than their parent metal [e.g., nickel oxides at 2090° C (3794° F)]. Despite the high heat of the welder's arc, some of these oxides may fail to melt and may then appear as discontinuities in the weld structure, as free graphite does in gray iron. For best results, oxides should be removed by grinding, abrasive blasting, machining, or pickling.
21. Sulfur and lead are particularly pernicious contaminants, their complete removal somewhat difficult, and their damaging effects serious.

22. Metal to be welded should be brought to about 21° C (70° F) prior to welding.
23. Stress relief may be required if parent metal is subject to hydrofluoric acid applications.
24. The higher the alloy content, the more important is joint design; that is, it should be opened up for easy weld distribution without excessive meltdown of parent metal. A test bead on scrap metal will quickly demonstrate the shallow penetration, sluggish arc action, and poor spreadability of the nickel electrodes. Increasing the amperage beyond the recommended range will avail the operator nothing except a flaking off of the electrode coating and overheating of the electrodes. Excessive puddling will cause a loss of deoxidizers. Butt joints of metals more than 2.5 mm in thickness should be beveled. (Maximum thickness may be increased to 5 mm if the joint is welded from both sides.) Thicknesses greater than 10 mm should be a double-V or double-U groove. GTAW gives best underbead contour on root passes where the weld cannot be made on both sides. V grooves should be 80°, and the root face should be kept to 2 mm. Single bevels on T joints should be 45°. Corner and lap joints should be avoided in high-temperature applications and under thermal and mechanical cycling conditions. Where corner joints are used, a full-thickness weld should be made, keyholing a melt-through or making an inside fillet weld. Grooved backup chill bars promote good penetration and aid in eliminating trapped gases and flux (use copper if arc-welding). If gas-welding, backup bars should be grooved 2 mm in depth and about 10 mm in width.
25. SMAW electrodes should be chosen to resemble closely the parent metal, with the exception that certain elements are sometimes added to satisfy weldability requirements. Electrodes may be left in their moisture-proof containers (if such is their packaging) in a dry storage area. Once opened, they should be kept in a cabinet containing a desiccant or provided with constant heat of a few degrees above the highest expected ambient temperature. If moisture pickup is suspected, electrodes should be rebaked at 315° C (600° F) for 1 hour or 260° C (500° F) for 2 hours in a vented oven.
26. In downhand and overhead welding, tilt the electrode about 20° in the line of travel and hold a short arc. If welding vertical-up, keep the electrode perpendicular to the parent metal. Because of the poor spread of the molten puddle, a slight weave is recommended to flatten somewhat the face of the deposit.
27. If using GTAW (TIG), use argon or helium or a mix. Avoid oxygen, carbon dioxide, and nitrogen. Five percent hydrogen can be tolerated in single-pass welds and gives a hotter arc and more uniform bead contour. If welding without a filler rod, helium gives greater welding speed and reduces porosity. Use DCEN, ACHF, thoriated tungsten, and a tapered tip. Avoid abrupt arc breaks. Reduce current and use side swing or forward swing.

Hold the torch as close to perpendicular as possible consistent with good visibility, and keep stick-out and arc length as short as possible. The size of the filler rod should be compatible with the size of the weld and the heat of the arc. The tip of the filler rod should be kept within the protection of the shielding gas envelope. Keep the weld puddle as quiet as possible to prevent burnout of deoxidizers. If touch starting is used, the arc is more difficult to start and maintain at currents below 60 A when helium is used.

28. With the GMAW process, spray, short arc, or pulsed arc may be used. Globular transfer can be used but tends to uneven bead contour and erratic penetration. As with stainless steels, the short-arc method is excellent on thin materials (to 3 mm). Use spray or pulsed arc for thicker materials. Pure argon shielding gives good results with spray and globular transfer, but as helium content is increased, wider and flatter beads are produced and penetration is reduced. Avoid oxygen and carbon dioxide. Helium alone produces an unsteady arc and excessive spatter. If using short arc, use argon with sufficient helium to flatten the weld contour and reduce cold lapping (lack of fusion). Since helium is much lighter than argon, the cfh rate must be increased as the helium percentage is increased. If using pulsed arc, use argon with helium additive up to 15 to 20%. Use DCEP (reverse polarity). Adjust wire feed consistent with stable spatterless arc. When using short arc, keep the arc in contact with the parent metal, not the weld puddle. When using pulsed arc, use a technique similar to that used for SMAW, giving particular attention to the edges of the puddle so as to eliminate any undercutting tendency.

29. If using the OAW process, use only bottled acetylene. Do not use generator-produced acetylene due to the sulfur content of calcium carbide. Use a slightly carburizing flame, not in excess of $1\frac{1}{2}$ x. An oxidizing flame will produce cuprous oxides in the Ni–Cu alloys and chromium oxides in the Ni–Cr alloys. Use a tip size that can accommodate a soft flame. Flux is required except for nickel 200, fluoride types are recommended, and borax should not be used.

30. For cleaning, Virgo descaling salt or sodium hydride (DuPont) or DGS oxidizing salt have been used effectively. If sandblasting must be used, be careful of embedding sand in the thin-gauge sections. Sandblasting also work-hardens the surfaces.

31. Most alloys can be machined using either HSS or carbide tips. On work-hardening types, the tip must be kept engaged at all times. If very close tolerances are required, finish by grinding.

Characteristics of Selected Alloys

The following treatment of several selected alloys is offered as an aid to the welder's understanding of the nickel-based alloys and provide some clue as to the probable identity of a metal whose factual identity is unknown and is not traceable back to its manufacturer.

Pure nickel. Commercially pure nickel is represented by nickel 200 and 201. Nickel 201 contains less carbon; thus any tendency toward graphitization is lessened and makes it preferable for applications at temperatures over 316° C (600° F). Nickel has exceptional resistance to alkalies (hot or cold) and finds use in caustic evaporator tubes. Its use in dilute acids, such as sulfuric, hydrochloric, and phosphoric is limited by such oxidizers as air or salts. Nickel is excellent for high-temperature chlorination and fluorination reactions because the formation of nickel chlorides or nickel fluorides inhibits the corrosive reaction. Nickel is likely to be found in food-handling and synthetic fibers applications, and in many applications where the special properties of the other nickel alloys are not essential.

MONEL alloy. MONEL is a trademark of Huntington Alloys, Inc., and actually embraces a series of alloys. The principal elements are nickel and copper. MONEL alloy 400 (67% nickel, 31% copper) retains most of the characteristics of the nickel 200 series and improves the corrosion resistance in such applications as seawater and brackish waters. It is likely to be found in propellers, propeller shafts, pump shafts, impellers, and condenser tubes; chemical processing equipment; gasoline and fresh water tanks; crude petroleum stills, process vessels, and piping; boiler feedwater heaters and other heat exchangers; and deaerating heaters.

The addition of 1.25% iron improves resistance to cavitation and erosion in condenser tubes. Antipitting is improved over nickel 200, but it will pit in stagnant seawater. Its resistance to corrosion by nonoxidizing acids such as sulfuric, hydrochloric, and phosphoric is improved over nickel 200, but the effects of the presence of oxidizers are the same for both. MONEL alloy 400 has excellent resistance to hydrofluoric acid (in the absence of aeration or oxidizing salts, of course).

MONEL alloy K-500 combines the excellent corrosion resistance of Alloy 400 with added strength due to its precipitation-hardening characteristic. Cold working prior to the age-hardening process adds further strength. K-500 maintains its strength from 650 to $-235°$ C (1200 to $-423°$ F). It is commonly used in pump shafts, impellers, doctor blades and scrapers, oil-well drill collars and instruments, and springs.

INCONEL alloy. INCONEL is a trademark of Huntington Alloys, Inc., and includes a series of nickel–chromium–iron alloys designated with 6xx and 7xx suffixes. INCONEL alloy 600 is an alloy of 76% nickel, 16% chromium, and 8% iron with traces of silicon, manganese, and copper. Its oxidation resistance is excellent up to 1100° C (2012° F) in air. Sulfur in the atmosphere will reduce the in-service temperature to about 315° C (600° F). Alloy 600 is likely to be found in furnace muffles, carburizing baskets, halogenic applications at elevated temperatures, and chlorination equipment up to 540° C (1000° F). Its corrosion resistance in high-purity water encourages its use in nuclear reactors, including steam generator tubing and primary water piping. Its absence of molybdenum restricts its usage where pitting is a factor, such as in reducing-acid solutions. Its resistance to ammonia solutions is excellent.

INCOLOY alloy. INCOLOY is a trademark of Huntington Alloys, Inc., and covers a series of alloys carrying suffixes 8xx. The principal elements are nickel, chromium, and iron. Alloy 800 is an alloy of 32% nickel, 20% chromium, and 46% iron. Its best use is for oxidation resistance at elevated temperatures. It is commonly used in the petrochemical industries; in heat exchangers, process piping, carburizing fixtures, and retorts; and in coal gasification processes when the environment is high in sulfur and oxygen. Its aqueous corrosion resistance is generally no greater than that of 316 stainless steel and, therefore, it is little likely to be found in aqueous service. INCOLOY alloy 825 retains the character of alloy 800 with improved aqueous corrosion resistance due to the addition of copper (1.8%) and molybdenum (3.0%) and an increase in the nickel content up to 42%. Alloy 825 is commonly used in phosphoric acid evaporators, pickling tank heaters, pickling equipment, chemical process equipment, propeller shafts, tank trucks, and spent nuclear fuel element recovery.

HASTELLOY. The tendency of welders to refer to HASTELLOY as a single alloy is very widespread. HASTELLOY is a registered trademark of the Cabot Corporation and applies to a series of alloys. We have chosen HASTELLOY alloys B-2 and G for discussion and refer the reader to Table 10.3 for others.

HASTELLOY alloy B-2 belongs to the nickel–molybdenum series. It is rich in nickel (around 70%), low in silicon and carbon (0.08 and 0.02% maximum, respectively), and contains less than 1% chromium. Molybdenum is its primary alloying element, hence its excellent resistance to pitting-type corrosion in reducing environments. It is likely to be found in applications in the distillation, condensation, and handling of hydrochloric acid and hydrogen chloride gas. It has excellent resistance to pure sulfuric acid at all concentrations below 60% acid and good resistance up to 100° C (212° F) above the 60% acid level. It is not likely to be found in applications above 760° C (1400° F), due to its absence of chromium, nor is it likely to be found in oxidizing environments, such as the oxidizing acids and salts.

HASTELLOY alloy G is a nickel-based alloy with 22% Cr, 19% Fe, 6.5% Mo, 2% Cu, and 2% CbTa. It has excellent resistance to reducing acids such as sulfuric and phosphoric. It also resists combinations of sulfuric and halides. It is commonly used in wet-process acid evaporators, agitator shafts, pumps, superphosphoric acid evaporators, the processing of organic chemicals, pollution-control equipment, garbage-incinerator systems, and the like.

Table 10.1 is offered as a guide in the selection of filler metals based on the practice of matching the filler metal as closely as possible to the parent metal.

Nickel-Based Alloys

TABLE 10.1 Filler Metals for Nickel Alloys

AWS[a]	Ni[b]	Cr[b]	Fe	Mn	Cu	Al	Ti	Cb	Mo	C	Si
ENi-1	96	—	0.05	0.30	—	0.25	2.50	—	—	0.03	0.60
ERNi-3	96	—	0.10	0.30	0.02	—	3.00	—	—	0.06	0.40
ERNiCu-7	65	—	0.20	3.50	27.0	—	2.20	—	—	0.03	1.00
ERNiCu-8	65	—	1.00	0.60	30.0	2.80	0.50	—	—	0.15	0.15
ERCuNi	31	—	0.50	0.75	67.5	—	0.30	—	—	0.02	0.10
ENiCuAl-1	64	—	1.00	2.50	30.0	1.80	0.75	—	—	0.20	0.30
ECuNi	32	—	0.60	2.00	65.0	—	—	—	—	0.02	0.15
ENiCu-2	65	—	0.30	3.10	30.5	0.15	0.55	—	—	0.01	0.75
ERNiCrFe-5	74	16	7.50	0.10	0.03	—	—	2.25	—	0.02	0.10
ERNiCrFe-7	73	15	6.50	0.55	0.05	0.70	0.25	0.85	—	0.04	0.30
ERNiCr-3	72	20	1.00	3.00	0.04	—	0.55	2.55	—	0.02	0.20
ERNiCrFe-6	71	16	6.60	2.30	0.04	—	3.20	—	—	0.03	0.10
ENiCrFe-1	73	15	8.50	0.75	0.04	—	—	2.10	—	0.04	0.20
ENiCrFe-3	67	14	7.50	7.75	0.10	—	0.40	1.75	—	0.05	0.50
EniCrFe-2	70	15	9.00	2.00	0.06	—	—	2.00	1.50	0.03	0.30
ERNiCrMo-3	61	21	2.50	0.25	—	0.20	0.20	3.65	9.00	0.05	0.25
ENiCrMo-3	61	22	4.00	0.30	—	—	—	3.60	9.00	0.05	0.40
ENiFeCr-1	42	21	30.00	0.70	1.70	—	0.10	—	3.00	0.03	0.30

[a]E, electrode for SMAW; ER, bare wire for GTAW and GMAW.
[b]Nickel and chromium rounded to the nearest 1%.

Many of these fillers have been classified by the American Welding Society (AWS) and are available from several sources. Others are proprietary, such as INCONEL and INCOLOY (Table 10.2) and HASTELLOY (Table 10.3). Table 10.4 lists the principal characteristics of nickel-based alloys.

Brazing Nickel-Based Alloys

Many brazing alloys for the nickel alloys are available commercially. The basic factors to consider in brazing are cleanliness, joint design, closely abutting bonding surfaces, avoidance of unfused areas, effective capillarity, avoidance of any movement during the plastic period, and effective expulsion of flux residue (see Chapters 2 and 9).

Silver brazing is usually restricted to in-service temperatures of 205° C (400° F). Most of the commercially available silver alloys (see Table 10.5) are suitable except those containing phosphorus (nickel phosphides form on the bonding surface). Fluxes for silver brazing are usually fluoride–borate mixtures that melt slightly below the melting point of the brazing alloy. Special fluxes are required for aluminum-bearing alloys, such as MONEL alloy K-500 and INCONEL alloy X-750. Any method of heating may be used that will meet the required bonding temperature within a reasonable time. Flux residues should

TABLE 10.2 Huntington Alloys Most Frequently Encountered by the Maintenance Welder[a]

Weldment	Characteristics (see Table 10.4)	SMAW	GTAW	GMAW	SAW	OAW	TB	FB
Nickel 200	1, 8a, 9, 11	141	61	61	61	61	c	c
Nickel 201	1, 8a, 9, 11, 15	141	61	61	61	61	c	c
MONEL[b]								
Alloy 400	1, 8a, 9, 22	190	60	60	60	60	c	c
Alloy 404	1, 8a, 9, 22	190	60	60	60	60	c	c
Alloy K-500	2, 4, 8a, 17, 19, 22	190	60	60	60	60	c	c
Alloy 502	2, 4, 8a, 13, 17, 19, 23	190	60	60	60	60	c	c
INCONEL[b]								
Alloy 600	1, 7, 8a, 11, 12, 17	182	62	62	62	62	c	c
Alloy 601	1, 7, 8a, 11, 12	182	601	82	82	601	c	c
Alloy 625	1, 7, 8a, 12, 17, 19	112	625	625	d	d	c	c
Alloy 617	6, 8b, 11	d	617	617	d	617	c	c
Alloy 671	8b, 11, 12	d	671	d	d	671	c	c
Alloy 706	2, 5, 7, 8b, 13, 17	135	718	d	d	718	c	c
Alloy 718	2, 5, 6, 7, 8a, 17	112	718	d	d	718	c	c
Alloy X-750	2, 7, 8b, 17, 19	d	718	d	d	718	c	c
INCOLOY[b]								
Alloy 800	1, 8a, 12, 19	112	82	82	82	82	c	c
Alloy 825	1, 8a, 12, 14, 16, 22	135	65	65	d	65	c	c
Alloy 801	1, 8a, 12	112	625	625	625	625	c	c
5 and 9% nickel steels	8a, 17	113	82	82	82	82	c	c
Cu-Ni 70/30	8a, 22, 23	187	67	67	67	67	c	c
Cu-Ni 80/20	8a, 22, 23	187	67	67	67	67	c	c
Cu-Ni 90/10	8a, 22, 23	187	67	67	67	67	c	c

[a] Consult Figure 10.1 and Appendix D for more complete coverage of alloys, electrodes, and filler metals.
[b] Registered trademark of Huntington Alloys, Inc.
[c] BAg-1, -1a, and -3 are preferred. BAg-3 and -4 are used for wide clearances. (Zinc and cadmium fumes are toxic; use adequate ventilation.) Precipitation-hardened alloys are aged after brazing. Therefore, melting points of brazing alloys must be above the aging temperature to be used. Alloys functioning in higher-temperature environments may be brazed in the range 1140° C (2050° F). Consult the manufacturer for methods and heating atmospheres.

TABLE 10.3 HASTELLOY[a] Alloys

Metal	Major Elements	Minor Elements	Characteristics (see Table 10.4)	Recommended Filler for Welding Process:			
				SMAW	GTAW	GMAW	SAW
Alloy B-2	Ni-Mo		1, 8a, 16, 18	B-2	B-2	B-2	[b]
Alloy C-4	Ni-Cr-Mo		1, 8a, 19	C-4	C-4	C-4	[b]
Alloy C-276	Ni-Cr-Mo	Fe-W	1, 8a, 19, 22	C-276	C-276	C-276	[b]
Alloy G	Ni-Cr-Fe	Mo-Cu	1, 8a, 16, 20	G-3	G	G	[b]
Alloy G-3	Ni-Cr-Fe	Mo-Cu	1, 8a, 16	G-3	G-3	G-3	[b]
Alloy N	Ni-Mo	Cr-Fe	1, 8a, 20	N	N	N	[b]
Alloy S	Ni-Cr-Mo	La	1, 7, 8a, 12	S	S	S	[b]
Alloy W	Ni-Mo	Fe-Cr	1, 7, 8a, 20, 21	W	W	W	[b]
Alloy X	Ni-Cr-Fe	Mo	1, 7, 8a, 12	X	X	X	[b]

[a]HASTELLOY is a registered trademark of the Cabot Corporation. The High Technology Materials Division, Cabot Corporation, will be pleased to provide detailed individualized data on any of these alloys.
[b]Process not recommended for use in corrosion-resistance applications.

TABLE 10.4 Characteristics of Nickel-Based Alloys

1. Solid-solution strengthened.
2. Precipitation (age) hardenable.
3. Strength proportional to Al–Ti content.
4. Larger amounts of Al–Ti decrease weldability.
5. Columbium improves strength *and* weldability.
6. Molybdenum and cobalt add strength, do not affect weldability.
7. Used in gas turbines, and in aircraft and spacecraft.
8. Weldability: a, good; b, TIG, PAW, EBW, brazing, soldering.
9. Acid, chlorine, and alkaline resistant.
10. Resistant to severe cyclic heating conditions.
11. Corrosion resistance good at elevated temperatures.
12. Good corrosion resistance at extremely high temperatures [up to 1100° C (2012° F)].
13. Excellent machinability.
14. Acid resistant.
15. Lower carbon or silicon.
16. Resistant to hot sulfuric, phosphoric, and nitric acids [up to 760° C (1400° F)].
17. Cryogenic applications (to −160° C).
18. Excellent resistance to hydrochloric acid.
19. Exceptional resistance to a wide variety of severely corrosive media.
20. Excellent resistance to halide salts.
21. Good welding of dissimilar high-temperature alloys.
22. Good corrosion resistance to seawater and brackish waters.
23. Strengthened by cold working.
24. High thermal and electrical conductivity.

be completely removed. The thermal shock method is generally adequate; that is, when the joint has cooled to about 370° C (700° F), it is plunged into water at room temperature.

Copper and nickel brazing are used for high in-service temperatures. Consult the manufacturer for proper procedures. These applications are usually carried out by furnace heating in controlled atmospheres.

Soldering of the nickel alloys approximates that of soldering stainless steel (see Chapter 9).

Cutting Nickel-Based Alloys

The cutting and grooving of nickel alloys are readily feasible using the air carbon-arc process (see Chapter 13). Gas tungsten-arc (TIG) is the better choice for cutting. Thicknesses up to 15 mm are most commonly cut by this method, but it may be applied to thicknesses up to 5 cm. Use DCEP, a gas flow of 33 1pm (70 cfh) (80% argon, 20% hydrogen), about a 4-mm thoriated-tungsten electrode, a constricted arc, 400 A, and 70 to 85 arc volts. To eliminate adherence of dross on the bottom of the cut, clamp a waster plate of 3 to 4 mm mild steel to the bottom of the plate being cut.

Where available, plasma arc is, of course, the best of all possible choices for cutting nickel alloys. Use nitrogen–hydrogen on material up to 12.5 cm thick and argon–hydrogen for greater thicknesses.

Nickel-Based Alloys

TABLE 10.5 Silver Brazing Alloys

Classification		Nominal Composition				Solidus[a]	Liquidus[a]
AWS	SAE	Ag	Cu	Zn	Other	[°C (F°)]	[°C (°F)]
BCuP-5		15	80		5 P	643 (1190)	802 (1475)
BAg-2	AMS-4678	35	26	21	18 Cd	607 (1125)	702 (1295)
BAg-4		40	30	28	2 Ni	671 (1240)	780 (1435)
BAg-1	AMS-4679	45	15	16	24 Cd	607 (1125)	618 (1145)
BAg-1a	AMS-4770	50	$15\frac{1}{2}$	$16\frac{1}{2}$	18 Cd	627 (1160)	635 (1175)
BAg-3	AMS-4771	50	$15\frac{1}{2}$	$15\frac{1}{2}$	16 Cd 3 Ni	632 (1170)	688 (1270)
	AMS-4774	63	$28\frac{1}{2}$		6 Sn 2.5 Ni	690 (1275)	802 (1475)
BAg-8		72	28			780 (1435)	780 (1435)

[a]Temperatures are approximated; consult manufacturers for precise data.

Gas-Welding Nickel-Based Alloys

The following guidelines should be observed when gas-welding nickel-based alloys.

1. Gas welding is not recommended for the HASTELLOY alloys.
2. When welding other nickel-based alloys, use only acetylene as fuel, because the other fuel gases do not provide the necessary high temperatures. Use only bottled acetylene; generator acetylene can contain up to 0.05% sulfur.
3. Use a slightly carburizing flame, not more than $1\frac{1}{4}$ ×, and follow the same procedures as those for gas-welding stainless steel (see Chapter 9). Keep in mind that when welding chromium-bearing alloys, excess oxygen produces chromium oxides that are not easily soluble, and when welding nickel–copper alloys, an excess of oxygen will produce copper oxides (also called cuprous oxides), which embrittle the weld and reduce corrosion resistance.
4. Flux is required for welding nickel–copper, nickel–chromium, and nickel–chromium–iron alloys but not for nickel 200 or 201. If fluxes are available commercially, they should be chosen to approximate the following as closely as possible:
   (a) *MONEL alloys 400 and 404:* Mix calcium fluoride 16%, sodium fluoride 4%, borium fluoride 60%, borium chloride 15%, and 5% gum arabic.
   (b) *INCONEL alloys 600 and 800:* Mix 1 part sodium fluoride with 2 parts calcium fluoride, and add 3% hematite and 0.05% Duponol ME.

(c) *MONEL alloys K-500 and 502:* Mix 2 parts of either of the fluxes listed above with 1 part lithium fluoride.

5. Fluxes should be mixed to a thin slurry (watery, insoluble consistency). Apply flux to both sides of joint and let dry before welding. *Do not use borax.*

COPPER-BASED ALLOYS

Copper and most copper-based alloys are readily welded by the maintenance welder, the choice of process depending on the equipment and filler metals available, and, of course, on the size, shape, and elemental analysis of the metal. The following sections provide guidelines for welding various copper-based alloys.

Tough Pitch Copper

1. Soft soldering and silver brazing are readily feasible by any method. Minimal fluxing is required if the metal is clean. The brazing temperature should not exceed 760° C (1400° F).
2. Fusion welding may result in a weakened zone near the weld which can reduce the strength to as low as 50% of the original when using the oxyacetylene flame (hydrogen is introduced). The weldment may then be restored to close to the original strength by reheating the weld area and forging it with a hammer.
3. Due to copper's high thermal conductivity, the oxyacetylene process will require a larger tip and perhaps a broad preheat before the welding joint may be melted. Two torches may be required. A good brazing flux should be used on the parent metal and filler rod when fusion welding with oxyacetylene.
4. In carbon or tungsten arc welding, paint the parent metal with a brazing flux. The parent metal should be preheated prior to the start of the weld and the arc should be very intense [up to 450 A for 12 mm ($\frac{1}{2}$ in.) plate].
5. In shielded metal-arc welding, electrodes should be deoxidized copper. Preheating is necessary due to the high thermal conductivity of the parent metal. Adequate preheating and continued heating permit a lower-amperage setting for the arc.

Deoxidized Copper

1. All methods are readily adaptable. Since there is no cuprous oxide in the metal, there will be no adverse effects from fusion welding by either arc or oxyacetylene. Procedures 3, 4, and 5 for tough pitch copper should be followed.
2. If TIG welding, use deoxidized copper filler; argon, helium, or mixed gas;

ac, high-frequency stabilized, with DCSP/DCEN; and a thoriated-tungsten electrode.

Aluminum-Copper Alloys (Aluminum Bronze)

1. Soft soldering of aluminum bronze has always been considered unsatisfactory due to the rapid formation of oxides at such low heats; but some specialized fluxes and solders are worthy of note, such as Marco, Eutectic, and others.
2. Silver brazing is satisfactory if done with care. Start with scrupulously clean metal, use a fluoride flux, and apply silver at the very moment the metal reaches bonding heat (similar to silver brazing of stainless steels and the nickel alloys).
3. Fusion welding by oxyacetylene is feasible, but only with a high-fluoride flux. At present the results must be classified as being less than excellent.
4. Fusion welding with the carbon arc is satisfactory when done with a fluoride flux (preferably suspended in alcohol). If a second pass is required, the welded area should be cooled and the oxide removed by grinding before application of the second pass and any subsequent passes.
5. Fusion welding with TIG is relatively easy using RCuAl-2 filler; argon gas; ac, high-frequency stabilized; and a thoriated-tungsten electrode. Adequate shielding, concentration of heat, and small weld pool will usually allow the metal to fuse before the oxides can interfere.
6. Fusion welding by the MIG process is also relatively easy, using ECuAl (A-1, A-2, or B, depending on the analysis of the parent metal).
7. Special electrodes for shielded metal-arc welding (stick) are also available.
8. The preheating recommendations outlined previously should be followed.

Beryllium-Copper Alloys

1. Soft soldering varies from difficult to unsatisfactory. Follow the practices outlined in "Aluminum–Copper Alloys."
2. Silver brazing is satisfactory. Follow the procedures outlined in "Aluminum–Copper Alloys."
3. Fusion welding by the oxyacetylene process is not recommended.
4. TIG welding will give the optimum results. Follow the procedures outlined in "Aluminum–Copper Alloys," using ac, high-frequency stabilized. If a flux is necessary, do not use a fluoride. The TIG arc will vaporize the fluoride, and the fumes are damaging to the lungs.
5. Strength after welding may be improved by cold or hot peening, or by heat treating.
6. Beryllium copper mist is toxic. Weld only under well-ventilated conditions. High-velocity ventilating is recommended to reduce dust, mist, and so on, to absolute zero.

Copper-Zinc Alloys (Brasses)

This group should be divided into two subgroups, with a dividing point of 20% zinc.

1. Those alloys, such as gilding brass (5% Zn), commercial bronze (10% Zn), red brass (15% Zn), and low brass (20% Zn) present no real problems but differ from copper in that the melting point is progressively lower as the zinc content is increased, and care in overheating must be exercised lest the zinc vaporizes. Serious vaporization of zinc occurs just a few degrees above the melting point of the alloy being heated.
   (a) These brasses are easily fusion-welded by the oxyacetylene process if a little care is exercised not to vaporize the zinc.
   (b) If TIG welding, use RCuZn (B, C, or D); argon gas (helium or mixtures, also); ac, high-frequency stabilized; and a thoriated-tungsten electrode. Preheating is recommended. Some concentration of the heat on the tip of the filler rod lessens the probability of zinc vaporization by heating the weld puddle indirectly (in this case, use a zinc-free filler rod).
   (c) If necessary, a zinc-free copper alloy electrode may be used to stick-weld, but the process is not recommended.
   (d) The MIG process is not recommended.
   (e) Soft soldering, silver brazing, and brazing are readily feasible by observing the ordinary precautions.
2. These copper-zinc alloys are the more common brasses. They contain up to about 40% zinc and include the *yellow* brasses, cartridge brass, and Muntz metal.*
   (a) These brasses are readily soft-soldered and silver-brazed. Brazing, however, should not exceed a bonding temperature of 815° C (1500° F).
   (b) These brasses may be fusion welded with a high-zinc bronze such as Tobin bronze. This type of filler rod is a braze on the lower-zinc brasses but a true fusion weld on brasses with more than 30% zinc. (Think "melting points.")
   (c) Observe welding recommendations for oxyacetylene and TIG welding as outlined for group 1.

Copper-Zinc-Lead Alloys

These alloys include the leaded brasses, forging brass, architectural bronze, the Muntz metals, and CDA class 3xx. These brasses may be soft-soldered, silver-

---

*Muntz metal contains some lead and therefore is treated with the next group of alloys, the leaded brasses.

Copper-Based Alloys

brazed, brazed, and fusion-welded. The procedures discussed previously for the brasses should be applied but with added precautions relative to vaporization of the zinc. The oxyacetylene process is best and although neutral flame is more generally used, an oxidizing flame tends to reduce or eliminate the vaporization of zinc. (The flame produces a film of zinc oxide over the weld pool, which inhibits the escape of the zinc vapor.) TIG welding may be feasible by careful indirect heating as described in "Copper-Zinc Alloys," group 1.

Copper-Zinc-Tin Alloys

These *tin* brasses include Admiralty and Naval brasses, and the CDA group 4xx. Follow the same procedures and observe the same precautions as in working the leaded brasses.

Copper-Tin Alloys (Phosphor Bronze)

The tin content affects the welding of these alloys by oxidizing rapidly and by giving the alloy a wide plastic range (i.e., a wide temperature range between liquid and solid). Phosphorus acts as a deoxidizer of tin.

These bronzes may be soft-soldered, silver-brazed, gas-welded, and arc-welded. If using the oxyacetylene process, use a high-zinc yellow-bronze filler metal if permissible. Metal electrodes are available for shielded metal-arc welding (stick). TIG welding should not concentrate the arc because the phosphorus will evaporate to an appreciable depth in the metal. To reduce this effect, draw a long arc, say 12 to 24 mm ($\frac{1}{2}$ to $1\frac{1}{2}$ in.).

Copper-Silicon Alloys

The copper-silicon alloys are the easiest of all the copper alloys to weld. They may be silver-brazed, brazed, metal-arc welded (SMA and GMA), and welded with TIG and carbon-arc processes. (If leaded, observe the precautions discussed previously for the leaded brasses.) Since the thermal and electrical conductivity is reduced to one-twelfth that of copper, preheating or concentration of heat is no problem.

In oxyacetylene fusion welding, use a slightly oxidizing, concentrated flame and a good brazing flux. If TIG welding, use argon gas, DCSP/DCEN, a RCuSi-A filler rod, and a thoriated-tungsten electrode. If soft-soldering, clean the metal thoroughly and immediately coat it with flux before applying any heat. If the metal becomes heat-tinted prior to the application of flux, the solder will not tin, and then the metal will have to be cooled and the process restarted. Use a zinc chloride flux or zinc chloride with ammonium chloride. The solder must be applied at the precise bonding temperature, and overheating will necessitate a restart. Powdered solder mixed with a paste flux ensures that the solder tins out at the proper bonding temperature.

NICKEL-SILVER ALLOYS

The *nickel-silvers* are, in reality, copper–nickel–zinc alloys. Nickel is used to strengthen, harden, and whiten the metal; and zinc appreciably reduces the amount of nickel required to obliterate the redness of the copper content. Nickel reduces the fusion weldability compared with the brasses because nickel oxide is refractory and resists fluxing reactions.

Due to the vaporization characteristic of zinc, arc welding is not recommended. Oxyacetylene fusion welds may be made by using copious amounts of flux. All of the nickel-silvers may be easily soft-soldered and silver-brazed. A few nickel-silvers, such as the Ambracs, contain so little zinc that they should be treated with the cupronickels.

CUPRONICKELS

With few exceptions, the copper–nickel alloys, CDA group 7xx, are readily weldable. Brazing, silver brazing, and soft soldering are no problem by observing only the ordinary precautions. Special flux-coated electrodes are available for easy SMAW (stick). TIG welding is relatively easy: Use argon gas, DCSP/DCEN, a RCuNi filler, and a thoriated-tungsten electrode. Argon gas is also recommended for MIG welding.

MANGANESE BRONZES

Observe practices for leaded brasses.

OTHER NONFERROUS METALS

The metals listed in Table 10.6 have been variously classified as *other nonferrous metals, difficult-to-weld metals,* or *exotic metals.* We have grouped them into reactive, refractory, low-melting, and precious metals.

TABLE 10.6 Other Nonferrous Metals

Metal	Reactive	Refractory	Melting Point [°C (°F)]
Beryllium	×		1376 (2330)
Columbium		×	2567 (4475)
Molybdenum		×	2610 (4730)
Tantalum		×	2995 (5425)
Titanium	×		1668 (3035)
Tungsten		×	3410 (6170)
Vanadium	×		1750 (3182)
Zirconium	×		1795 (3365)

Low-Melting Metals

The reactive and refractory metals are described as follows:

1. *Reactive:* The metal has a high affinity for oxygen and other gases (other than inert gases) at elevated temperatures. Some gas contamination is possible well below the melting point of the metal. Reactive metals have oxides whose melting points are considerably higher than the melting point of the metal itself. Unmelted oxides will appear in the weld deposit as discontinuities and weaken the metal. (Compare with the graphite flakes in gray cast iron.) At ordinary temperatures, the oxides of reactive metals resist further oxidation.

2. *Refractory:* These metals have extremely high melting points. At elevated temperatures, they are easily contaminated when exposed to air. Molybdenum and tungsten, for example, have a high oxidation rate at temperatures above 800° C (1500° F). The oxides of refractory metals do not afford much protection against further oxidation due to their lack of cohesion with the parent metal.

Since the offending gases are oxygen, nitrogen, and hydrogen, fusion welding of these metals by the oxyacetylene or oxy-fuel gas processes should not be contemplated. SMA welding does not provide adequate shielding.

The TIG and MIG processes may be used provided that shielding is adequate. Since any part of the metal heated above 535° C (about 1000° F) must be protected from contamination by the ambient air, a much wider shielded area must be provided. Use larger nozzles and a trailing shield. Backups should be used to provide a gas shield for the root side of the weld. If trailing-gas nozzles and suitable backups are not practical, use chill bars on the back side (root) and face sides as close to the weld seam as possible. Absolute cleanliness is essential (even to the point of removing fingerprints). *Recommendation:* Use GTAW; DCEN, high-frequency stabilized; a pointed thoriated-tungsten electrode, and the shortest possible stick-out. Argon gas is generally preferred, except in the case of tantalum, which is usually welded without filler wire, and the deeper penetration characteristic of helium is preferred.

Beryllium is a toxic metal and should not be welded without special precautions in ventilation and handling.

The foregoing discussion of these metals does not apply to the elements as they are used in alloying with other metals, even in the relatively high content of molybdenum in some of the HASTELLOY alloys.

LOW-MELTING METALS

Lead is generally welded with the oxyacetylene process. Use a 000 tip (75 drill size) with about a $1\frac{1}{2}$x flame. Flux should not be used. Metal should be clean and free of oxidation. Filler metal is usually sheared from sheet lead. If TIG welding is used, draw a relatively long arc. No special precautions are necessary except an awareness of the extremely low melting point [328° C (620° F)]. *Caution:* Since lead fumes are toxic, some form of active ventilation should be provided.

Zinc die-cast (sometimes called *white metal* or *pot metal*) is generally welded with the oxyacetylene process, observing the recommendations for lead welding. The melting point of the average alloy is just slightly higher than that of lead [385° C (725° F)]. Due to the lower wettability, however, the weld joint should be opened to about a 90° root opening. The filler metal and the parent metal should be heated equally, and the filler rod should be pressed gently into the melting parent metal. This pressure is necessary to overcome the skin resistance of the parent metal. *Puddling* with the filler rod or a spatula will be necessary only if the welder has melted filler into the weld joint without fusing it. The filler metal must fairly approximate the elemental analysis of the parent metal; therefore, the better practice is to melt down a like metal and cast one's own filler rods. This may be done by melting irreparable parts in a crucible of sorts and pouring the melt into a suitably positioned angle iron that has been dammed at both ends (similar to a tinker's dam). Zinc die-cast is easily distinguishable from aluminum and magnesium by its weight, which approximates that of steel.

As an added precaution against collapse of the weld area during heating, a backing of asbestos sheet may be used or a mold may be fashioned from commercially obtainable wet asbestos under such trade names as Form-Eze (Marco Weld Products) or Form-a-Jig (Eutectic), among others.

PRECIOUS METALS

It is not probable that many maintenance welders will encounter such metals as silver, gold, and platinum, but these metals are easily welded by oxyacetylene and TIG.

The welding of silver may be compared with the welding of copper. As a general rule, the silver encountered will be alloyed with copper, and silver brazing alloys are available that will meet or approximate almost any weldment encountered. Silver, like copper, has a low affinity for oxygen, and no special precautions are necessary. Silver is easily fused, brazed, or soldered.

Gold is generally welded with the oxyacetylene process. It is readily brazed or soldered. It is probably the metal most easily cold-welded. Although the melting point of gold is 1060° C (1945° F), the weldment will be of such minute section that the 000 or 00 tip (75 and 72 drill size) will be required.

Platinum may also be welded with the oxyacetylene process and TIG. The melting point of platinum is 1770° C (3218° F).

DISSIMILAR METALS

When welding dissimilar metals (such as stainless steel to steel, cast iron to steel, etc.), both the similarities and dissimilarities of the metals should be taken into

Dissimilar Metals

account. Steel and cast iron, for example, are similarly iron–carbon alloys, as the equilibrium diagram (Figure 5.10) will attest. The same diagram, however, shows them to have entirely different structures. Therefore, before welding cast iron to steel, the maintenance welder must understand the steel (low to high carbon, or tool-and-die) as discussed in Chapters 6 and 7, and must also understand cast iron (Chapter 8). A review of those chapters will show immediately that the greatest problem will be the cast iron.

A review of the nickel alloys will show that in the welding of MONEL to INCONEL, the nickel content of each metal is so similar that, certainly, a high-nickel alloy filler would be a good choice. But what of the dissimilarity of the other major alloying elements, copper and chromium? Should the welder use a pure nickel filler and justify the decision by arguing: "Both are predominantly nickel; therefore, the solubility on both sides of my weld will be good; and if I open up the joint and distribute my filler carefully, the copper in the MONEL will not be mixing with the chromium in the INCONEL"? (The welding of MONEL to INCONEL is, or should be, a most unlikely event, but the marriage serves to exemplify the problem.) Buttering each metal with the nickel filler before joining would, of course, reduce any admixture of copper and chromium virtually to zero.

Cast iron to steel:

1. Observe all of the precautions for welding cast iron outlined in Chapter 8.
2. Any of the processes and procedures of Chapter 8 may be used.
3. Since the tensile strength of steel is approximately twice that of gray cast iron, the weld surface on the cast iron should be approximately twice the area on the steel. For example, the leg of a fillet on the cast iron would be twice the dimension of the leg on the steel.
4. If studding is used, do not stud the steel.

Cast iron to stainless steel. Many welders will be tempted to use an ordinary stainless steel electrode, such as E-308 or E-309, but due to the high chromium content of such electrodes, a better choice would be either a pure nickel or a Ni-Fe electrode, or to butter the cast iron with either the nickel or the nickel–iron electrode and fill with the E-308 or E-309. (The only advantage of the pure nickel electrode is its greater ductility and the machinability of the fusion zone on the cast iron.)

Observe the precautions for welding cast iron (Chapter 8). If brazing and braze-welding, review those processes as applied to stainless steel (Chapter 9) and cast iron (Chapter 8).

Cast iron to other alloys. Use filler metal of the other alloy as it relates to steel and observe the precautions pertinent to cast iron. If brazing or braze-welding, refer to the problems of the other alloys and observe uniform heating of the cast iron.

Steel to MONEL. Use MONEL. (See "Nickel-Based Alloys," this chapter.)

Steel to INCONEL. Use INCONEL. (See "Nickel-Based Alloys," this chapter.)

MONEL to stainless steel. Use MONEL. (See "Nickel-Based Alloys," this chapter.)

INCONEL to stainless steel. Use INCONEL. (See "Nickel-Based Alloys," this chapter.)

Copper to steel. Use copper. If GTAW, use pulsed arc and concentrate the heat on the copper as much as possible to avoid picking up iron, which will embrittle the weld deposit. Or butter the copper with nickel (two layers preferred), being careful to minimize dilution with the copper. Heavy copper sections should be preheated prior to buttering [540° C (1000° F)].

Copper to stainless steel. Use an aluminum-bronze electrode or rod.

Brass to steel. Use aluminum bronze.

Aluminum bronze to steel. Use aluminum bronze.

Aluminum to steel. Dip-coat the steel with aluminum (or a compatible metal) and TIG-weld with a suitable aluminum alloy or use a special insert.

Aluminum to stainless steel. Same as aluminum to steel.

Aluminum to copper. Use an insert.

Aluminum to other alloys. Soft soldering is relatively easy, attaining tensile strengths up to 17.6 kg/mm^2 (25,000 psi). These soft solders are available commercially (Marco, Eutectic, etc.).

11

ALUMINUM AND ITS ALLOYS

UNDERSTANDING ALUMINUM

Aluminum is a white, relatively soft metal somewhat resembling tin in appearance. It is only slightly magnetic. In its pure state, it is highly corrosion resistant, due to the formation of a superficial film of oxides (alumina, or Al_2O_3) on its surface that strongly resists further oxidation. This film gives aluminum a frosted appearance, and it must be broken down and floated out of the weld puddle if a sound weld is to be produced. Aluminum is strongly electropositive and thus is susceptible to corrosion when in contact with other metals. Although aluminum is one of the most abundant constituents of the earth's crust, it was not discovered until 1825 and was first introduced in 1852 at a cost of $545 per pound (a goodly sum in the 1850s).

The principal source of aluminum is from the ore bauxite: first in the form of aluminum oxide and subsequently converted to a pure metal by electrolysis of the oxide. In its pure form (99.5 to 99.9%) aluminum is very soft and ductile, having a tensile strength from 9000 psi in its single-crystal state to 25,000 psi when fully work-hardened. Aluminum may be readily rolled, drawn, and extruded. It is generally alloyed with other metals when casting. These alloys are cast in sand or metal molds and are easily die-cast.

The constituents most commonly alloyed with aluminum (Table 11.1) are copper (Cu), silicon (Si), manganese (Mn), zinc (Zn), and magnesium (Mg), generally used as follows.

Si: for corrosion resistance, fluidity, precipitation hardening at high temperatures

Mn: to increase tensile strength and aid in work hardening
Mg: for precipitation hardening at room temperature
Cu: for corrosion resistance and age hardening
Cr: to increase tensile strength and corrosion resistance
Zn: principally as an aid in die casting, its expansion during the cool-down making sharp corners possible

TABLE 11.1 Aluminum Alloys

Number Series	Major Alloying Constituent[a]
1xxx	None (99.0% minimum aluminum)
2xxx	Copper
3xxx	Manganese
4xxx	Silicon
5xxx	Magnesium
6xxx	Magnesium-silicon
7xxx	Zinc
8xxx	Other elements

[a]Consult manufacturers' data sheets for more detail.

Wrought Aluminum Alloys

These metals develop higher physical properties due to the beneficial effects of rolling, forging, and extruding. As in the case of steels, hot working refines the grain structure and cold working distorts the grains, both of which strengthen the metal. Smaller amounts of alloying ingredient are used in wrought metals than in castings.

Hot working is done between 480 and 315° C (900° F). It is much more difficult than the hot working of steel.

Cold working is done as follows:

1. The aluminum is cast in the ingot form.
2. The ingot is reduced by hot working to the appropriate size and shape for cold working.
3. The cold working begins where the hot working leaves off.

The degree of cold-worked hardness is often referred to as: 0, $\frac{1}{4}$ H, $\frac{1}{2}$ H, $\frac{3}{4}$ H, H; with 0 meaning soft (annealed) and H meaning fully hardened.

Aluminum-Copper Alloys

The aluminum-copper alloys, diagrammed in Figure 11.1, are mixtures of solid solution (alpha) and a eutectic that is composed of a mechanical mixture of solid solution plus the compound $CuAl_2$. The eutectic increases in amount until

Understanding Aluminum

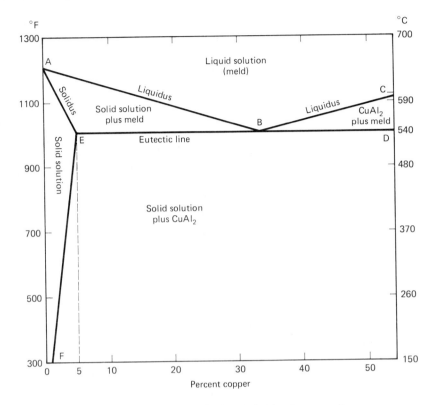

Figure 11.1   Equilibrium diagram of aluminum–copper alloys.

100% eutectic is found at 33% copper and 67% aluminum.*

At about 540° C (1005° F) aluminum will hold 5% copper in solution and, upon cooling, this solid solution will precipitate its copper in the form of $CuAl_2$ until, at 150° C (300° F), aluminum will hold only 0.5% copper in solution (see line FE). The precipitated $CuAl_2$ remains within the grains.

Thus if we take an alloy of 95% aluminum and 5% copper, we see that it freezes as a solid-solution alloy. Upon slow cooling of this solid solution, the copper precipitates out of solution along the FE line. This condition we can consider as the normal or annealed condition: a structure of rather hard $CuAl_2$ constituent embedded in a soft aluminum matrix.

Heat Treatment of Aluminum

To anneal commercially pure aluminum after it has been cold-worked, it is heated slightly above 345° C (650° F) (softening is instantaneous) and allowed

*This straight-line diagram is for illustrative purposes only and should not be used where absolute accuracy is required.

to cool down to room temperature. The heating is not critical but must slightly exceed the recrystallization temperature. Wrought aluminum alloys may be annealed in the same way, except that a very slow cool-down is required if the heating exceeds 345° C (650° F) by more than 10 degrees.

The aluminum–copper diagram (Figure 11.1) shows that heating the aluminum to above 345° C (650° F) will recrystallize the aluminum without appreciably affecting the $CuAl_2$ component. The rate of cooling is not important if the $CuAl_2$ constituent does not go into solution. It can be seen, then, that if the $CuAl_2$ goes into solution and is cooled too rapidly, the $CuAl_2$ will be trapped in the solid solution similarly to the entrapment of carbon in austenite in the iron–carbon diagram.

If the alloy is to be completely annealed for severe cold-forming operations, it should be heated to a temperature from 400 to 425° C (750 to 795° F) and soaked for about 2 hours; then cooled 20 degrees per hour down to 260° C (500° F).

Solution heat treatment. It can be seen by the *FE* line in the diagram that if we heat a normally annealed 4% copper alloy to 480° C (900° F), the $CuAl_2$ component will go into solution. [Complete solution will take about 15 hours at 480° C (900° F)]. If we then cool the solid solution rapidly as in a water quench, most of the $CuAl_2$ will be trapped in the solid solution. Oil should be used if a slower quench is desirable. This treatment enhances the corrosion resistance and strength. Cold working will then further enhance the strength without affecting the corrosion resistance.

After 30 minutes to 1 hour at room temperature, the strength is then further increased by a natural phenomenon called "precipitation hardening" or "age hardening," similar to the decomposition of residual austenite in martensitic steels. This precipitation of $CuAl_2$ takes place at the aluminum grain boundaries and along the slip lines of the aluminum crystals, the result being submicroscopic particles that resist slippage (as in the usual analogy of sand interfering with the sliding of a deck of cards). This resistance is considerable. A 17S alloy, for example, heat-treated as above, has a tensile strength of 31 kg/mm^2 (44,000 psi), and the precipitation hardening increases the tensile strength to 43.5 kg/mm^2 (62,000 psi) without reducing the ductility.

It should be noted that this precipitation hardening is possible in 17S aluminum due to its containing magnesium. Alloys without magnesium age-harden much less. The addition of silicon does little for age hardening at room temperatures but greatly enhances age hardening at elevated temperatures. The procedure for AlMgSi alloys is to quench in water from 515° C (960° F) and age-harden at 155 to 170° C (310 to 340° F) for up to 18 hours for maximum results.

The heat treating of aluminum castings requires more time at the suggested temperatures than does the heat treatment of wrought alloys.

WELDING ALUMINUM

Aluminum is one of our most weldable of metals and may be welded by most established welding processes. Choosing the process to be used will depend on:

1. The availability of equipment and the skill of the welder in the various processes
2. The type of aluminum in question
3. The size, shape, and gauge of the aluminum
4. The availability of ancillary equipment such as heat-treating processes

Characteristics that affect weldability are:

1. Low melting point [657° C down to 482° C (1215 to 900° F)]
2. High thermal conductivity, expansion, and contraction
3. High oxidation susceptibility
4. Gas porosity
5. No change of color when heated (even to the melting point)
6. Melting point of oxides is three times that of the metal itself

General information for best results when welding aluminum is as follows:

1. Non-heat-treatable alloys (1xxx, 3xxx, and 5xxx), when overheated, lose the valuable properties induced into the metal by cold working (rolling, forging, extruding, etc.). The strength, however, will not be lower than in the same metal in the annealed condition. The more concentrated the heat, the narrower the heat-affected zone.
2. Pure aluminum (1xxx) and magnesium alloys (3xxx) are readily weldable by TIG, MIG, stick, or gas.
3. Aluminum–magnesium series (5xxx) have good weldability because of their high strength in the annealed condition. Those with more than 2.5% Mg should not be gas- or stick-welded.
4. Most of the heat-treatable alloys (2xxx, 4xxx, 6xxx, and 7xxx) are considerably strengthened by their heat treatment and therefore stand to lose the most from the annealing effects of welding.
5. Heat-treatable alloys in the 2xxx, 4xxx, and 7xxx series will give the greatest problem in welding due to the very high heat necessary to penetrate them. This resistance to penetration virtually eliminates gas welding, brazing, and soldering. Arc welding is difficult but feasible.
6. Aluminum should be welded at the highest possible speed: MIG, TIG, and stick are preferable.
7. The oxides that form on the surface of aluminum have a melting point

of 2040° C (3700° F). The welder should not attempt to "burn through" this film and float it out, because the aluminum may melt beneath the film, with the result of poor fusion. The surface should be wire-brushed or sanded prior to welding. Chemolysis may also be used, and is preferable when surface is anodized.

8. Preheating is always desirable and is a necessity when gas-welding. In TIG welding, preheating often saves time. In stick welding preheating permits an easy flow of metal from the first electrode and the use of lower current. (The proper heat for aluminum stick-welding electrodes is about the same as for the comparable size of E-6010.)

9. Avoid overwelding and excessive penetration. Use the minimum weld deposit and minimum penetration required to do the job.

10. Heat-treatable alloys are prone to considerable precipitation of their alloying ingredients during welding. The precipitated elements migrate to the grain boundaries, and the effect is discernible by a change of color adjacent to the weld. If appearance is of critical importance, the weld must be solution heat-treated and re-anodized.

11. When appearance is a factor, it must be remembered that the weld is essentially a casting process and cannot have the same appearance as the wrought aluminum. The best solution is to use a filler metal of the higher alloy content.

12. Aluminum castings present the same problems as wrought aluminum except that they are more prone to cracking. Preheating should be higher [around 315° C (600° F), never above 370° C (700° F)]. Arc welding is best, and the faster, the better. If stick welding is used, a 5% silicon electrode is best.

Gas-Welding Aluminum

Gas welding of aluminum is far from satisfactory when compared with the results obtainable by other processes, especially TIG, but is often a necessary skill for the maintenance welder where the preferred equipment is not available.

One of the problems encountered by the gas welder of aluminum is that the melting point of the heated surface is much less apparent. The gas flame does not produce the bright, shiny, wet-surface melt that is characteristic of a TIG puddle. Part of this problem lies in the fact the oxyacetylene flame does not concentrate the heat as restrictively as does the TIG arc, and the heat is dispersed over a wider area. A second part of the problem is that the shielding of the puddle by using a flux that must be melted and floated away is not as effective as inert-gas shielding and does not permit as clear a view of the molten puddle.

If we ask an old-timer how he knows when the aluminum is near the melting point, he is apt to say, "by guess and by golly." Many welders first blacken the surface with a pure acetylene flame and then heat the surface broadly with

Welding Aluminum

a neutral flame until the surface is cleared of all traces of the carburized film. This should be about 235° C (445° F). Their next step is to apply flux to the weld area and heat until the flux melts and flows freely. The flux used should have a melting point slightly below the melting point of the aluminum. When the flux is running freely, the welder touches the filler rod to the parent metal with a slight wiping action. At the fusion temperature, the filler rod will wipe away some of the parent metal and with the flame cone directed at the filler rod, the filler will melt and fuse with the parent metal. The welder then proceeds with the fusion weld in a manner similar to the gas welding of steel, except that he must add flux to break up and float away the oxides that form while welding.

Although the neutral flame may be considered ideal, most welders use a slightly reducing flame (about 2 ×). This flame is used because of aluminum's great affinity for oxygen, and a creeping regulator is not so noticeable when it creeps toward the excess-oxygen side. A slight excess of acetylene does not seem to have any serious ill effects.

Gas welding of aluminum is considered one of the "dirtiest" of all welds and complete cleansing is necessary after welding to minimize postweld corrosion. Even with the best of cleaning, anodizing of the gas weld is unsatisfactory.

Aluminum castings present an additional problem in that they are often impregnated with foreign substances such as oil or grease in transmission cases or food acids in the case of cooking vessels. Although a good soak in an oven will boil out more of these substances, good preheating with a torch will usually suffice. When a weld joint is exceptionally stubborn, the welder may elect to use a spatula for raking the joint down to cleaner metal. (An E-6010 electrode flattened at its bare end will make an excellent puddler.)

Alloys containing high percentages of magnesium (more than 2.5%) should not be gas- or stick-welded. Lap joints should not be used when gas welding due to the entrapment of residual flux between the two surfaces. If lap joints are unavoidable, weld both sides and boxweld the ends of the joints.

TIG-Welding Aluminum

Choice of gas shielding. Argon and helium are the two gases used most commonly for shielding with TIG welding of aluminum. The distinct characteristics of the two gases are:

1. Argon is heavier than air and thus permits more stick-out of the electrode per unit of gas flow.
2. Helium is less expensive, but more gas escapes into the air because of its being lighter than air.
3. Argon gives a quieter, smoother arc with ac welding.
4. Helium gives a hotter arc and deeper penetrating power, thus giving greater welding speeds on heavier sections.

5. Argon maintains the arc better when the arc length is varied.
6. The two gases may be mixed in any desired proportions in order to compromise their differences.
7. Argon with high-frequency-stabilized ac is by far the preferred method of welding aluminum.

Choice of current. Aluminum may be welded by dc straight polarity, dc reverse polarity, or ac. The distinct characteristics of the three methods are:

1. DCSP heats the parent metal rapidly because of the electron stream impinging on the parent metal at a high velocity, thus giving the weldment the greater part of the heat of the arc. The positive gas ions are attracted to the negative electrode with a resultant cooling effect. Thus with straight polarity, higher currents may be used on electrodes of the same size. A deeper, narrower bead is possible: narrow due to the smaller electrode diameter and deep due to the greater penetration. DCSP is preferred on the heavier welds [12 mm ($\frac{1}{2}$ in.) and above].
2. DCRP reverses the electron-ion behavior described above; that is, the work is cooler and the electrode hotter. The result is that a much larger electrode is needed for a comparable current setting. The disadvantages are that the higher heat consumes the so-called "nonconsumable" electrode at a much higher rate, and the larger electrode size inhibits the view of the weld puddle. The advantages are:
   (a) The shallow penetration makes reverse polarity preferable on very thin sections.
   (b) It produces better cleaning action in the puddle. The cleaning action may be either the electrons leaving the weldment (attracted to the positive electrode) or the gas ions impinging on the weldment (ions, being positive, are attracted to the negative weldment).
3. In the case of ac, since the current is alternately negative or positive, it would seem that we could say that ac welding is a combination of DCSP and DCRP welding. In effect, however, a form of rectification takes place because the oxides, moisture, and scale on the weldment tend to prevent the flow of current in the reverse direction. This is made apparent by the fact that we must use a high-frequency unit (high voltage, high frequency, low current) to keep the arc from extinguishing on the reverse cycle. Since the heat is evenly distributed between electrode and weldment, the penetration is less than that of DCSP but greater than that of DCRP, and the contour of the weld shapes itself somewhere between the wide shallow bead of DCRP and the deep narrow bead of DCSP. High-frequency-stabilized current has the following advantages:
   (a) All-position welding is easier.
   (b) Less current is needed.
   (c) There exists a wider current range for the size of the electrode.

(d) The electrode does not get as hot (less cooling required).
(e) A longer arc is possible (very advantageous in the sweating technique of wearfacing and in braze-welding high-zinc brasses).
(f) It results in better arc stability.
(g) Most important, it permits starting the arc without momentarily touching the weldment.

Choice of electrode. Tungsten electrodes are used because their melting point is the highest of the known metals [3400° C (6150° F)]. The heat of the arc is rated at 2760° C (5000° F) and does not melt the electrode. Although it is called a nonconsumable electrode, it does waste away slowly, similarly to the wasting away of carbon electrodes.

Tungsten electrodes are presently available in pure tungsten, thoriated tungsten (1 to 2% thorium), and tungsten with 1% zirconium. Pure tungsten is generally used with ac welding. Avoid thorium. The zirconium type is also excellent for ac welding, and has certain advantages:

1. High resistance to contamination
2. More stable arc
3. Higher current-carrying capacity

Thoriated tungsten, for DCSP welding, has a lower rate of consumption, runs cooler, has better touch starting, and has improved arc stability. The "striped" electrode is a thoriated-tungsten type with the thorium (2%) added to the electrode in the shape of a wedge inserted the full length of the electrode. The striped electrode has the following advantages:

1. Smaller-diameter electrodes may be used
2. Improved arc stability
3. Less spitting
4. Excellent balling
5. Easy starting
6. Compatibility with all metals

Choice of filler rod. It is generally not necessary to match the filler material exactly with the weldment. For the maintenance welder to do so would require an inventory of hundreds of different welding rods. Table 11.2 is offered as a general guide. For additional information, consult aluminum manufacturers' literature.

Welding techniques. The welder who is skilled in gas-welding aluminum will have no problem adapting his skill to the TIG welding of aluminum. Actually, he will find the experience so gratifying that he may well forsake the oxyacetylene method forever, except when TIG-welding equipment is not available.

Setting the gas flow, time of after-flow, and electrode stick-out are the same as for welding other metals. The length of stick-out and size of electrode determine the cfh setting [$\frac{1}{8}$-in. stick-out, 9.4 lpm (20 cfm) should handle most jobs for the $\frac{3}{32}$-in. electrode]. The after-flow should be limited to the several seconds necessary to shield the solidifying weld puddle from contamination by the ambient air. The stick-out should be as short as possible, consistent with good vision of the weld puddle. If a long stick-out is required to reach a hard-to-get-at corner, the cfh rate must be increased adequately.

TABLE 11.2 Filler Selection

Sheet or Plate	TIG	MIG
EC	EC, 1100	EC, 1100
1100	1100, 4043	1100, 4043
3003	1100, 4043	1100, 4043
3004	4043, 5154, 5356	4043, 5154, 5356
5005	4043, 5356	4043, 5154, 5356
5050	4043, 5356	4043, 5154, 5356
5052	5154, 5356, 4043	5154, 5356, 4043
5056	5183, 5356	5183, 5356, 5154
5154	5183, 5386	5183, 5356, 5154
5083	5183, 5356	5183, 5356
5086	5183, 5356[a]	5183, 5356[a]
6061	5356, 4043[b]	5356, 4043[b]
6062	5356, 4043	5356, 4043
6063	5356, 4043[b]	5356, 4043[b]

[a]Do not use 5154 filler on 5086 plate, due to transverse cracking susceptibility.

[b]Use only 4043 filler if 6061 or 6063 is used as the electrical conductor.

When using high-frequency-stabilized ac, the welder should find the "ball" point method to be the most satisfactory. An electrode operating on such a current will ultimately ball-up spontaneously, but the best practice is to start the weld with a balled point. Balling the point is a relatively simple procedure.

1. Set the machine to deliver about 40 A ($\frac{3}{32}$ electrode).
2. Set the polarity switch on "reverse" (dc+).
3. Establish an arc on a practice piece (copper preferred, such as a penny) and maintain the arc until a shiny ball tip appears on the end of the electrode.

The diameter of the ball should barely exceed the electrode diameter. Reset the machine to ac and the proper welding current for the job at hand.

When using DCSP, the electrode point is best prepared by grinding it to a relatively sharp point. Welders often call it a "needle point," although that may be an overstatement.

When using a high-frequency unit with ac, the step-by-step procedure is as follows

1. Touching the weldment to establish the arc is not necessary. Place the torch parallel to the weldment with the cone touching the metal. Depress the foot pedal or fingertip control and raise the torch toward a perpendicular position, being careful not to touch the weldment with the electrode tip.
2. Hold the torch about 15 to 20° backward from the line of travel (similar to forehand gas welding).
3. Heat a confined area until the metal sweats. If the aluminum is clean, the heated spot should show a very shiny, wet surface melt. If so, the spot is ready for fusion. (Time can be saved on heavy sections by preheating with a torch.)
4. Add a droplet of filler as in gas-welding steel, and if it flattens out with a shiny surface, your weld start is proper.
5. Proceed as in gas welding. The torch may be moved steadily ahead, may be oscillated very slightly, or may be moved back and forth in the line of travel, whichever suits the job or the whim of the welder. The size, shape, and appearance of the weld will, of course, depend on the rhythm of dipping the filler rod and the consistent moving of the torch.
6. Postweld cleaning is not necessary.

Precautionary Notes:

1. Heavy sections should be preheated.
2. Touching the weldment with the electrode or touching the electrode with the filler rod will have dire, or at least exasperating, results. Evidence of such touching is seen immediately as a black smudge on the weldment, and examination of the electrode will show aluminum adhered to the electrode tip. The welder now has two choices:
   (a) He may elect to conserve the preheating process and fight through the contamination until the electrode cleans itself.
   (b) He can dissemble the torch, regrind and re-ball the point, and clean the contaminated weldment.
   The smarter welder will know that inadvertent touching does happen regardless of degree of skill, and he will have a second, and even a third electrode ready for a quick change.
3. The longer the stick-out, the more likely is the accidental touching.
4. Accidental touching with the filler rod can be nearly eliminated by holding the filler rod almost perpendicular to the weldment. (Pushing the filler rod toward the puddle at about a 45° angle is asking for trouble.)
5. If shrinkage cracks occur in the crater, these cracks may be eliminated

by foldback, forward swing, or side swing; or by gradually lengthening the arc while adding filler metal, by making and breaking the arc several times while adding filler metal, or by using the foot pedal to reduce the weld current before breaking the arc.

6. If the electrode becomes contaminated with aluminum, the contamination must be removed. Minor contamination can be removed by increasing the current while maintaining an arc on a piece of scrap metal. Excessive contamination can be removed on a grinding wheel, but such gumming of the wheel is frowned on in most shops and schools. Contamination may also be removed by breaking off the faulty portion and forming a new contour on the tip. (Because of the high cost of tungsten electrodes, the operator should exercise care that a minimum of electrode is wasted.)
7. Never strike an arc on a piece of carbon.
8. If DCSP must be used, a high-frequency start should be used to avoid touch starting. The torch movement should be steadily forward, and the filler wire should be fed into the leading edge of the weld puddle (see also item 5 above). Use a thoriated-tungsten electrode. Since there is no cleaning action characteristic of the positive cycle of ac welding, precleaning the weld area is necessary. Preheating is not necessary, even on heavy sections. Do not expect to see the shiny, clean weld surface during welding. The surface oxides characteristic of DCSP welding are easily removed by gentle wire brushing.

Stick-Welding Aluminum

Because of the high welding speeds obtainable, stick welding of aluminum may prove advantageous on castings and heavy sections when the more desirable MIG equipment is not available. Sound welds can be produced that approximate the E-6010 and E-6011 welds in appearance.

Stick welding of aluminum presents no real problem for the welder who is skilled with the E-6010 electrode, except that verticals and overheads should be avoided if at all possible. A few practice beads will quickly acquaint the welder with the difference between the E-6010 and the aluminum stick.

Many welders believe that they must "set the heat up" when welding aluminum with a stick electrode. Actually, the current setting should be about the same as for a comparably sized E-6010 that has been set for a good arc action and may even be reduced as the weldment grows hotter.

The angle of the electrode should be at right angles to the weld bead—do not tilt the electrode out of the perpendicular. The arc should be so tight that the hand can feel the coating rubbing the metal.

After the arc is established, there is little danger of the electrode sticking, due to the rapid melt-off of the core.

The electrode must be fed into the puddle much more rapidly than with the E-6010. A long arc will result in excessive spatter and a lack of fusion.

Prepare the joint so as to minimize the need for meltdown of the parent metal. As in the gas and TIG welding of aluminum, the stick welding puddle is not as fluid as it is with steel.

Because of the high thermal conductivity of aluminum, the weldment should be preheated to at least 205° C (400° F). If the weldment is cold, the parent metal will "steal the heat" from the arc so rapidly that the first, or first several electrodes will not fuse but will spatter away as the arc preheats the weldment.

There is relatively little difference between the two electrodes, but most welders prefer an extruded coating to the dipped type. The electrode should have a 5% silicon content to enhance the fluidity of the melt.

Because of the minimal cleaning action of the arc, each bead should be brushed clean before depositing a subsequent bead. The weldment should be thoroughly cleaned after welding. Postweld anodizing is generally unsatisfactory and is not recommended.

SOLDERING ALUMINUM

For simplification, we may look at soldering as being a very low temperature brazing process. The essential difference is the temperature range [205 to 400° C (400 to 750° F)]. Choice of melting and bonding temperatures will depend on the strength desired and the temperatures to which the joint will be subjected when put in service.

There are four groups of solders for aluminum:

1. *Zinc-based solders:* Give shear strengths upward of 10.5 kg/mm^2 (15,000 psi); good corrosion resistance; bonding temperatures vary from 315 to 425° C (600 to 800° F).

2. *Zinc-cadmium-based solders:* Give shear strengths above 7 kg/mm^2 (10,000 psi); bond in the range 265° to 400° C (510 to 750° F).

3. *Tin-zinc-based solders:* Give shear strengths above 5 kg/mm^2 (7000 psi); have good corrosion resistance; bond at 290° C (555° F) and upward. One manufacturer of zinc-tin solder [melting point 265° C (510° F)] claims a shear strength of 14 kg/mm (20,000 psi). The same manufacturer also markets a solder that claims a tensile strength of 17.5 kg/mm^2 (25,000 psi) and melts at 370° C (700° F).

4. *Tin-lead solders (containing cadmium or zinc):* Give shear strengths above 3.5 kg/mm^2 (5000 psi); good corrosion resistance; bond at 232° C (450° F) and higher.

Table 11.3 lists types of aluminum according to solderability.

Fluxes for aluminum soldering must necessarily be of the corrosive type, and the joints made with such fluxes must be cleaned thoroughly after solder-

TABLE 11.3 Solderability of Aluminum Types

Excellent		Good	Poor (Difficult)	
1030	1175	3004	2011	7178
1050	1180	5005	2014	7277
1060	1187	5357	2017	4032
1070	1197	6053	2018	4043
1075	1230	6061	2024	4045
1080	1235	6062	2025	4343
1085	3003	6063	2117	5055
1090		6151	2214	5056
1095		6253	2218	5083
1099		6951	2225	5086
1100		7072	5050	5154
1130		8112	5052	5254
1145			5652	5356
1160			7075	

ing. They are generally termed "special purpose." Some manufacturers of aluminum solders have developed fluxes that perform as temperature indicators for their solders.

BRAZING ALUMINUM

The brazing method of joining aluminum is especially effective on sections too thin for welding, and for complex assemblies. Brazing rods giving shear strengths and tensile strengths up to 23 kg/mm^2 (33,000 psi) are manufactured by several companies. Many of these alloys may be used for fusion welding with TIG or oxyacetylene. Although brazed, some of these alloys, when cut through for inspection, give the appearance of having been fusion welded. The brazing technique is the same for aluminum as for other metals. Therefore, study or review of "Brazing," Chapter 2, is recommended, as well as the following guidelines.

1. Butt joints should be avoided; lock, lap, T, and upset joints are preferable.
2. As in all brazing, the filler rod must melt at a temperature below the melting point of the parent metal. If oxyacetylene is used, the torch flame should be feathered to about 3×.
3. Since aluminum does not change color under heat, a flux that acts as a temperature indicator should be chosen (generally available under the same numbered designation as the filler rod). The flux should melt and flow out thin and transparent at the proper bonding temperature.
4. When the flux is distributed, fluid, and transparent, the filler rod should

be applied as rapidly as possible in a manner similar to that used in silver brazing.
5. If the assembly is complex, that is, composed of heavy and thin sections, care must be taken in heating. The thin sections can obtain their heat by absorption from the heavy sections and from the outer periphery of the heat radiation. Again, the heating is the same as for other metals.
6. If the joint is a long one, the flux may be distributed along the entire length, and the filler applied with a sweeping motion or continuous travel.

Caution: Anodizing of brazed joints is unsatisfactory and is not recommended if a flux has been used. It is important that the capillary action carry the filler through and completely around the joint, so that the filler will displace any flux that might otherwise be trapped within the joint.

Aluminums that are easily solderable will usually be easily brazable (see Table 11.3).

CHARACTERISTICS OF ALUMINUM WELDMENTS

1. The weld metal, as deposited, is essentially a cast structure, influenced by its composition and rate of solidification. Generally, the higher rates of solidification produce finer microstructures and greater strengths.
2. The more concentrated the welding heat, the narrower the heat-affected zone.
3. The alloying constituents in heat-treatable aluminum alloys are dissolved in the aluminum at high temperatures by a process called *solution heat treatment*, and they are kept in solution by rapid quenching. Further hardening occurs by precipitation out of solution of the alloying constituents at room temperatures. (See "age hardening" and "precipitation hardening" in the Glossary.)
4. The welding heat exceeds the annealing temperature of strain-hardened alloys, resulting in lowered strength and increased ductility. These alloys (not heat-treatable) require very little time at the annealing temperature to soften them substantially.
5. Heat-treatable alloys require substantially more time at the annealing temperature to soften the metal fully. Thus the greater the speed of welding, the less the microstructural change, such as eutectic melting, grain boundary precipitation, or grain growth.
6. Heat-treatable alloys may be re-heat-treated after welding. All weld failures will then occur in the weld metal unless there has been some overweld or reinforcing weld. If overweld is left intact, failures usually occur in the fusion zone.
7. Where complete heat treatment is not practical, solution-heat-treated alloys may be welded as is and subsequently artificially age-hardened.

8. Various weld defects may occur.
    (a) Crater cracks may be avoided by proper welding technique. See "side swing," "forward swing," and "foldback" in the Glossary, as well as item 5 of the "Precautionary Notes" regarding welding techniques, earlier in this chapter.
    (b) Although they occur rarely, axial or longitudinal cracks are caused by incorrect filler alloy, too high or too low a rate of welding, incorrect edge preparation, or incorrect joint spacing.
    (c) Cold cracks can be avoided by putting in heavy enough welds to withstand cooling stresses occurring during solidification.
    (d) Incomplete penetration is caused by improper joint preparation, too low a welding current, welding travel too fast for heat input used, and because weld beads do not interpenetrate when welding is done on both sides. In fillet welds, lack of penetration is due to the bridging tendency and can be eliminated by proper welding technique, such as notching the filler wire in the weld puddle. There is some substance to the argument that interpenetration of two-sided butt welds is not desirable since the high heat required will broaden the heat-affected zone and even the slightest overweld makes facial fracturing difficult.
    (e) Incomplete fusion is due to failure to dissolve the surface oxides completely. This phenomenon is difficult to see in the weld puddle. Since the oxides of aluminum melt at approximately three times the melting point of the alloy itself, these oxides may not dissolve in the weld puddle and may function as a film between the weld deposit and the parent metal. The result is a superficial adhesion that may easily separate under loading or vibration.
    (f) Porosity is caused by entrapment of hydrogen in the weld deposit. Aluminum in the molten state can hold nearly 20 times as much hydrogen as it can in the solid state. This hydrogen is generally frozen out upon solidification but can become entrapped if the cooling rate is too rapid. This defect is not nearly as prevalent as it should be to warrant the attention it gets. Hydrogen entrapment is more likely to occur in GMAW than in other welding processes.
    (g) Inclusions may be metallic or nonmetallic. These inclusions occur in GTAW because an excessive current for a particular electrode size will result in tungsten being deposited in the weld metal. Tungsten inclusions are also caused by touching the electrode to the weldment and by touching the filler rod to the electrode. Tungsten inclusions are not a serious defect except in the very strict code-welding applications. Copper inclusions may occur because of burnback of the electrode to the contact tube. Copper inclusions cause brittleness and corrosion. Improper brushing with metallic wire brushes may result in parts of the bristles being incorporated in the weld metal. Nonmetallic inclusions occur because of improper precleaning of the weld area and entrapment of residual flux.

12

AUSTENITIC MANGANESE STEEL

UNDERSTANDING MANGANESE STEEL

Characteristically, manganese is a gray-white metal, relatively hard and very brittle. It is almost indispensable in the making of steel and is used as an alloying ingredient with the iron-carbon alloys as a toughening and hardening agent. In steels of less than 1% manganese, the manganese is used solely to combine with the sulfur and/or phosphorus to reduce embrittlement and hot-shortness. It is used in percentages up to 14% to toughen and improve hardenability, but, phenomenally, at 3 to 4%, it causes embrittlement.

Late in the nineteenth century, Sir Robert Hadfield discovered that a combination of carbon and manganese in steel in a ratio of 1:10 produced a steel that improved in toughness and durability by in-service abuse, such as hammering, battering, and applied stress. This steel became known as *Hadfield steel*.

Sir Robert has become all but forgotten by the modern welder, who calls it manganese steel and, sometimes, simply *manganese*. Engineers often call it austenitic manganese steel because of its structure, which is gamma iron with carbon and manganese in solution. The 10 to 14% manganese content prevents the transformation of austenite, as does the manganese in the 200 series of stainless steels (see Chapters 5 and 9). We can, therefore, deduce that austenitic manganese steel and austenitic stainless steel should have some similarities. They both work harden, they are both nonmagnetic, and neither is hardenable by heat treatment. The usual ratio of carbon to manganese in this steel is 1.0 to 1.4% carbon and 12 to 14% manganese.

The toughness of austenitic manganese steel is produced through the phenomenon we call *work hardening*. The steel comes from the mill at a soft, ductile 10 Rockwell C. It may then be cold worked up to +50 Rockwell C. It may, of course, be placed in service without such cold working, with the prospect that the in-service abuse by impact will increase its hardness. The steel maintains its ductility because only the surface is hardened, the depth of the hardening depending on the extent of the applied stress.

A railroad rail is an excellent example of austenitic manganese steel in service. As placed in service, the rail will be relatively soft, and the first freight car passing over it will deform the rail slightly. The steel hardens immediately, and the pounding of the last car of a 100-car train will have much less effect than did the first car. The rail remains flexible because only the surface has hardened.

This surface hardening may be likened to the case hardening of carbon steels, which we studied in Chapters 5 and 7. The hard outer surface of the case-hardened steel, however, will ultimately wear away, necessitating heat treatment to obtain a new case. The austenitic manganese steel rail, on the other hand, maintains its outer case, because, as the hard steel wears away, the pounding of the wheels hardens the steel that is close enough to the surface to be affected by the work-hardening phenomenon.

Therefore, wherever a steel may be found subjected to constant abuse from impact, the maintenance welder may well suspect it to be manganese steel. If that steel is somewhat gray, with an almost imaginary blue nuance and is somewhat shiny smooth, the steel is almost undoubtedly manganese steel. The welder who can discern the yellow and blue that make up the color of *pure* green will have no trouble distinguishing Hadfield steel from carbon steel and both from stainless steel.

Many maintenance welders who must rely on their personal knowledge and expertise maintain a sample board of metals. When they have identified a specific metal, and a piece of it is available, they center-punch an identity on it and file it on their sample board for reference when time has dimmed their memory. Such sample boards are especially helpful in discerning between such metals as MONEL and INCONEL.

CHARACTERISTICS THAT AFFECT WELDING

For successful welding of manganese steel, the welder should note the following characteristics.

1. The carbon and manganese are in solution in an almost pure austenitic structure, the degree of purity depending on the heat treatment and the success of the water quench.
2. Heating the steel to 540 °C (1000 °F) will cause carbides to precipitate out of solution and migrate to the grain boundaries, thereby embrittling and weakening the metal.
3. The addition of nickel (about 5%) prevents the precipitation of the car-

bides, but since the problem is encountered in welding, the obvious preventive measure is to add the nickel to the electrodes.

WELDING PROCEDURES AND RECOMMENDATIONS

Since we are dealing with a somewhat unstable structure, the less trauma we inflict on the parent metal, the more successful the weld will be. Use minimal heat, shallow penetration, and avoid excessive meltdown of the parent metal structure. More specifically:

1. Do not use oxyacetylene welding unless you are certain that the parent metal is of the 5% nickel type.
2. For high-strength joining, use a nickel-bearing electrode (about 5%). These electrodes, usually patented, contain 3 to 5% nickel, up to plus 1% carbon, and as much as 2% silicon. Their manganese content is usually 12 to 14%.
3. Keep the parent metal as cool as possible, using the skip-weld technique. Do not deposit additional weld metal until your bare hand can be held on the parent metal 15 cm (about 6 in.) from the weld zone.
4. When deep-groove welding, use the backstep block method: that is, weld about 5 to 6 cm (2 in.), completing each section to the top of the groove before laying the root bead of the next section.
5. When welding a partial crack, drilling holes at the ends of the crack will help eliminate further cracking.
6. When breaking the arc, swing the arc to one side so as not to leave crater cracks in the throat of the bead. The foldback technique will also help eliminate crater cracks but will make a restart more troublesome, especially with bare electrodes.
7. If welds fail or peel away, and your welding technique has been proper, weld with the higher alloys, such as E-308, E-309, and E-310, bearing in mind that the weld deposit cannot then be cut with the oxyacetylene torch.
8. In general buildup work, overweld slightly and hammer into shape while hot. This procedure hardens the metal and reduces expensive grinding time. Excessive hammering will embrittle the metal. Embrittlement is evident by a flaking off of particles of metal.
9. Use the Shuler (see the Glossary), crescent, or figure-8 weaves. The Shuler weave is especially good for retracing on a deposit, thereby working the puddle sufficiently to allow gases to escape before the puddle solidifies.
10. Metal cut or otherwise prepared by oxyacetylene torch should be ground clean before welding.
11. When building up carbon steel with austenitic manganese steel, remember that the austenitic manganese steel will shrink 50% more than the carbon steel. If contraction becomes a problem, it can be alleviated by buttering

the carbon steel with a nickel-bearing electrode that has the needed elongation, such as E-308, E-309, or E-310 (8%, 12%, and 20% nickel, respectively). The least expensive E-308 will usually suffice.

12. Bare electrodes give a heavier, thicker bead with less heat input to the parent metal than do the coated electrodes. The bare electrode is somewhat more difficult to control, but if the operator uses the proper heat and can oscillate the weave uniformly, the result will be satisfactory. The bare electrode does necessitate a longer arc, and working out the gas pockets requires more time in the puddle (see item 9). Here we are referring only to an austenitic manganese steel electrode.

13

WELDING TECHNIQUES

Although the maintenance welder is often called on to fabricate a part, to alter an existing design, or to reinforce a structure, the great preponderance of his expertise is directed toward putting back together something that has failed in service. Luckily, all of these welding jobs are interrelated and experience in one can be carried over to another.

JOINT DESIGN

Aside from an ability to weld, joint design is probably the most important aspect of maintenance welding. The main factors to consider in joint design are:

1. Strength required
2. Accessibility
3. Control of stress and distortion
4. Type of metal
5. Welding processes available

Strength. The joint design should provide only as much strength as is needed. Overwelding does nothing for the weldment except increase the cost and contribute to strain and displacement.

Accessibility. The joints and the welding sequence should be designed so as to make the welding as easy as possible. If the welder is not highly skilled in the overhead position, the welding should be planned so as to be welded in the flat position. The welder should never design a sequence that leaves a weld in an inaccessible position, such as an inside fillet in a tank too small to get into or a weld calculated to seal the welder inside the tank.

Distortion. Joints should be designed so as to reduce instead of promote distortion. The basics of stress and distortion are dealt with in the following pages.

Type of metal. The physical characteristics of a metal often dictate the process to be used and the desirable welding position, both of which, in turn, affect the choice of joint design. For example, lap joints should not be used to join dissimilar metals having dissimilar electrical properties, due to possible electrolysis.

Welding process. In preparing the welding procedure, the welder should ask the following questions:

1. How shall I distribute the deposit to minimize the shrinkage effect?
2. How many beads will be needed, and how should I arrange them to avoid possible slag entrapment between the beads?
3. Which direction will the stresses take?
4. How much angular or axial displacement will occur, and how can I offset that displacement?
5. Will there be a hinging or tearing effect at the root of the weld?
6. Is the joint open enough to get adequate weld depth without excessive meltdown of the weldment?

The answers to some of these questions are given in the following pages. An understanding of stress, strain, and distortion can best be gained by factual experimentation, as shown in Figure 13.7 to 13.20. Joint design may be selected from Figures 13.25 to 13.54.

Stress and Strain

Stress, strain, and distortion are the inevitable results of welding, due to the natural law that metal will expand and contract at variance with heat. Metals expand as we heat them, and they contract as they cool (with an exception or two, such as zinc). If metal expands and contracts uniformly over its entire volume, we have no problem. That is why we sometimes preheat an entire weldment prior to welding it. Unfortunately, most of welding is done without preheating, and the weldment is subjected to uneven expansion and contraction. We have a choice to make: We can let the metal dictate to us just what shape and structure it will take, or we can understand stress and strain and make these forces work for us instead of against us.

These forces are defined as follows:

Distortion: As applied to a weldment, distortion is the twisting out of the natural, normal, or original shape or condition of the welded material.

Stress: Stress is the force that tends to produce a deformation or fracture of a metal.

Strain: Strain is a condition in a weldment wherein the metal structure is stretched to its maximum extension and tautness due to stress.

Cause. The cause of a strain or distortion is the nonuniform expansion and contraction of both the filler metal and the proximate parent metal during the heating and cooling periods of the welding process.

Relief and prevention of stress. Relief of strain and prevention of built-in stresses become more important as the carbon content of the steel is increased. Many steels, as manufactured and treated, have inherent strains, such as the unstable martensitic structures of some tool-and-die steels (see Chapters 6 and 7). These steels should be preheated and postheated, or special techniques and materials must be used in welding them. If the design of the weldment is such that distortion cannot take place, the weldment will have a locked-in stress which, if added to an external stress such as impact, vibration, or fatigue, may cause premature failure in service.

Controlling distortion. If we put into a weldment all the welding and heat-treatment techniques encompassed in this book, we could conceivably have a weldment that would be completely undistorted and unstrained, and completely free of built-in stress. But not only is that result impractical, it is also unnecessary: most weldments can accommodate some distortion. Just as the unstable martensite is a desirable feature in many tool steels, the strain produced by welding some softer, weaker steels makes the weld-affected area stronger than the parent metal. Thus we have a decision to make: How much stress, strain, and distortion can a weldment tolerate? Quite a bit! Such is the ductility of low-carbon steel, which is the greater percentage of our welding, that being stressed beyond its yield strength does not seriously impair its rupturing strength. We will, however, have to concern ourselves with alignment. Warping out of alignment may relieve some strain, but wreaks havoc with appearance and fit.

By understanding the stresses caused by expansion and contraction, we can often play them against each other or make them work in a direction favorable to our design. There are three types of stress: compressive, tensile, and shear. Figures 13.1 to 13.3 illustrate their direction.

The effect of a deformation in a right-angle bend is shown in Figure 13.4. The tensile stress shows that the metal has elongated (stretched). The compressive stress has compacted the metal. Figure 13.5 shows the lateral and axial

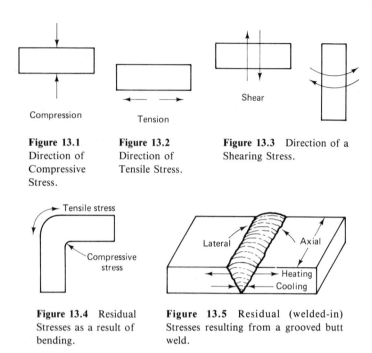

Figure 13.1 Direction of Compressive Stress.

Figure 13.2 Direction of Tensile Stress.

Figure 13.3 Direction of a Shearing Stress.

Figure 13.4 Residual Stresses as a result of bending.

Figure 13.5 Residual (welded-in) Stresses resulting from a grooved butt weld.

stresses, and the reversing compressive and tensile stresses during welding and cool-down.

In Figure 13.6 the lateral and axial stresses in a T assembly are illustrated. If the welds are evenly distributed on alternate sides and welding sequence plays one stress against the other, no angular displacement occurs. Axial displacement is restrained by the *strongback* or *stiffener* effect of the vertical section. The stresses are therefore locked in. As mentioned, if the steel is of low carbon content, these stresses are not serious. Some steels, however, would require stress-relief treatment.

Figure 13.7 shows an axial displacement. This deformation can be alleviated by prebending as shown in Figure 13.17, or by offsetting as shown in Figure 13.16.

The angular displacement inherent in a single-V butt weld is shown in Figure 13.8. This displacement is due to the shrinkage of unequal amounts of filler metal between the root and face of the weld. Such angular displacement can be eliminated or mitigated by redesign of the joint as shown by Figures 13.9 to 13.13.

The single V gives the most displacement with or without backup due to unequal shrinkage at the root and face (Figure 13.9). The open square butt gives equal shrinkage but is difficult to weld except on thin sections (Figure 13.10). The square butt with strapping provides equal shrinkage but requires planning of weld distribution in order to avoid slag entrapment (Figure 13.11). The U joint gives equal shrinkage but is expensive to prepare (Figure 13.12). Finally,

Joint Design

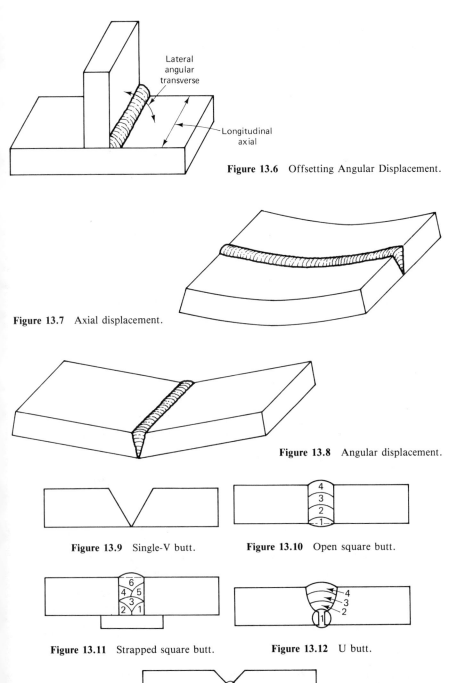

Figure 13.6 Offsetting Angular Displacement.

Figure 13.7 Axial displacement.

Figure 13.8 Angular displacement.

Figure 13.9 Single-V butt.

Figure 13.10 Open square butt.

Figure 13.11 Strapped square butt.

Figure 13.12 U butt.

Figure 13.13 Double-V butt.

the double-V butt, welded on alternate sides, keeps section in alignment according to the welder's planning (Figure 13.13).

Since most undesirable deformation is caused by the shrinking effect, we should make every effort to control or distribute that shrinkage. Perhaps one or more of the following practices will apply to the job at hand.

1. Keep the shrinkage force as low as possible by using only the amount of weld necessary to provide the strength required.
2. Increase the welding speed using larger electrodes, MIG, flux core, or submerged arc.
3. Use fewer weld passes.
4. Skip-weld (balance the force of one weld against the force of another).
5. Weld from the point of maximum stress toward points of maximum freedom.
6. Make all welds near a neutral axis.
7. On a complicated assembly, weld the lighter gauges and sections first.
8. Use intermittent welding. Many weldments do not require a continuous weld but can do with one-half to one-fourth of the length of the assembly (stiffeners, for example).
9. Backstep-weld toward a free end, such as beginning at the center of a seam and moving toward the ends.
10. Make the shrinkage forces work in the desired direction. [See out-of-position settings (Figure 13.16) and prebending (Figure 13.17)].
11. Balance the shrinkage forces with opposing forces. When restraints are removed, there will be some distortion but measurably less than if no restraints were used.

Following are various specific welding techniques for deformation prevention.

1. Refer to Figure 13.14. Assemble the T and tack at end X on both sides so that the perpendicular is squared. Mark end Y with the center punch as shown. Do not tack end Y. Backstep-weld from X to Y on alternate sides. Adjust the welding to maintain alignment.
2. Refer to Figure 13.15. As shown in (a), assemble a T and tack both ends so that the vertical section is plumb. Weld a single-pass fillet on one side only and observe the degree of angular displacement. As shown in (b), assemble a second T but offset the vertical. Observe the amount of weld required to bring the vertical section into proper alignment.
3. Refer to Figure 13.16. Assemble two sheets (16 to 12 gauge) as in (a). Begin to weld at X and continue toward Z. If the plates are thin enough, they will begin to overlap at about Y. If previously tacked at Z, they cannot overlap and the angular displacement shown in part (c) will occur. Alterna-

Joint Design

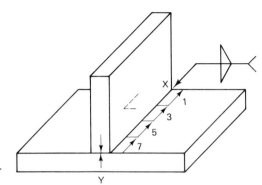

Figure 13.14 Offsetting welds.

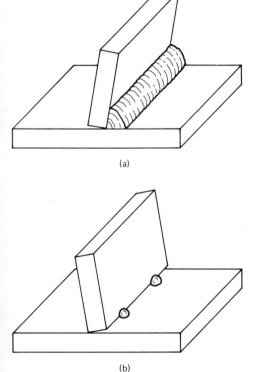

Figure 13.15 Offsetting angular displacement.

tively, assemble two plates as in (d) and try to offset the transverse stress of (b). If welding speed, amount of weld deposit, and width of gap are proper, no displacement will occur.

4. Refer to Figure 13.17. In *prebending*, first align the plates and tack the weld sufficiently to withstand bending. Tack-weld the corners A and B to the layout plate. Bend the weldment over the spacer and tack-weld corners C and D. Begin the weld at the center and backstep to the ends.

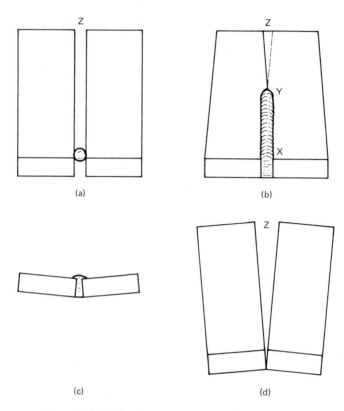

Figure 13.16 Offsetting angular and axial displacement.

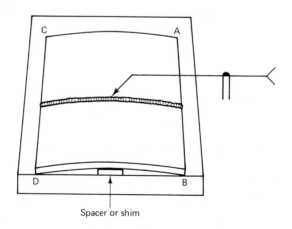

Figure 13.17 Prebending.

Joint Design

Tacks can be cut out with a disk grinder. If adequate clamps are available, use them. The symbol means: "Weld arrow side, squared open butt, burn through."

5. Refer to Figure 13.18. Shown here is the *chain intermittent technique* (sometimes called, erroneously, "skip welding"). On light gauges, weld all perpendiculars *vertical-down*. The symbol means: "Fillet weld, both sides; 6-unit welds with 15-unit centers." Always place welds at both ends.

6. Refer to Figure 13.19. In the *staggered intermittent technique*, the process is the same as that described for the chain intermittent procedure. The symbol means: "Weld both sides, 6-unit welds with 12-unit centers." Al-

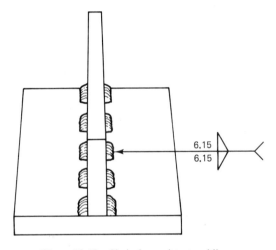

Figure 13.18 Chain intermittent welding.

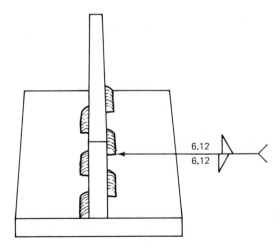

Figure 13.19 Staggered intermittent welding.

ways place a weld at the ends. Note the position of the fillet symbols.
7. Refer to Figure 13.20. To *offset* the angular displacement occurring in Figure 13.8, position the weldment as shown by the solid lines prior to welding. The contraction of the filler metal should then shrink the weldment into alignment as shown by the dashed lines. Such offsets, in maintenance welding at least, are educated guesses based on the welder's experience.

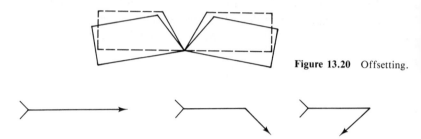

Figure 13.20 Offsetting.

Figure 13.21 Base of the welding symbol.

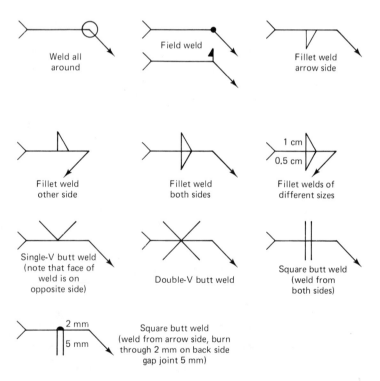

Figure 13.22 Welding instructions.

Joint Design

Welding Symbols

As shown in Figure 13.21, the base of the welding symbol is an arrow that may be bent as the drafter may see fit for convenience or for symmetry. The welding instructions are then draped along the arrow as shown in Figure 13.22.

As shown in Figure 13.23, the welding symbol is read more easily if we realize that the symbol approximates very closely the actual shape of the joint. The fletcher is used to carry specific instructions (Figure 13.24).

Figures 13.25 to 13.54 are examples of joint design diagrams. In Figures 13.29 to 13.32, numbers such as 2-4, 3-10, and so on, refer to the ratio of weld length to the unwelded portion. The first number is the length of the weld. The second number is the distance from the *center* of one weld to the *center* of the next weld.

The upset joint shown in Figure 13.41 is particularly useful in sheet-metal work. It permits the use of larger electrodes, speeds up welding, and adds rigidity to the assembly. An upset of 2 mm and a 6-mm backstep weld will produce a neat-appearing weld with a minimum of deformation.

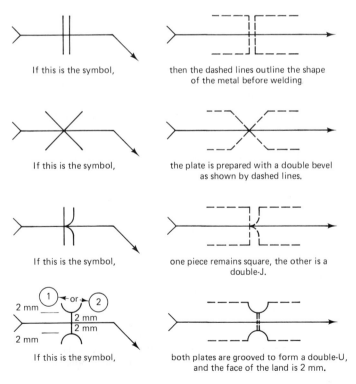

Figure 13.23 Reading the welding symbols.

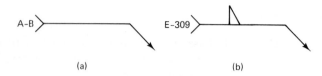

Figure 13.24 Fletchers: (a) find the instructions A and B before welding (they are on the sketch); (b) weld with E-309 electrode.

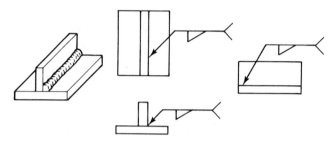

Figure 13.25 Joint design diagram.

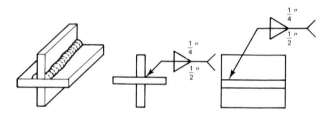

Figure 13.26 Joint design diagram showing arrow side with $\frac{1}{2}$-in. fillet and other side with $\frac{1}{4}$-in. fillet.

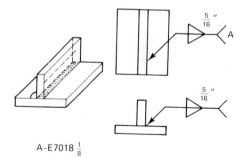

Figure 13.27 Joint design diagram with fletcher. The draftsman may show the symbol on one or both views.

Joint Design

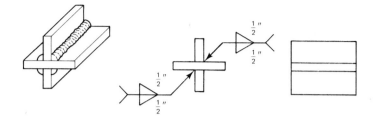

Figure 13.28 Joint design diagram showing both sides equal.

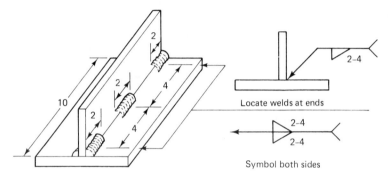

Figure 13.29 Chain intermittent diagram.

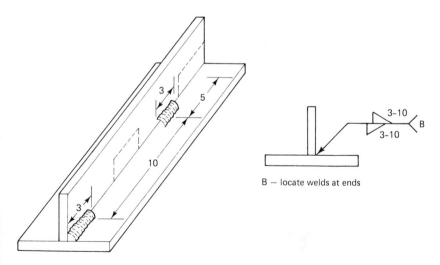

Figure 13.30 Staggered intermittent diagram.

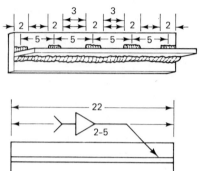

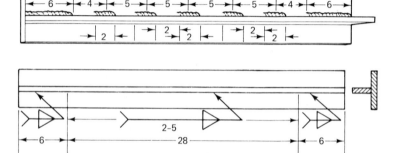

Figure 13.31 Weld arrow side; 2-inch welds with 5-in. centers. Chain Intermittent; Weld other side continuous.

Figure 13.32 Diagram for combined intermittent and continuous welding.

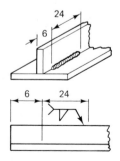

Figure 13.33 Weld specific areas.

Figure 13.34 Weld specific length.

Joint Design

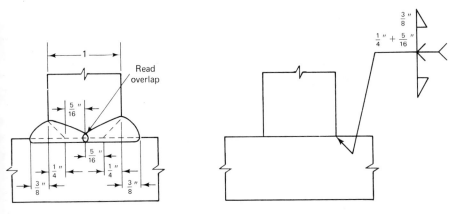

Figure 13.35 Size of weld and root penetration.

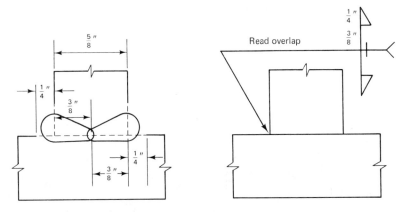

Figure 13.36 Size and penetration specified.

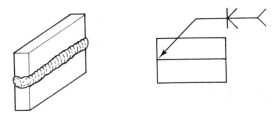

Figure 13.37 Double-bevel one edge only.

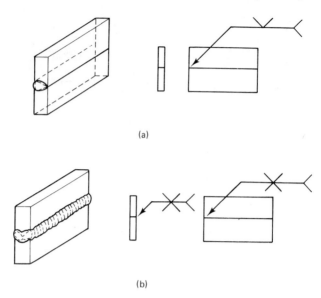

Figure 13.38 V grooves: (a) single; (b) double.

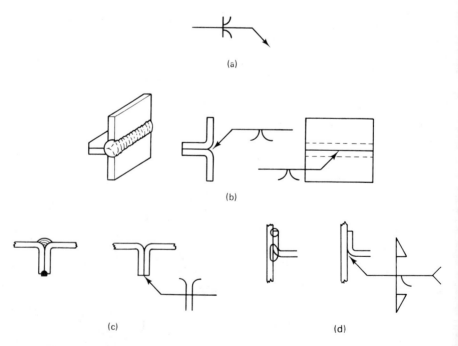

Figure 13.39 (a) J grooves; (b) flare V grooves; (c) flare V and square grooves; (d) flanged-T.

Joint Design

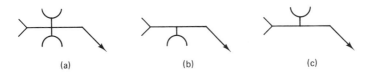

Figure 13.40 U-groove symbols: (a) both sides; (b) arrow side; (c) other side.

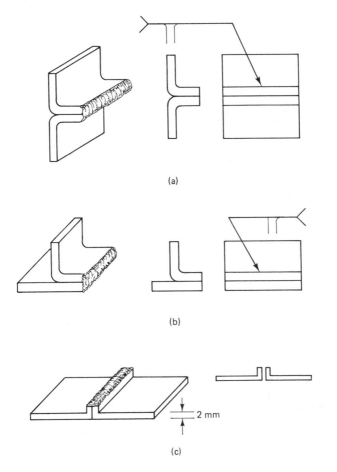

Figure 13.41 (a) Edge flange; (b) corner flange; (c) upset joint.

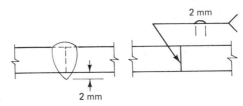

Figure 13.42 Melt-through.

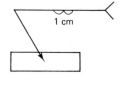

Figure 13.43 Buildup of entire surface. Thin layers are sometimes called overlays or cladding.

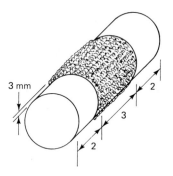 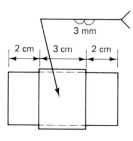

Figure 13.44 Buildup of specified area.

Figure 13.45 Backweld.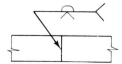

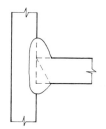 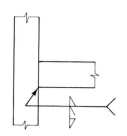

Figure 13.46 Single bevel groove and double fillet.

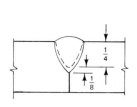 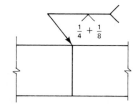

Figure 13.47 Read root penetration.

Joint Design

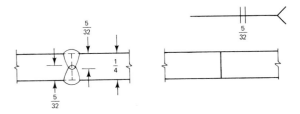

Figure 13.48 Read penetration and overlap.

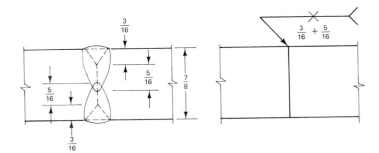

Figure 13.49 Bevel plus penetration.

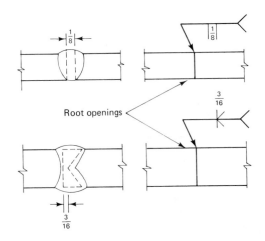

Figure 13.50 Root openings.

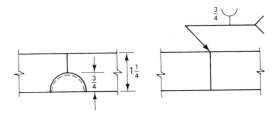

Figure 13.51 No root opening.

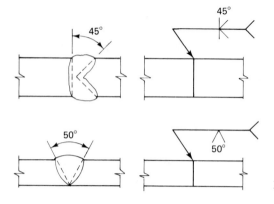

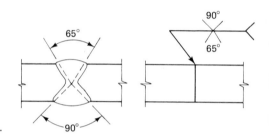

Figure 13.52 Read groove angle.

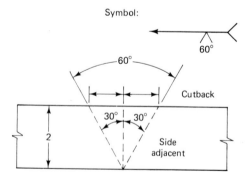

Figure 13.53 Different bevel angles.

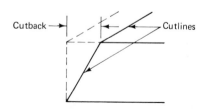

Figure 13.54 Cutback: angle of cut = 60°/2 = 30°; cutback = tangent × side adjacent = 0.57735 × 2 = 1.1547.

METAL WORKING

Oxyacetylene Cutting

The cutting or scarfing of steel by the oxyacetylene method is accomplished by controlled burning of the steel. Steel will burn in pure oxygen at a temperature of approximately 870° C (1600° F). The steel is heated to its ignition temperature by the preheat flames of the torch, and a jet of pure oxygen is directed at the heated spot. The steel burns but instead of giving off carbon dioxide into the air, the oxygen combines with the iron to produce iron oxide. The melting point of iron oxide is much lower than that of the melting point of steel and is easily blown out of the way by the force of the oxygen jet. As the iron oxide is forced out of the way, more steel is exposed to the oxygen jet.

Many mechanics inexperienced with the oxyacetylene torch erroneously believe that the steel is melted and blown away. To change this belief, the mechanic is encouraged to try an experiment: Take a steel plate about 8 to 12 mm thick, heat the edge to the ignition point (bright red to early orange), depress the jet lever, and establish the cutting kerf. When the kerf is well established, gradually raise the torch tip above the cut. With a steady hand, the operator will find that the torch can be raised up to 4 or more centimeters above the cut without losing the cut. The molten iron oxide, with very little help from the preheat flames, appears to provide ignition. The distance that the torch can be raised in this experiment will depend on the pressure of the jet stream. Such an extreme distance is not recommended as a general practice.

Although cast iron and high-alloy steels such as the stainless steels can be cut by oxyacetylene, the practice is not recommended due to the special techniques involved and the less-than-satisfactory results obtainable.

Procedures. If unfamiliar with a welding torch, read Chapter 2.

In Figures 13.55 to 13.57, Richard (Dick) Barnes of the San Francisco Community College District demonstrates oxyacetylene flame-cutting techniques.

General guidelines for oxyacetylene cutting are as follows:

1. Select the proper tip from Table 13.1 (shown on p. 203) according to the thickness of the steel.
2. Insert the tip in the head of the cutting attachment and tighten securely, bearing in mind that the torch head is undoubtedly brass and the cutting tip is copper. If the tip does not screw in easily by hand, inspect the threads. Do not try to use the tip as a self-threading screw. Use a wrench lightly for the final tightening.
3. Set the regulator for proper pressures for the tip and thickness of the steel, as shown in Table 13.1.
4. Open completely the oxygen valve on the welding torch.
5. Open the preheat oxygen control valve on the cutting attachment and adjust the regulator to the desired delivery pressure.

Figure 13.55 Operator demonstrates the pulling technique, which lends itself to short runs.

(a)

(b) **Figure 13.56** (a) Operator demonstrates the arm–elbow–wrist pivotal technique, which permits cuts of up to 30 cm (3 ft) nonstop and is also preferred for circular cutting. Note the distance from face to torch that is characteristic of the skilled operator. (b) Another view of (a).

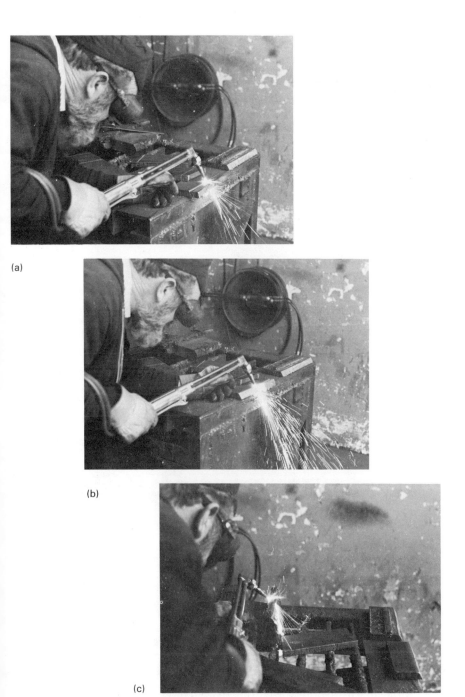

Figure 13.57 Operator (a) bevels a plate using the pushing technique; (b) has disposed of a portion of the unwanted metal in order to restart with good vision; (c) demonstrates the inside-out technique of beveling a plate. This method allows control of the width of a root face (land) and finds good use for the in-position beveling of the bottom quadrant of a pipe in a horizontal run.

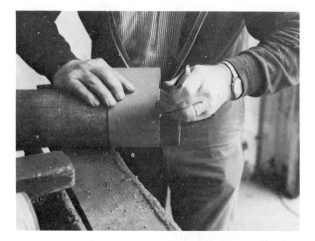

(d)

(e)

Figure 13-57 *(continued)* (d) Wraparound is used to mark a radial cut on a pipe. (e) Operator demonstrates the simultaneous bevel-and-cut technique. When working with long sections of pipe, the pipe welder will straddle the pipe and roll it with the thighs and buttocks.

6. Close the oxygen preheat control valve.
7. Open the fuel valve on the welding torch handle and adjust the fuel regulator delivery range.
8. Close the fuel control valve on the torch handle.
9. Momentarily depress the cutting oxygen lever to purge the high-pressure oxygen jet passage.
10. Open the fuel control valve one-half turn and ignite the gas with a spark-type lighter. (*Caution:* Although the light from the oxyacetylene flame may appear harmless to the eyes, it is not. Use a number 5 or 6 protective

TABLE 13.1 Cutting Specifications[a,b]

Metal Thickness [mm (in.)][c]	Cutting- Tip Size	Oxygen Pressure (psi)	Acetylene Pressure (psi)
3 ($\frac{1}{8}$)	000	20–25	3–5
6 ($\frac{1}{4}$)	00	20–25	3–5
10 ($\frac{3}{8}$)	0	25–30	3–5
12 ($\frac{1}{2}$)	0	30–35	3–5
20 ($\frac{3}{4}$)	1	30–35	3–5
25 (1)	2	35–40	3–6
4 ($1\frac{1}{2}$)	2	40–45	3–7
5 (2)	3	40–45	4–9
6 ($2\frac{1}{2}$)	3	45–50	4–10
8 (3)	4	40–50	5–10
10 (4)	5	45–55	5–12
13 (5)	5	45–55	5–13
15 (6)	6	45–55	7–13
20 (8)	6	45–55	7–14
25 (10)	7	45–55	10–15
30 (12)	8	45–55	10–15

[a]The tip sizes and respective pressure settings are arbitrary, but the ranges shown should accommodate most torches under most conditions. The welder should consult the recommendations of the manufacturer of the particular torch.

[b]*Caution:* Use a 9.5-mm ($\frac{3}{8}$-in.) hose for tip size 6 and larger. Never use a withdrawal rate of greater than one-seventh of the fuel cylinder's capacity per hour. If a higher gas flow is necessary, use several cylinders in a manifold system (as many cylinders as are needed based on the maximum withdrawal rate shown here).

[c]Do not use the metric/English approximations to obtain equivalents. The figures are very close approximations only.

filter lens. Goggles should fit snugly against the skin completely around the eye sockets.)

11. Increase fuel until the flame shows a 3- to 4-mm gap between the flame and the tip.
12. Reduce the fuel until the flame returns to just touch the tip.
13. Slowly open the oxygen valve on the cutting attachment until the preheat flames establish a sharp inner cone (a neutral flame).
14. Depress the oxygen-jet lever.
15. The preheat flame should change to a slightly carburizing flame. If it does not, increase the fuel slightly until the flame is just barely carburizing.
16. Increase the oxygen until the oxygen and fuel gas are again neutral.

Note: The oxygen-jet control valve is depressed during the procedures in items 15 and 16.

17. Hold the cutting torch comfortably in both hands as shown in Figure 13.55. Note that the operator is using his elbow and wrist as pivotal points. Note also that he is not crowding the cut; that is, his cut is some distance from his face. The operator shown is comfortable and relaxed. The right-to-left cut is preferred by right-handers, and left-handers prefer the reverse hold and direction of travel. An experienced operator should be able to cut 30 or more centimeters without stopping.
18. While cutting, observe several things.
    (a) The *cut line:* Always mark a cut line in chalk, and if the metal is rusty, painted, lacquered, or otherwise obscured, center-punch the line. Many welders who are experienced with a cutting torch fill the center-punch marks with soapstone for better vision.
    (b) The *angle of the cut:* This factor is especially important when beveling.
    (c) The *kerf:* Make sure that the cut goes through cleanly.
19. Figure 13.58 shows the proper preheating angle and tip angle for the usual right-angle cut. Note position 1. Much preheating time can be saved if the operator learns to heat a sharp point instead of a rounded one and learns to pick a sharp point when restarting a lost kerf. The object is *ignition*, not *melt*. Study the quality of cuts and the causes shown in Figure 13.59.

There is a right tip size, a proper pressure, a proper distance between tip and steel, and a correct travel speed for every cut. All of these factors assume that the welder has at hand all of the proper equipment and that the equipment is in good condition. More often than not, however, the maintenance welder will have to compromise. The ability to compromise comes with experience. A few hints may speed up the learning process.

1. If the tip size is too large for the job, set the fuel pressure as low as possible without pop-outs (this setting will be the low end of the range in Table 13.1).
2. If the flame is still too great, speed up the cut.
3. If the tip is too small, reverse the procedure in item 1 and slow down the cut.
4. If the cut calls for a 000 tip and only a number 1 tip is available, tilt the tip away from the line of travel and sweep the cutting jet into the kerf at a speed that keeps the preheat flames about 4 mm from the burning steel.

Mechanics who are somewhat fastidious about their work jealously guard their *private* selection of cutting and scarfing tips (especially if they have laid out an hour's wages for each of them).

Metal Working

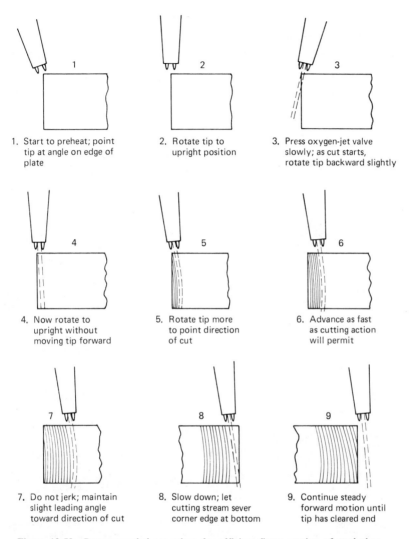

Figure 13.58 Recommended procedure for efficient flame cutting of steel plate. (Courtesy of Victor Equipment Company.)

Special uses of the torch:

Piercing (and Cutting a Hole). The procedure is as follows:

1. Hold the torch as shown in Figure 13.55.
2. Preheat the steel with the cutting tip about 5 mm from the plate.
3. When ignition heat is apparent, tilt the torch so that the tip is about 10° from perpendicular and touch the oxygen-jet lever just enough to test the

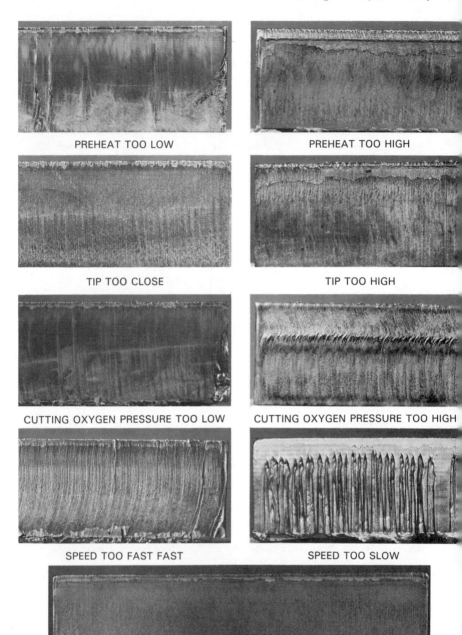

Figure 13.59 Effects on cut quality of cutting techniques.

metal. If it starts to burn, maintain the angle, raise the tip about 1 cm, and depress the oxygen-jet lever. When the initial pierce is made, lower the torch to the proper height, return it to the perpendicular, and proceed with the cut (circular, ellipitical, etc.).
4. If cutting circular, elliptical, or other shapes and the cutting must be of good quality, drill a starting hole, especially on thicker sections.

Scarfing. Removing old welds, grooving, and rough machining are possible with an oxyacetylene cutting torch. Experienced welders readily use the ordinary cutting tip, but specially angled tips make the operation somewhat less difficult. The important factors are to keep the preheat flames relatively hot, the oxygen jet suited to the washing away of slag, and careful control of the width and depth of burning. The force of the jet should neither burn the steel to a greater depth than that desired nor burn it so far ahead as to be out of control. The force of the jet should be just enough to wash the slag sufficiently out of the way of the scarf. Experienced maintenance welders remove sprocket gears from shafts without damaging the shaft, and keys from keyways without marring the keyway.

Stack Cutting. The maintenance welder will rarely be called on to cut several plates simultaneously, but it is relatively easy with an oxyacetylene torch and has the advantage of leaving slag only on the bottom plate.

1. Select a tip and pressures in accordance with the total thickness of the stack.
2. Make certain that the plates are clamped together tightly and evenly. (Air gaps between the plates will produce a disaster.)
3. Proceed as if cutting a single plate, but bear in mind that the flow of the molten slag across the tiniest air space between plates is important.
4. Do not attempt to pierce a hole in a stack of plates as described above. Drill a starting hole.

Other Uses for the Torch:

1. To remove oil, grease, paint, and other debris from metal prior to welding. Adjust the flame as for cutting, depress the oxygen jet lever, and heat and sweep.
2. As a preheat for arc welding. Do not open the oxygen jet; use exactly as you would use a welding nozzle.
3. As a temperature indicator for preheat. Blacken the surface of the metal with a pure acetylene flame, return the flame to neutral, and heat broadly until all traces of carbon have cleared. The temperature should be about 230° C (450° F). This relatively low preheating is especially useful when stick-welding metals having a very high thermal conductivity, such as aluminum, copper, brass, or bronze.
4. To provide for a welder's comfort in extreme cold. Fashion a heating stove

by blanking off one end of a short length of pipe and inserting the tip inside the other end, with the neutral flame deflecting obliquely off the pipe wall.

Caution: Never use the oxygen jet to dust off clothing. Aside from being an expensive brush, it is dangerous under certain conditions. For instance, oil and grease on clothing will ignite and burn in pure oxygen at a very low temperature.

Oxygen-Arc Cutting

The primary advantage of oxygen-arc equipment is its ability to cut metals that cannot be cut by the oxy-fuel process.

The additional equipment needed for this operation is relatively inexpensive, since the welder's present welding machine and oxygen regulator may be used. An oxygen-arc electrode holder and oxy-arc electrodes are all that is needed.

The electrodes are composed of a ferrous metal tube with a nonmagnetic coating. The arc melts the metal and the metal is blown away by the pressure of the oxygen jet. The electrode is kept in contact with the metal, since the ferrous metal tube melts before the coating, and this recessing of the tubular core material maintains the proper arc length. There is no problem of fusion of core material with the metal being cut, because the core material is completely oxidized in the arc. The angle of the electrode is important, especially in grooving and gouging. (See "Air Arc" and "Chamfering Electrodes," this chapter.)

Arc Cutting

Although less satisfactory than more sophisticated methods, metal cutting is also possible with the arc alone. Arc cutting is usually done with carbon electrodes, available in round, square, and rectangular shapes.

The process, however, is limited to cuts where the molten metal will fall away from the cut by gravity alone. Heavy sections may be cut with a sawing technique, may be positioned vertically, or may be cut from the underside. Basically, the technique is one of melting the metal and chasing it out of the cut before it can resolidify. The position and thickness of the cut will determine the manipulation of the arc.

Coated solid-core electrodes are available that will arc-cut somewhat less cleanly than will the carbon arc. With these electrodes, the cutting technique is quite similar to cutting with a carbon electrode. As in the oxygen-arc electrode, the core melts well up inside the coating (as much as 6 mm), creating a gas pocket that inhibits fusion and provides a slight jet action.

Coated solid-core electrodes are useful to the maintenance welder working out of the shop, since he can merely add a few of them to his selection of electrodes for the several jobs on his list of repairs to be made.

Metal Working

Chamfering electrodes. Common uses of these electrodes are:

1. Scarf off high spots, lumps, and so on, prior to machining.
2. Rout out old or defective welds.
3. Groove out cracks and cracked welds.
4. Scarf out for fill-in welding when hardfacing layers peel.
5. Make alignment of broken parts easy. Align and fit together exactly as broken, tack-weld in place, and groove-out and fill, grooving-out tacked areas last.
6. Cut any metal.

Figure 13.60 shows the proper welding positions. Position 1 is for grooving, routing out, and so on, 10° or less from parallel. Position 2 is for cutting heavy (thick) sections. Vary the perpendicular to suit, using a sawing motion. For cutting thin-gauge material, use almost parallel, as in position 1.

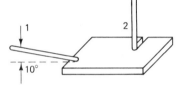

Figure 13.60 Cutting positions for the chamfering electrode.

The welding procedure is as follows:

1. Use dc straight polarity (electrode negative) or ac.
2. Set the current approximately double that of equivalent-diameter E-6010 (or check the manufacturer's recommendations).
3. Point the electrode in the line of travel, holding the electrode almost parallel to the metal's surface and pushing forward so that the tip of the electrode is in positive contact with the metal. The operator should actually feel the electrode in contact in front of and under the arc.
4. Do not attempt to gouge too deeply when grooving; widen the groove with a second run, then groove a second layer.
5. Practiced welders sometimes widen the grooving by oscillating the arc sideways but do not try for more than twice the diameter of the electrode (the coating included).
6. To restart the arc with a partially consumed electrode, simply rest the tip against the metal. The arc will restart within 1 second.

Air-Arc Cutting

In the air carbon-arc metalworking process, commonly called *air arc*, the metal is melted by the electrode arc and is chased away by a jet stream of compressed air. Although the molten metal is not oxidized as in the case of oxyacetylene burning, it does not adhere stubbornly to the parent metal, due to its being ionized.

The air-arc process has found wide acceptance in industry for a variety of purposes, such as plate-edge preparation, back-gouging for backwelding, rough machining, removal of old or defective welds, undermining of hardfacing layers that have peeled away, removal of old hardfacings, gouging out cracks for welding, and removing worn tips from dipper teeth. Its use in U-groove preparation is excellent, as shown in Figure 13.61.

The techniques for using air arc are:

1. The arc length should be the shortest arc that does not sputter. The noise level is influenced greatly by the arc length and uniform rate of travel.
2. The air pressure should be 414kPa to 690kPa (60 to 100 psi).
3. Stick-out is not critical, but important. The shorter the stick-out, the more effective the jet force.
4. The welding machine should be rated at 100% duty cycle for the size of electrode used.
5. Machines may be paralleled to obtain the higher current values. Consult the manufacturer for best procedure.
6. Use dc electrode positive (DCEP). Special ac electrodes have been developed for foundry use on cast irons.
7. If the job is of relatively long duration (more than 15 minutes), consider using a darker filter lens; wear ear protectors and breathing filter.
8. Keep the electrodes dry. If exposed to moisture, rebake for 10 hours at 180° C (about 300° F).

For manual operation, the necessary equipment is minimal. A specially designed electrode holder feeds the air jet parallel to and alongside the carbon electrode. The air jet is fed into the electrode holder from a source of compressed air [414kPa to 690kPa kg/cm^2 (60 to 100 psi)] and exits through tiny apertures in the copper electrode clamps, clearly visible in Figures 13.62 and 13.63. Table 13.2 gives the current ranges required for the Arcair* torches shown in these figures.

TABLE 13.2 Arcair Torch Current Ranges[a]

K-3	$\frac{5}{32}$	$\frac{3}{16}$	$\frac{1}{4}$	$\frac{5}{16}$	$\frac{3}{8}$		
Amperage	90–150	150–200	200–400	250–450	350–600	600–1000	800–1200
K-5				$\frac{5}{16}$	$\frac{3}{8}$	$\frac{1}{2}$	$\frac{5}{8}$

[a]Air pressure 5.6 to 7.03 kg/cm^2 (80 to 100 psi).

*Arcair and Arcair-Matic are registered trademarks of the Arcair Company, Lancaster, Ohio.

Metal Working

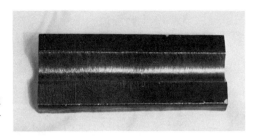

Figure 13.61 Typical U groove prepared by the air-arcing process. (Courtesy of Arcair Company, Lancaster, Ohio.)

Figure 13.62 Arcair Model K-3 torch for general use. (Courtesy of Arcair Company, Lancaster, Ohio.)

Figure 13.63 Arcair Tri-Arc torch for foundry applications. (Courtesy of Arcair Company, Lancaster, Ohio.)

Automatic equipment. Almost any metal can be worked with speed and precision using Arcair automatic equipment (Figure 13.64), and a relatively high degree of precision is obtainable with the manual equipment in the hands of a skilled operator. The Arcair-Matic equipment shown in Figure 13.64 will produce uniform U grooves up to 19 mm ($\frac{3}{4}$ in.) in depth in a single pass. With *arc voltage control*, the grooves can be controlled within ±0.635 mm (0.025 in.). Continuous electrode feed is possible by use of jointed electrodes. Accurate groove depth is controlled by automatic compensation for waviness of plate or out-of-roundness for tanks and pressure vessels. Groove depth and width may be altered by changing forward travel speed or by changing electrode size, or a combination of the two. Fully automatic Arcair torches require a variable-voltage power source; do not use constant-voltage equipment (MIG-FCAW).

Flame Straightening

Flat steel plate structures such as bulkheads and decks will sometimes buckle or warp from welded-in stresses. These distortions may be completely removed by flame straightening.

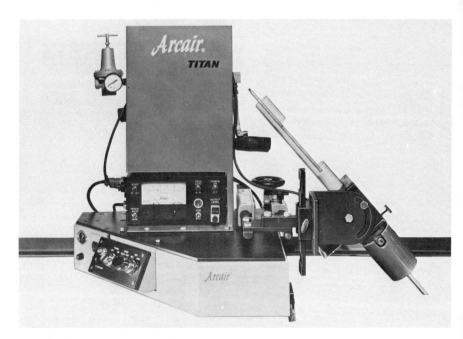

Figure 13.64 Arcair-Matic N torch mounted on Arcair Titan machine carriage. (Courtesy of Arcair Company, Lancaster, Ohio.)

Using the oxyacetylene torch, small circles [about 75 mm (3 in.)] are heated to a temperature of 650° C (1200° F) (barely red) and water quenched. These circular spots will first upset, then shrink in all directions. A sufficient number of these heated circles and a proper distribution or sequence of them will remove the distortion completely.

Flame straightening is, of course, an engineering feat of sorts, and the layout of the heat-and-cool pattern is performed by a heat-treatment expert. The maintenance welder, however, can put it to use on something less than the expert level by observing a few precautions and by using the try-and-fail, try-again-and-succeed method may attain a respectable expertise.

The procedure for flame straightening is as follows:

1. To straighten steel, never go above a temperature of 650° C (barely red) and restrict the process to low-carbon steels. Do not attempt to straighten medium- and high-carbon steels.
2. To straighten stainless steels, heat to 425° C (about 800° F).
3. To straighten aluminum, heat to about 205° C (400° F).
4. Use a tip size and flame that will heat the spot quickly and not allow the heat to disperse too widely.

5. Whether to air- or water-quench will depend on the ambient air temperature. Neither will affect the steel adversely. Fast cooling reduces the shrinkage time.
6. The heated side (torch side) will react to the expansion–contraction cycle more substantially than will the opposite side.
7. On thicker sections, the heating is best observed from the opposite side. A worker stationed on the underside of a deck plate, for instance, can signal for the quench by means of a hammer tap.
8. On larger jobs, or where the procedure is called for, a straight-pipe type of torch may be fitted with a parallel waterline, which will enable the operator to quench a heated spot while heating the next spot.

FABRICATED SHOP FACILITIES

Figures 13.65 to 13.70 illustrate several facilities for smaller shops which can be fabricated from surplus or scrap materials.

Figure 13.65 Cutting bin that prevents damage to concrete or other vulnerable floors. Note the access door for cleanout, clamping devices for restraining metals being cut, and removable slats whose top edges have been overlaid with two layers of stainless steel (some surplus E-308-15) for the purpose of resisting the bite of the oxygen jet as the kerf passes over them.

Figure 13.66 Layout and assembly table using H-beams adequately spaced to accommodate the use of C-clamps.

Figure 13.67 Smaller cutting bin constructed similarly to the one shown in Figure 13.65.

Figure 13.68 Gas welding–brazing booth with active ventilation for toxic fumes.

Fabricated Shop Facilities

Figure 13.69 Use for a scrap wheel rim.

Figure 13.70 Adjustable stanchions fabricated from a $1\frac{1}{4} \times 10$ in. stud, its mating nuts, and various sizes of pipe, a *must* for the pipe welder.

14

PIPE WELDING

The shortage of practice pipe made available to students of welding schools soon reminds instructors that pipe welding is essentially the same as plate welding. The welder merely welds in a circle instead of in the usual straight line.

The welding of a veed-out butted pipe joint requires the same welding procedure as does the weld of a butted plate. If the joint is fitted with a backup ring or a consumable insert, the welding is identical. The only difference with the respective open-butted joints is that the pipe-weld burn-through of the root pass must approximate a backweld on its own—it cannot be routed out and backwelded, as is the custom on high-quality welds on plate. The horizontal weld of a butted joint on a vertically positioned pipe is precisely the same as a horizontal weld on plate.

A *rolling weld* may be welded as a flat-position weld by maintaining the weld puddle at the top of the pipe, or it may be made at the 1 or 11 o'clock position, as preferred by most experienced welders. In this instance the welder may elect a slightly uphill or downhill travel direction, depending on the width of the gap.

The in-position weld on a horizontal run is for the most part a vertical-up, there being very little of the overhead position except on a pipe of extremely large diameter. (Some overland and subcritical pipe is welded vertical-down.)

Therefore, a student using this book as a guide to pipe welding and fitting should not forgo available practice hours simply because he has no pipe. The mastery of beveling and welding pipe can be adequately simulated by using steel plate of a thickness comparable to the wall thickness of standard pipe, which is 0.95 cm ($\frac{3}{8}$ in.) for blackwall, schedule 40, 15 to 16 cm (6 in.).

Taking the Pipe Test

Any experienced pipe welder will attest to the fact that it is much more difficult to get a burn-through that is the equivalent of a backweld on an overhead positioned plate than it is on a pipe positioned horizontally.

When no practice pipe is available, imagine the beveling, fitting, and welding experience that can be had by mocking up a hexagonal pipe from 12 pieces of steel plate. There will be 24 bevels (these should be beveled by hand, not using "automatic"), one flat weld, one overhead weld, four horizontal welds (one a little more difficult than the other), and a "circumferential" weld that approximates the in-position weld of a horizontally positioned pipe. The experience gained by the assembly of two such mock-ups should be sufficient practice to enable a good plate welder to "pass the pipe test" (and the pressure-vessel test).

Pipe is everywhere. It may be encountered by the welder as a new run, a rerouting or upgrading of an existing run, a tap into a line with a lateral or backstub, or a modification to accommodate a larger valve or a manifold.

Pipe is fabricated from most weldable metals, from the very ordinary to the exotic. But, as before, pipe may be treated as plate that is formed into a circular shape. Therefore, when welding steel pipe, the procedures and precautions discussed in Chapter 6 apply. For chrome-nickel pipe, refer to "Welding Stainless Steel," Chapter 9; MONEL, INCONEL, and the others are covered in Chapter 10. The exceptions are a few electrodes especially developed for pipe, such as the E-6010P.

We have selected the oil refinery as being representative of pipe welding for our purposes. An oil refinery contains more pipe per unit of area than does any other industry. And no other industry spends more welding hours upgrading and rerouting pipe lines.

The welding procedure for in-position welds may best be illustrated by the standard blackwall schedule 40 pipe test given by a leading refinery.

TAKING THE PIPE TEST

Take with you a ball-peen hammer, good chisel, slag hammer and brush, and a piece of saw blade with a taped handle (not a thin hacksaw or bandsaw, but one of the heavier, thicker types).

You will be given two pieces of pipe [usually 15.2 cm (6 in.)] with one end of each "milled." If the test is given without a backup ring (as is common in oil refineries), set the two pieces as shown in Figure 14.1b, with the two milled edges together. Set the gap you wish [2.5 mm ($\frac{3}{32}$ in.) recommended] by using a wire, two $\frac{3}{32}$ welding rods with the coating removed, or three pennies.

Make sure that the pipe is in perfect alignment. Use your fingers to feel all around the pipe joint; use a straightedge (your level or square) and "eyeball" down the inside of the pipe. When aligned, tack-weld in three places, equidistant around the pipe. Make tacks of root-pass quality, bearing in mind that when you make the root pass, you will have to blend into and through them so as not to have slag pockets and to have a uniform deposit on the inside. Make tack welds of sufficient strength to allow handling of the weldment when positioning it in whatever positioning device has been provided. Remove the spacers.

Clamp the pipe in the most comfortable welding position, preferably so that you can rest an elbow on something. Calling the top of the pipe 12 o'clock, position the pipe so that the tacks are at 4, 8, and 12 o'clock and you will not have to start your root pass on one of the tacks. The tester will center-punch his mark at the top of the pipe and you must always have the pipe in that position while making the vertical-overhead weld. You may take the pipe down for cleaning.

Refer to Figure 14.2. You will be welding from 6 o'clock up to 2 o'clock and from 6 up to 10 in a vertical-overhead position. Mark the pipe so that you will know where to stop. You will weld the portion from 10 to 2 o'clock in a horizontal position (see Figure 14.1b).

Start the root pass at 7 o'clock and weld nonstop (if possible) up to 2 o'clock, being careful to burn into and through the tack at 4 o'clock. Leave a crater and keyhole at two o'clock. When making the restart at 7 o'clock, do not strike an arc directly in the joint. Start it on the side of the pipe near the desired juncture, immediately draw a long arc so that no metal is deposited on the pipe, and when the arc is as you want it, keyhole the gap and burn well into the weld deposit. Weld up to 10 o'clock, being careful with the tack at 8 o'clock. Leave the keyhole at 10 o'clock.

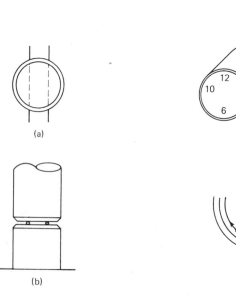

Figure 14.1 Setting up for the pipe test: (a) plan; (b) front view.

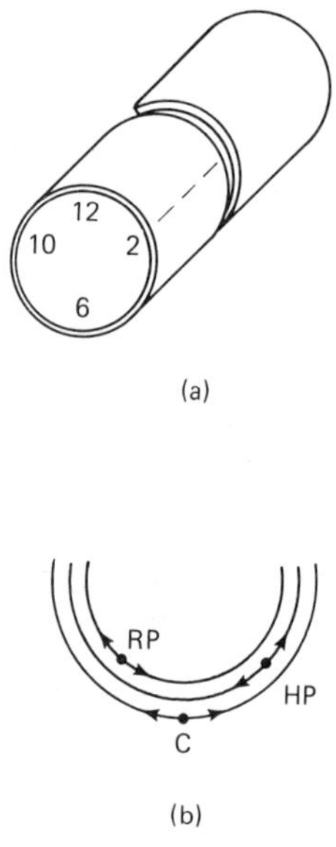

Figure 14.2 Beginning the weld. RP, root pass; HP, hot pass; C, cover.

Remove the pipe, stand it on end (as in Figure 14.1b), and make a horizontal root pass. It should give no trouble if you have left good keyholes. Inspect the root pass inside the pipe. If it appears to you to be satisfactory, call the tester to look at it. A failure of the root pass is failure in the test, so there is no need to continue unless you want the practice. Do not try to cover it up unless you are absolutely certain that you have flunked and have nothing to lose.

Clean the root pass thoroughly. Look for spots that may cause slag pockets because you cannot burn in the hot pass well enough to melt through and blend in the high spots. Use a hammer and chisel to knock out the high spots and use a saw blade to clean down into the edge of the bead, which should blend smoothly with the milled edge of the pipe. Brush briskly until the metal shines.

Start the hot pass (second layer) at 5 o'clock and go to 10, then from 5 o'clock to 2. Remove the pipe and hot-pass the horizontal. Clean thoroughly as before. (A *filler* layer may be needed.)

Start the cover pass at 6 o'clock. Most welders make a thin cover about 2 to 2.5 cm wide, using a very flat side-down figure 8 or a simple right–left–right oscillation. Since it is difficult to see and keep the width uniform, develop the "pendulum" effect in your hands: a natural, relaxed swinging of the hands becomes like the pendulum of a clock and always swings the same distance to the right and left of your eye's center.

The horizontal cover is done with stringer beads one on top of the other. Try to avoid overhang on the lower bead and an undercutting tendency on the top bead. Try to blend the beads so as to present a uniform surface. Some examiners will want the "rounded-out" look of a convex weld.

Dress up the cover. Using a saw blade, clean the edges of the deposit. This will make the cover look more uniform, will eliminate scale (which sometimes looks like undercut), and the shine will reflect in the eye and confuse it a little—perhaps enough to influence the inspector your way.

The welder–testee will generally be required to cut his own test coupons with an oxyacetylene torch and to grind the face- and root-bend specimens flush with the pipe wall; do not grind off the face overweld or root-pass burn-through on the nick-break specimens (see Figures 14.3 and 14.4). Note that in Figure 14.3 coupons are numbered for illustrative purposes only and do not indicate selection by the tester. [*Caution:* To eliminate any possible notch effect, finish grinding with the wheel marks linear to the bend, not transverse. The average width of coupons required will be about 2.5 cm (1 in.).]

MATHEMATICS OF PIPE FITTING

The mathematics of pipe fitting can be quite extensive for the civil engineer, the architect, and the like, but the welder can function more than adequately with a basic understanding of the horizontal and vertical planes (the words *level* and *plumb*), circles, spheres, squares, rectangles, polygons, and the right triangle.

An understanding of the right triangle is of paramount importance, since

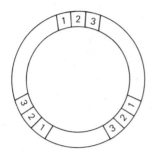 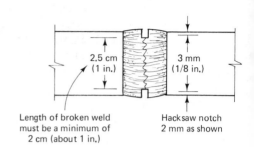

Figure 14.3 Coupons: 1, root bend; 2, nick break; 3, face bend.

Figure 14.4 Relative positions of dimensions in the right triangle.

any pipe making a turn must make that turn at a predetermined angle. All laterals and offsets are calculated from the right triangle. At the heart of the right triangle are the square root and trigonometry.

Since the term *trigonometry* is a composite of (Greek) *treis*, three; *gonia*, angle; and *metron*, measure, we may look at trigonometric functions as a measure of the interrelationships of the three angles of a right triangle.

Such formulas as "the square of the hypotenuse is equal to the sum of the squares of the other two sides" may sound formidable to the welder who has been confined to the ratio 3-4-5, but pocket calculators now give squares and square roots at the touch of a button. More expensive calculators will provide the trigonometric functions *sine, cosine,* and *tangent* of the angle, which are the functions most often used. Spending a little time "doing it the hard way" will, however, give a better understanding of the triangle and greater skill with a calculator.

Fractions

Converting fractions to decimals:

(a) $\frac{1}{4}$ (b) $\frac{1}{8}$ (c) $\frac{1}{16}$ (d) $\frac{1}{32}$ (e) $\frac{1}{64}$

$$\begin{array}{r} .25 \\ 4 \overline{)1.00} \\ \underline{8} \\ 20 \\ \underline{20} \end{array}$$

Converting compound fractions to decimals:

(a) $3\frac{3}{4}$ (b) $4\frac{1}{8}$ (c) $25\frac{1}{16}$ (d) $21\frac{5}{8}$

Mathematics of Pipe Fitting

```
    .75
4 ) 3.00
    28
    ‾‾
     20
     20
     ‾‾
```
or 3.75

Square Root

Square root of whole numbers. The procedure is as follows:

1. Start at the right and mark pairs of numbers toward the left. Each pair of numbers yields one number in the answer. If, when marking off to the left, a single number remains, treat it as a pair of numbers; that is, you may add a zero to the left if it will help clarify the situation.
2. Select a number that multiplied by itself equals or nearly equals the first pair of numbers. As in the example below, 6 × 6 = 36 (7 × 7 is 49 and exceeds the pair 42).
3. For the next number in the answer, double the first digit (6 × 2 = 12) and place it to divide into the second pair of numbers (the 12 in the figure 125 below).
4. Select a number to add to the right of the doubled number which will multiply the entire number into the new dividend (such as 5 × 125 = 625).

```
              6 5
(a)        √4225              (b) √15625              (c) √1262325
              36
       125   625
             625
```

Square root of decimals. Mark off numbers in pairs from left to right, then proceed as described for whole numbers.

```
                    . 3 9 5 2 8
         (a)  √0.15625000 00              (b) √0.50024
                    9
          6 9     662
                  621
         78 5    4150
                 3925
                 ‾‾‾‾‾
                 22500
        790 2    15804
                 ‾‾‾‾‾‾‾
                 6696 00
       7184 8    5747 84
                 ‾‾‾‾‾‾‾
                  948 16

         (c) √0.84262                     (d) √0.025097213
```

Square root of compound numbers. Mark off numbers in pairs to the left of the decimal point and to the right of the point, as shown here:

$$\overset{\frown}{25}\overset{\frown}{10}\overset{\frown}{6}.\overset{\frown}{13}\overset{\frown}{45}\overset{\frown}{6}$$
$$\longleftarrow \cdot \longrightarrow$$

Now proceed as before.

    (a) $\sqrt{306.5625}$ (b) $\sqrt{19.474636}$ (c) $\sqrt{23.87654}$

Square root of fractions. Change fractions to decimals and proceed as above.

(a) $\sqrt{\frac{1}{4}}$ (b) $\sqrt{\frac{1}{128}}$ (c) $\sqrt{3\frac{1}{8}}$

$1 \div 4 = 0.25$ $1 \div 128 = 0.0078125$ $1 \div 8 = 0.125$

```
         .  5                           .  0 8 8 3   8   8                      1.  7 6   7   7
     √0.25                          √0.00781250 00 00                       √3.1250 00 00
         25                                  64                                 1
        ___                          16     1412                            27  2 12
                                            1344                                1 89
                                     176    6850                            34 6  2350
                                            5289                                   2076
                                            1561 00                         352 7 274 00
                                     1766 8 1413 44                                246 89
                                            147 56 00                              27 11 00
                                     17676 8 141 41 44                      3534 7 24 74 29
                                             6 04 56                                2 36 71
```

(d) $\sqrt{\frac{1}{8}}$ (e) $\sqrt{\frac{1}{16}}$ (f) $\sqrt{\frac{1}{32}}$

(g) $\sqrt{\frac{1}{64}}$ (h) $\sqrt{4\frac{3}{16}}$ (i) $\sqrt{5\frac{7}{8}}$

Answers:

    (d) $\sqrt{\frac{1}{8}} = 0.35355$ (e) $\sqrt{\frac{1}{16}} = 0.25$

    (f) $\sqrt{\frac{1}{32}} = 0.17677$ (g) $\sqrt{\frac{1}{64}} = 0.125$

    (h) $\sqrt{4\frac{3}{16}} = 2.04633$ (i) $\sqrt{5\frac{7}{8}} = 2.42383$

Calculator Example. To find the square root of $4\frac{3}{16}$, first convert $\frac{3}{16}$ to decimal form, using a calculator to divide. Add the whole number 4 to the quotient, then apply the square-root function.

    $3 \div 16 = 0.1875 + 4 = 4.1875$ (F) $\boxed{\sqrt{X^2}} = 2.04633$

Mathematics of Pipe Fitting

Right Triangles

Solving right triangles by square root:

$$A^2 + B^2 = C^2$$
$$A = \sqrt{C^2 - B^2}$$
$$B = \sqrt{C^2 - A^2}$$
$$C = \sqrt{A^2 + B^2}$$

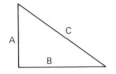

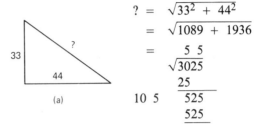

(a)

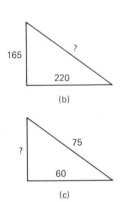

(b)

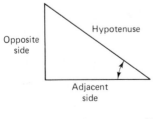

(c)

Trigonometry of right triangles:

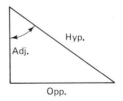

$$\text{sine of angle} = \frac{\text{opp.}}{\text{hyp.}} \qquad \text{cotangent} = \frac{\text{adj.}}{\text{opp.}}$$

$$\text{cosine of angle} = \frac{\text{adj.}}{\text{hyp.}} \qquad \text{secant} = \frac{\text{hyp.}}{\text{adj.}}$$

224 Pipe Welding   Chap. 14

$$\text{tangent of angle} = \frac{\text{opp.}}{\text{adj.}} \qquad \text{cosecant} = \frac{\text{hyp.}}{\text{opp.}}$$

The problems below require that the unknown dimension be found before finding the function. Use the formula for right triangles. If investing in a calculator, pay the extra cost of getting one that handles the trigonometric functions sine, cosine, and tangent.

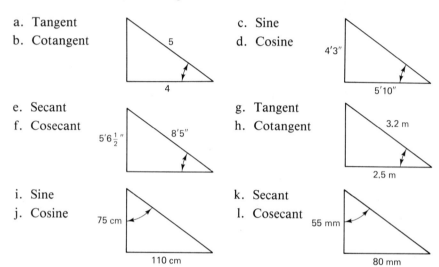

a. Tangent
b. Cotangent

c. Sine
d. Cosine

e. Secant
f. Cosecant

g. Tangent
h. Cotangent

i. Sine
j. Cosine

k. Secant
l. Cosecant

The Circle

Since pipe is round, and the majority of tanks are cylindrical, elliptical, or spherical, it is important to understand the circle in order to meet the problems that arise in working as a welder–fitter, and especially so for the steamfitter-welder. The basic elements to understand are (1) the diameter, (2) the radius, (3) the chord, (4) the segment, (5) the circumference, and (6) that "sine qua non," our friend pi (π), 3.1416 (see Figures 14.5 to 14.7). Some relationships among the parts shown in these figures are as follows:

$$\text{circumference} = D \times 3.1416$$
$$\text{area of circle} = D^2 \times 0.7854$$
$$\text{area of circle} = \frac{D}{2} \times \frac{C}{2}$$
$$\text{length of chord} = 2\sqrt{AB}$$
$$\text{area of sector} = \frac{\text{arc}}{2} \times R$$
$$\text{area of segment} = \text{area of sector minus area of triangle}$$
$$\text{area of triangle} = \text{base} \times \frac{\text{height}}{2}$$

Mathematics of Pipe Fitting

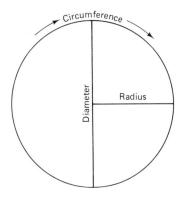
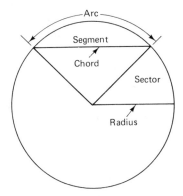

Figure 14.5 Dimensions of the circle   Figure 14.6 Portions of the circle

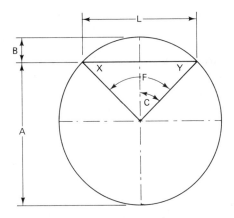

Figure 14.7 Mathematics of a sector

length of arc = number of degrees in arc × radius × 0.01745

The number of degrees in the arc is found as follows: (see Figure 14.7):

1. Find the sine of angle C.

$$\frac{L}{2} \div R = \text{sine}$$

2. Obtain the angle from trigonometry tables.
3. Double the number of degrees in the angle for angle F.

"Pi" is important to remember because so many formulas dealing with circles and spheres use 3.1416 or fractions thereof. If you know 3.1416 and that usually odd numbers such as 1.5708 and 0.7854 are a fraction of pi, you can recall the number by going back to pi and dividing by 2 or 4 or 8 and recognizing it when you see it—bearing in mind that if you are working with spheres, you may also run into $\frac{1}{3}$ or $\frac{1}{6}$ of pi.

In school, you may have learned to divide a circle into parts by using a compass and scribing on a piece of paper. What kind of compass would you use to divide a 16-in. pipe into four parts—not to mention a tank 150 feet in diameter? Are you on the inside or outside of the tank? Are you at the top, middle, or bottom of the tank?

You are a fitter working inside (Figure 14.8a) or outside (Figure 14.8b) of a cylindrical shape (tank, reactor, whatever). You have forgotten all the formulas but remember our friend π and that the cylindrical "whatever" is about 20 to 40 meters in diameter, but you do not know the diameter. How would you proceed to find a given spot on the tank at which to install a fitting or cut a hole? X is your given reference point. In part (a), the distance from X to point A is the farthest you can stretch your tape. Find A and mark it. Point B is where chords X-B and A-B are equal. Point C is found in the same manner. Use chords and arcs for other points. When working outside the tank, if the diameter is known, use the formula $C = D \times 3.1416$ to find the circumference. If the diameter is unknown, find the circumference by using your tape. Then the distance from X to A is one-half the circumference, and from X to B or C is one-fourth the circumference.

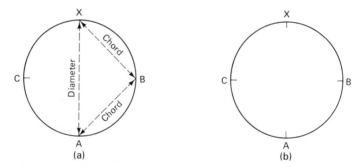

Figure 14.8 Measuring (a) inside a cylindrical tank and (b) outside the tank.

Capacities of Common Tanks

Figure 14.9 illustrates the most common tank configurations, and the following formulas apply.

Cylindrical tanks:

area of surface = $D \times$ pi $\times$ length + area of two bases

area of base = $D \times D \times 0.7854$

Mathematics of Pipe Fitting

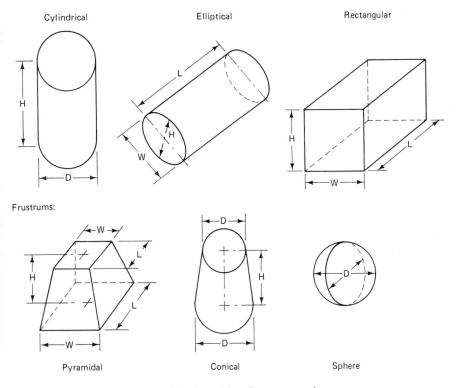

Figure 14.9 Capacities of common tanks.

$$\text{volume} = L \times \text{area of base}$$
$$\text{capacity in kiloliters} = \text{volume in cubic meters}$$
$$\text{capacity in gallons} = \text{volume in cubic inches} \div 231$$
$$= \text{volume in cubic feet} \times 7.48$$

To find the volume of a cylindrical tank with a convex head, add the volume of the head. If the head is concave, deduct the volume of the head.

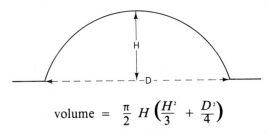

$$\text{volume} = \frac{\pi}{2} H \left(\frac{H^2}{3} + \frac{D^2}{4} \right)$$

The formula above may be used for any dome or dish-shaped volume.

Elliptical tanks:

area of end = short D × long D × 0.7854

volume = $W \times H \times 0.7854 \times L$

capacity in kiloliters = volume in cubic meters

capacity in gallons = volume in cubic inches ÷ 231

= volume in cubic feet × 7.48

Rectangular tanks:

volume = $L \times W \times H$

Spherical tanks:

area of surface = $D \times D \times 3.1416$

volume = $D \times D \times D \times 0.5236$

Cones:

volume = area of base × height × 0.333

Pyramids:

volume = area of base × height × 0.333

Frustrums (pyramidal or conical):

$V = \dfrac{H}{3}$ × (area of upper base × area of lower base + $\sqrt{AB}$

PIPE FITTING

Joining Pipe Sections

Double ending. While a crane or cherry-picker is available, put on wooden blocks as many sections of pipe as there is room for (see Figure 14.10). Of course, you may not have blocks; an experienced pipe welder can align and join pipe on the runway, on the ground, in a ditch, or even up and down a hillside.

The procedure for double ending is as follows:

1. Line up pipe as shown in Figure 14.10, or in the most efficient manner available.
2. Set the gap as desired. New pipe sections have milled edges, and evenness of gap indicates fairly close alignment.
3. Check for "high-low" on outer walls. Use fingers, level, straightedge, welding rod, or whatever. It seems that no matter what a welder uses, he usually ends it all up with a finger-check.

Pipe Fitting

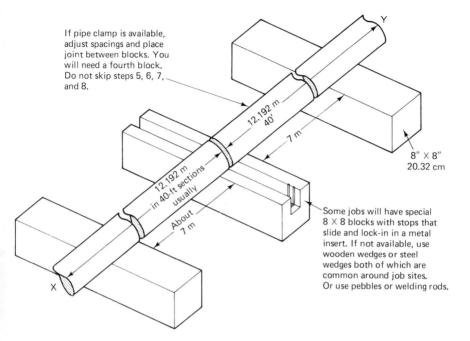

Figure 14.10 Double-ending.

4. If possible, level up the pipe.
5. Align the pipe by eye, standing at end X about 10 to 20 feet away. Go to end Y and "eyeball" for a recheck.
6. Tack-weld the pipe, about 1 cm long, of root-pass quality, at 3 and 9 o'clock.
7. Roll the pipe 90°. Repeat alignment as in step 5. *Do not skip this step!* Tacks made in step 6 should now be on top and bottom.
8. Again, tack-weld at 3 and 9 o'clock.
9. Recheck the alignment, then weld according to specifications.

Joining a pipe turn. The squaring process is illustrated in Figure 14.11. When "squaring" a pipe turn with pipe, the distances A and B must be equal. (If a fitting or Tube-Turn* is put on an existing run or there are other fittings on the pipe, the Tube-Turn must be "leveled" at right angles at point Y.)

As an alternative method of "squaring," level up the pipe at X and level the tube turn at Y (or "plumb" the Tube-Turn at Z).

The joining procedure is as follows:

1. Rest the Tube-Turn between your legs and butt it against the pipe end.
2. Feel the sides and bottom for "high-low."

*Tube-Turn is a registered trade name.

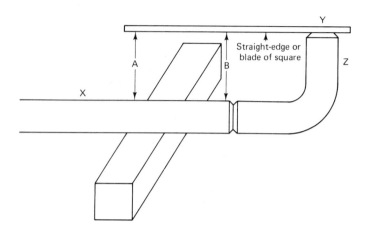

Figure 14.11 Joining a pipe turn.

3. Letting the edges touch at the bottom (6 o'clock) and feeling high–low with one finger, cock the Tube-Turn back toward you to set the proper gap. Keep feeling with your fingers along the sides of the pipe as high up as you can stretch.
4. If you are working in pairs, the welder tack-welds at the top (12 o'clock), of root-pass quality.
5. Do not let go of the Tube-Turn. Insert a tapered steel wedge at the 6 o'clock point and wedge open so that the gap is even all around the pipe. But square up the pipe with Tube-Turn regardless of gap!
6. The welder then tack-welds on the bottom on both sides of the wedge at approximately 5 and 7 o'clock.
7. Weld as convenient. If the pipe assembly is easy to handle, push it back so that the Tube-Turn can lie on the block or workbench and "flop-weld"; that is, weld from 3 o'clock up to 12 o'clock and from 9 o'clock up to 12 o'clock, then flop the Tube-Turn and repeat.
8. If working alone, lay the Tube-Turn on its side and use shims, wedges, spacers, or whatever to make alignment.

Flanges

There are three common types of flanges: slip-on, weld-neck, and socket (Figure 14.12). At A and B, assemble as shown in Figure 14.12 and weld as shown on the blueprint, as directed, or as is common practice by (1) multipass stringer beads, (2) multipass weave beads, or (3) multipass stringers with a lacing cover. Practice all three.

Be careful not to let the weld spill out beyond the face of the flange. Do not scar the face of the flange. The preferred shape of the weld is shown at C.

Pipe Fitting

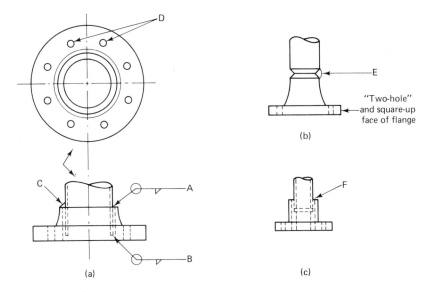

Figure 14.12 Flanges: (a) slip-on; (b) weld-neck; (c) socket.

Alignment of holes (D) is important when flange is welded in position on an existing pipe run and when there are other fittings on the section of pipe you are working on. (See the *two-holing* of flanges and valves in Figure 14.14.)

Weld-neck flanges (E) may be treated like a piece of pipe. Square up the face as shown in Figure 14.13. The flange must be "squared" two ways: roll square or pipe 90° and repeat measurements. Shape the weld convex. Old-timers say: "Burn it in, round it out, and make it wide."

Socket-type flanges (F) present no problems but since there is some looseness in the socket, align as in Figure 14.13 by two-holing.

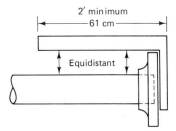

Figure 14.13 Squaring a flange.

Two-holing a spool:

1. Assemble and weld the first flange as shown in Figure 14.13. Have no concern about the alignment of holes.
2. Place the second flange on the end of the pipe (Figure 14.14a).
3. Now two-hole the first flange. Place the level as shown in Figure 14.14b.

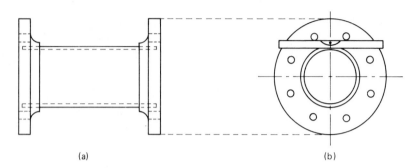

Figure 14.14 Two-holing a spool.

With thumbs pressed against the level, grip the flange and level between fingers and thumbs. Roll the spool until the two holes are level (you can use the tops or bottoms of the holes as guides).* When level, chock-up (block) the spool so that it cannot roll out of level.

4. Two-hole the second flange and tack-weld at the top or bottom (12 o'clock or 6 o'clock). Square-up the flange face as shown on page 234. (Always be sure to square-up in two directions, 90° apart.)
5. Recheck the entire assembly and weld.

Flanging a 90° vertical turn:

1. Find the centerline of a horizontal pipe run and mark as shown (Figure 14.15).

Alignment pins for two-holing of flanges

2. Find the centers between two holes (Figure 14.15b).
3. Set the flange on the pipe and adjust the flange so that the four chalk marks line up (or by sighting along the level and side of the pipe run).
4. Place the level on the pipe flange as shown in Figure 14.15b. Chalk-mark centers as shown by the arrows and "eyeball" down the centerline of the pipe. Line up the four marks. Alternatively, place the level at two holes, as shown, and sight along the side of the pipe. Use both where possible.
5. Adjust the flange so that the pipe end is 1 cm below the face of the flange at point A. Allow the binding effect of the pipe and flange to absorb most of the weight. This binding will cause point B to be slightly more than

---

*Alignment pins are used by some welders. Two are needed. Pins slide through the holes until the right size is in the hole. You can then rest the level on top of the part of the pin that sticks out. Most welders do not use them: They are expensive, and you will be lucky if they are not lost or stolen during the first month you have them.

Pipe Fitting

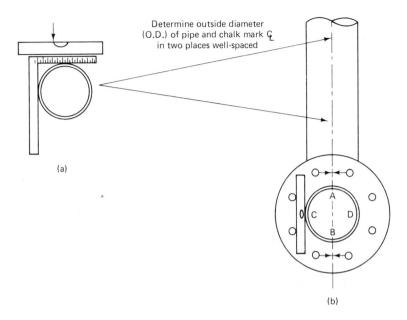

Figure 14.15 Flanging a 90° turn.

1 cm below the face of the flange. Allow that to happen. Just be sure that the chalk marks line up and that point A is correct.

6. The welder tack-welds at point A.
7. Lift the flange so that the face of the flange shows level.
8. Tack-weld at point B.
9. Turn the level at right angles to the pipe run (crosswise on the flange face) and rock the flange to show that it is level.
10. Tack-weld at points C and D.
11. Recheck the two-hole alignment. On an existing line, level two directions regardless of the run.

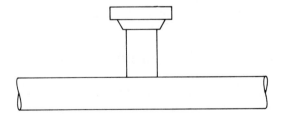

Note: This two-holing and alignment may be used for squaring up a "tee" (T):

In squaring the flange parallel to a pipe run, roll straight edge 90° and square it at right angles (see Figure 14.16). If working on an existing run and

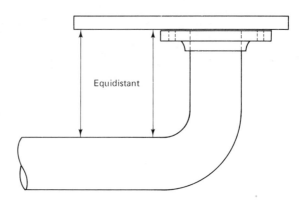

Figure 14.16 Squaring the flange parallel to the pipe turn.

if the pipe run is out of proper alignment (not level or plumb), make the flange face level lengthwise and at right angles.*

Cracking a Flange:

1. Learn to use two wrenches: one box-end and one open-end. Use the box-end to hold the nut while you twist off the other nut with the open-end wrench. If both nuts turn on the stud, cock the box-end wrench so that it binds the nut on the stud.
2. Loosen the two bottom nuts first. Do not twist off all the way unless the stud is so short that you cannot open the joint. Proceed toward the top of the pipe flange, loosening each nut a little less than the one below it. The idea is to allow the joint to seal at the top and open at the bottom so that any substance in the pipe will spew out at the bottom and not all over you. Circumstances alter cases; sometimes you will want a pipe joint to open sideways. At any rate, always open a joint so that you will not be directly in front of a spill-out. "Empty" pipes have been known to flow oil for hours.

Installing a blind. Usually, the purpose in "cracking" a flanged joint is to insert a blind or remove a blind. Observe the following procedures:

1. Do not remove all the studs. Remove only enough studs to allow you to remove and insert the blind. Leaving the other studs helps you in aligning the gasket and/or the blind.
2. If you are removing a ring-type blind to open a line, invert the blind; that is, insert the ring part of the blind. If the gasket is a standard asbestos type, throw it away, scrape clean the face of the flanges, and put in a

---

*Standards are necessary because miles away a welder-fitter in the shop may be putting together a pipe assembly that will be connected to yours. If his flange does not face your flange properly, and his is right and yours is not, you are in trouble!

Pipe Fitting

new gasket. If the gasket is the expensive metallic type, inspect it for flaws and reuse it if it is satisfactory.

3. If the blind is just a piece of metal with a stub on it, place the removed blind nearby for future use.

The ring-type blind is probably the more popular, for although they are more expensive to make, they have several advantages: (1) the pump-station operator can look out over an area and see immediately if a line is open or shut; (2) the blind is always available to him, no one can borrow it; and (3) when reconnecting the flanges, the blind always fills the space, and you do not have to force the joint apart to insert it or pull it together after removing the blind. Some pipe joints just do not have room to be forced.

Laying Out Angles Using a Square

The procedure for using a square to lay out angles is as follows:

1. Find the tangent of the angle using trigonometry tables. (See the appendix to this chapter.)
2. Move the decimal in the tangent one digit to the right. The resulting number is then read in inches (or the units of measure that you are using).
3. Draw a line from the "10" mark on the blade to the "tangent" on the tongue. For example, for a 30° angle, the tangent is 0.57735. Move the decimal to 5.7735. Then 5.7735 = 5 and $\frac{25}{32}$ in. (if using inches). [*Note:* To make a larger layout, you could use 100 on the blade and move the decimal of the tangent to 57.735 for the distance on tongue. If using a metric rule, you would use 10 cm on the blade and 5.7735 cm on the tongue, or 100 cm (1 meter) and 57.735 cm (see Figure 14.17).]

Since trigonometry tables are based on a right triangle, angles of more than 45° must be obtained on the other side of the midtangent. For a 70° angle, 90° − 70° = 20°. The tangent of a 20° angle is 0.36397. 3.6397 = $3\frac{35}{8}$ in. (see Figure 14.18).

Refer to Figure 14.19. The 12 is a constant. Lay off dimension A on a straightedge. Obtain dimension A by multiplying 12 times the cotangent of angle. 12 × 1.7320 = 20.784 = $20\frac{3}{4}$ in. (or metric measure).

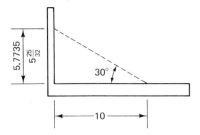

Figure 14.17 Laying out a 30° angle.

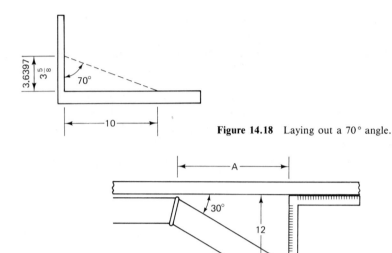

Figure 14.18 Laying out a 70° angle.

Figure 14.19 Laying out an angle with T-square.

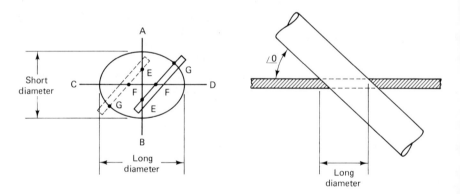

Figure 14.20 Laying out an elliptical hole.

Elliptical hole for angular travel.   Refer to Figure 14.20.

short diameter = outside diameter (O.D.) of pipe

long diameter = O.D. of pipe × cosecant of angle of pipe to plate

$F - G = \frac{1}{2}$ of short diameter

$E - G = \frac{1}{2}$ of long diameter

1. Lay out lines *A-B* and *C-D* at centerlines of opening.
2. Lay out *F-G* and *E-G* on a flat piece of wood (or substitute).

3. Draw the curved line at point G, keeping E and F on lines A-B and C-D as the piece of wood is rotated.

Note: When cutting, keep the angle of the tip at the same angle as the pipe will be when installed.

Laying Out Angles Using a Protractor

The usual procedure for laying out angles using a protractor is shown in Figure 14.21a; this procedure may also be followed without using a level. To use the protractor, set and lock the arm, and use the arm and block to scribe the angle (Figure 14.21a).

To get the most out of your investment, you should know the relationships of angles in the full 180° range and know that there are complementary right triangles abutting, a total of 4 to the 360° circle (see Figure 14.21b-d).

Laying Out Miter Cuts for Pipe Turns

Figures 14.22 and 14.23 illustrate miter cuts for tube turns. Note that the procedure and formulas are the same regardless of the degree of the turn and the number of pieces in the turn.

$$\frac{\text{number of degrees of turn}}{\text{number of welds} \times 2} = \text{angle of cut}$$

(Count only the welds *in* the turn.)

$$\text{dimensions of } A \text{ and } B = \frac{\text{O.D. of pipe} \times \text{tangent* of the angle}}{\text{number of welds} \times 2}$$

From Figure 14.23a:

$$\frac{90°}{2} = 45°$$

$$\text{tangent} = 1.0000$$

$$A \text{ and } B \text{ of Figure 14.22} = \frac{22 \times 1.0000}{2} = 11 \text{ cm}$$

From Figure 14.23b:

$$\text{angle} = \frac{\text{number of degrees of turn}}{\text{number of welds} \times 2} = \frac{90}{4} = 22\frac{1}{2}°$$

$$C = \text{radius} \times \text{tangent} = 27 \times 0.41421 = 11\frac{1}{8}$$

$$D = A \times 2 = 22\frac{1}{4} \text{ in.}$$

Lay out the cuts as shown in Figure 14.22.

*Many pipe fitters' handbooks give this figure as the "factor" of cut in a table giving degrees and "factors," but it is the tangent you have in your trigonometry tables.

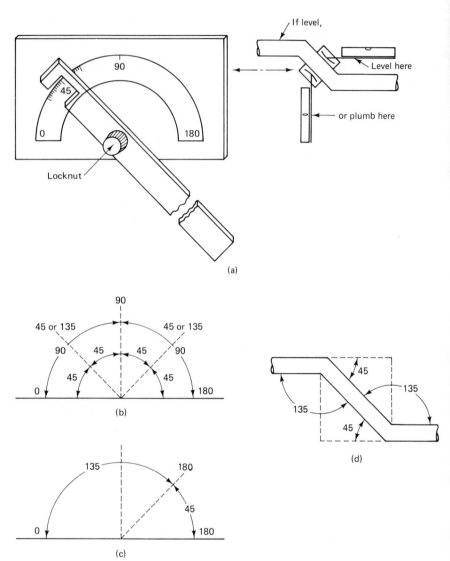

Figure 14.21 Using a protractor.

The procedure for laying out miter cuts is as follows:

1. Divide pipe into four equal parts.
   (a) *Mode 1:* Wrap a strip of paper around the pipe and tear off the excess circumference. The ends of paper should barely touch. Fold the paper back on itself (in halves) and crease the paper well. Fold again (quarters) and crease. Rewrap the paper around the pipe and make a longitudinal chalk line at each crease and where the two ends meet.

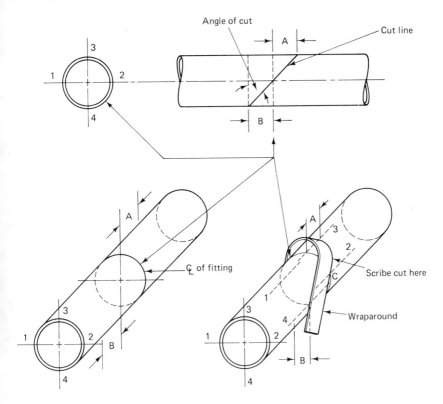

Figure 14.22 Laying out miter cuts for tube turns.

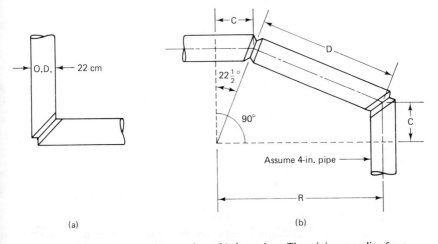

Figure 14.23 90° turns: (a) two-piece; (b) three-piece. The minimum radius for a three-piece turn is six times the outside diameter of the pipe.

(b) *Mode 2:* Find the circumference of the pipe and divide by 4. Mark off these quadrants on your wraparound and use the wraparound to mark the pipe. Assume a 20-cm pipe with an O.D. of 22 cm.

$$\text{quadrant} = 22 \times 3.1416 = \frac{69.152}{4} = 17.28 \text{ cm}$$

(This method is necessary on larger pipe. How often on the job can you find a piece of paper to go around a 6-in. or larger pipe?)

2. Find the dimensions for A and B (see Figure 14.22). Note that A is on chalk line 3 and B is on line 4, and that they go in opposite directions from the centerline of the fitting.
3. Shape the wraparound carefully around one-half of the pipe so that its edge is held at chalk lines 1 and 2 and the edge at the loop crosses chalk line 3 at dimension A. Draw a line from line 1 through line 3 and end the line at 2. This is the cut line.
4. Rotate the pipe 180° (if movable) and repeat. Run a chalk line from point 1 through line 4 and end the cut line at 2.

Notes: For best-quality miter cuts, always center-punch the cut line. Learn to center-punch so that you can cut away half the center-punch mark and leave half of it on the piece you are going to use. Center punching is mandatory on rusty, painted, or lacquered pipe. In making miter cuts, always keep the torch tip aimed at the line on the other side of the pipe. Bevel *after* you know that the cut is correct.

Branches and Headers

Full-size tee. This tee is diagrammed in Figure 14.24. Following is the procedure for laying out the header.

1. Center-punch the header at the exact centerline of the branch. Wrap the wraparound around the pipe and soapstone a circumference of the same as that for making a cut.
2. Divide the circumference into four equal parts. At each quadrant, draw a longitudinal line about 10 in. (a little more than the O.D. of the branch) with its center at the circumference line. Number these lines 1, 2, 3, and 4, as shown in Figure 14.22, with 3 on the top, 4 on the bottom, and 1 and 2 on the sides.
3. Mark off points A and B on line 3 on each side of the circumferential line. The distance from A to the centerline and B to the centerline will be one-half the O.D. of the branch. (*Note:* Point C will be located at the intersection of the circumferential line and lines 1 and 2.)
4. Place the wraparound on the pipe and line it up with point A and points C on each side of header. Draw a line from C through A to C (see Figures 14.22 and 14.24). Draw a similar line from C through B to C.

Pipe Fitting

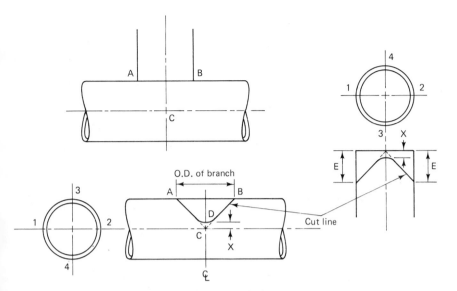

Figure 14.24 Branch and header for a full-size T.

5. Pointed cuts at point C are usually undesirable. Round the point as shown in Figure 14.23. The distance x should be twice the wall thickness of the pipe.
6. Make a radial cut. Bevel 45° after the cut is determined to be correct.

For laying out the branch, proceed as follows:

1. Divide the pipe into four equal parts at or near the end. Draw a straight line longitudinally from the end of the pipe at each of the quarter marks. Number the lines 1, 2, 3, and 4, as above.
2. Mark off points A and B on lines 1 and 2. The distance from A to the end of the pipe should be one-half the O.D. of the branch, and the same for B to the end of the pipe. Line up the wraparound on point A and extending around the pipe to both points C on lines 3 and 4 at the end of the pipe. Do the same with point B. You should now have a line from the end of the pipe at C, extending through A, and connecting with C again at the end of the pipe. And you should have an identical line on the other half of the circumference of the pipe.
3. If the header has a rounded point at C, the branch *must be* rounded in the same manner.
4. Make a radial cut. *Do not bevel the branch.*

Note: Most welders in the field lay out and cut the branch, then set the branch on the header, and chalk a line inside the branch (unless the branch is too tall), outlining its shape on the header.

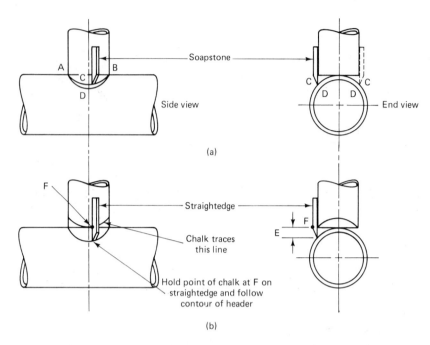

Figure 14.25 Reducing Tee: (a) laying out the header; (b) laying out the branch.

Reducing Tee. Follow the diagrams in Figure 14.25 for performing the following procedures. The steps for laying out the header are as follows:

1. Set the branch pipe on the header. With a soapstone pencil (sharpen to a flat edge on one end) held flat against the branch, draw a smooth curved line on the header. Points A, B, C, and C will be located on this line.
2. Mark off points D. The distance from C to D is twice the wall thickness of the header. Connect points A, B, D, and D freehand with a smooth curved line: D through A to D, and D through B to D. This is the cut line for the opening in the header.
3. Make a radial cut, pointing the torch tip to the center of the pipe at all times. Bevel the edge of the opening 45°.

Laying out the branch is done as follows:

1. Obtain a flat piece of wood (or reasonable substitute) and sharpen it to a flat edge on one end. This piece of wood will be used as a straightedge. Lay off dimension E from the sharpened end. (One could notch the piece of wood to aid in holding the soapstone or pencil or marker.)
2. Set the branch in position on the header (some little distance from the hole in the header). Place the straightedge flat against the branch with the sharpened end resting on the header, and hold the soapstone pencil

at F on the straightedge. Move the straightedge around the branch to draw the line. Be sure that soapstone is always on the proper mark on the straightedge.
3. Make a radial cut. Do not bevel the edge of the branch.

Note: Since the branch must be held level (right and left of header) and you need two hands to hold the straightedge and soapstone, do not attempt the near impossible and try it alone. The situation becomes even more difficult if the branch is already flanged and you have to two-hole it. You can, of course, tack-weld the branch to the header at points A and B. If you can hold the branch level with one hand, flip your shield down without losing level, and weld with the other.

Fabricating Fittings from 90° Pipe Turns

See Figure 14.26, which illustrates the dimensions used in the following procedure for fabricating fittings from 90° pipe turns. The formulas for the dimensions labeled in the figure are:

$$A = \text{radius of turn}$$
$$B = A - \tfrac{1}{2}\,\text{O.D. of pipe turn}$$
$$C = A + \tfrac{1}{2}\,\text{O.D. of pipe turn}$$
$$D = \text{degrees of fitting desired}$$
$$E = D \times B \times 0.01745$$
$$F = D \times A \times 0.01745$$
$$G = D \times C \times 0.01745$$

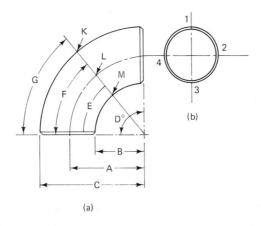

Figure 14.26 Fabricating fittings from 90° pipe turns.

The procedure for fabricating fittings is as follows:

1. Divide the Tube-Turn into four equal parts at both ends (Figure 14.26b).
2. Draw longitudinal lines connecting the numbers at both ends.
3. Using the formulas above, find the dimensions of arcs *E*, *F*, and *G* (Figure 14.26a).
4. From the face of the fitting, lay off arc *G* on line 1 and mark the fitting at *K*. Lay off arc *F* on lines 2 and 4 and mark at *L*. Lay off arc *E* on line 3 and mark at *M*.
5. Using the wraparound (or other guide), connect points *L*, *K*, and *L* on lines 1, 2, and 4 (soapstone); then connect *L*, *M*, and *L* on lines 2, 3, and 4.
6. Make a miter cut and bevel the edges. Save the other piece, you may need a fitting for which it will be right.

Alternative methods. Many welders (fitters) use a simpler method.

1. Lay out the desired angle of cut on any clean flat surface.
2. Place the fitting on the angle and freehand a cut line that corresponds to the line on the triangle.

Note: If trig tables are available (you should have them with you), use 10 on your tape and the tangent of the angle. (If working on a very long radius, use 100 and move the decimal of the tangent two digits. On pipe of very large diameter, this method loses accuracy.) This method is shown in Figure 14.27. The tangent of a 50° angle is 1.1917. Move the decimal point one place to the right to obtain 11.917. (The scale shown is 1 cm = 2 in.)

A field welder-fitter often finds the work area unsuitable for either of the two methods shown. This next method can be used by a welder sitting on a grassy hillside and holding the fitting on his lap.

1. Measure the long and short arcs with a tape.

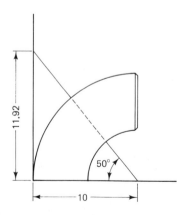

Figure 14.27 Freehanded cutting line.

Pipe Fitting

2. 90° ÷ length of arc = degrees per unit (cm or in.) of length.
3. Mark the pipe accordingly, draw a cut line freehand, and cut.

Assume that you have measured the fitting (Figure 14.26) and the long arc is 60 cm, the short arc is 40 cm, and the angle you want is 50°. You are making a 50° fitting from a 90° fitting.

$$90° \div 60 = 1.5°$$ per centimeter in the long arc

$$90° \div 40 = 2.25°$$ per centimeter in the short arc

Work backward to the first method (Figure 14.26). 90° − 50° = 40°, which you don't want.

$$40° \div 1.5° = 26.666 \text{ cm}$$

$$40° \div 2.25° = 17.777 \text{ cm}$$

Start at the face of the fitting (as at top of Figure 14.26b) and locate K as being 26.66 cm from face. Locate E at 17.77 cm from the face. Draw a cut line from K to M on both sides of pipe. Miter the cut and bevel.

Branch pipe back from the pipe turn. Refer to Figure 14.28 (*Note:* A back stub having the bottom of the stub even with the bottom of the fitting may be used as a support for a descending pipe run that falls between runway stanchions.) Following is the procedure for laying out the branch pipe.

1. Mark, with a line on the side, the centerline of the forward run and the centerline of the backward run. Unless otherwise instructed, if the backward run is of a lesser diameter, line up its centerline with the centerline of the larger run.
2. Sight down the side of the pipe and when properly aligned, tack-weld the pipe to the fitting as shown in Figure 14.28. Tack-weld the top of the brace to the fitting. Adjust the brace and pipe in correct alignment and tack-brace to the pipe.

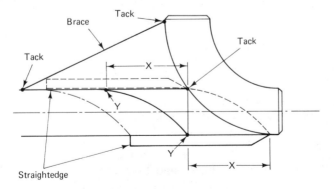

Figure 14.28 Branch pipe back from pipe turn.

3. Sharpen a flat piece of wood to make a straightedge. Measure distance *X* and lay it off on the wooden straightedge, point *Y*. Holding the chalk at point *Y*, move the straightedge and chalk up the side of the pipe to its position at the top of the pipe (dashed outline). The two curved-dashed lines show the path of the straightedge, and the solid line between them will be the cutline. Duplicate the marking on the other half of the pipe. Make sure that the sharpened end of the straightedge contacts the fitting at all times. When the backward pipe is the same size as the fitting, keeping contact can be a problem.

For laying out the hole in the fitting, the procedure is as follows:

1. Set the branch in position against the fitting and scribe the cut line for the hole by using the end of the branch as a guide.
2. Keep the cutting tip toward the center of the fitting at all times. Cut *inside* the chalk line (leave the chalk) because the hole should be the size of the inside diameter of the branch. Do not bevel.

Welding Offsets

Following are the formulas that are commonly used when welding offsets (see Figure 14.29):

$$\text{angle} = \frac{\text{side opposite (set)}}{\text{side adjacent (run)}}$$
$$= \text{tangent}$$
$$\text{set} = \text{travel} \times \text{sine}$$
$$= \text{run} \times \text{tangent}$$
$$\text{run} = \text{travel} \times \text{cosine}$$
$$= \text{set} \times \text{cotangent}$$
$$\text{travel} = \text{set} \times \text{cosecant}$$
$$= \text{run} \times \text{secant}$$
$$\text{angle of cut} = \text{one-half of angle of turn}$$
$$\text{cut-back dimension} = \text{O.D. of pipe} \times \text{tangent of angle of cut}$$

Refer to Figure 14.30 and observe the following procedure.

1. Locate the centerline of the turn. With the wraparound, draw a line around the pipe. Divide the pipe into quadrants.
2. Extend these points lengthwise on the pipe; number the lines 1 to 4.
3. Lay out dimension *C* on lines 1 and 3 from the centerline as shown. The intersection of the plane between lines 2 and 4 with the centerline will be the center of the fitting.

Pipe Fitting

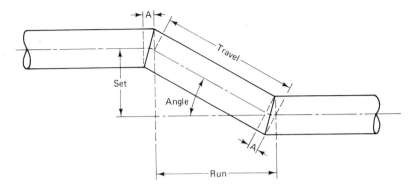

Figure 14.29 Welded offsets.

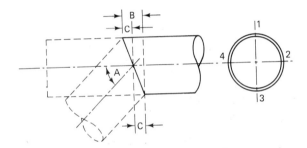

Figure 14.30 Preparing a miter cut.

4. Connect these points with the wraparound to lay out the cutline.
5. Miter the cut. Bevel the edges.

Note: Travel, set, and run can also be solved by the square-root formulas covered previously. (See also "Laying Out Miter Cuts for Pipe Turns," this chapter.)

Installing a Steam Line

A steam line is run alongside a pipe run to reduce the viscosity of the substance in the pipe so that it will flow better. The two lines are shown in Figure 14.31. Note that the fittings at *A* are socket-type fittings, and are the same as those on the other side of the flange. The steam line should be kept as tight against the pipe run as possible (B), as laggers must cover both runs with insulation. Make up the two sections separately where welding conditions are good; then field-weld at C. When you stud a flange, make the stud flush with the nut on one side. On new lines or when upgrading, use new studs of the proper size.

Although Figure 14.31 is for the purpose of showing how a steam line is run, it can be used to point out the following:

1. Learn to plan your pipe runs so that field welds (in-position welds) fall

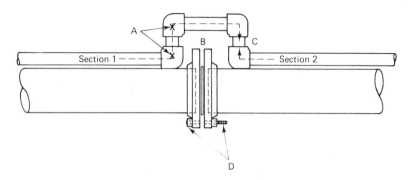

Figure 14.31 Installing a steam line.

where you want them, that is, where they are easily accessible. Even if you are the welder and you have a fitter for a partner, do not hesitate to tell him where you would like to make the weld. He should know. A bad fitter can make your day miserable if he does not know what he is doing. As you look at his sketch or plan, talk with him about the best place to make the in-position weld. If he has no sketch, and is going "from here to there," survey the run and advise him.

2. Plan where to put studs in a flanged joint. Even though the line is way out "nowhere," appearance does count. Joints made up so that some studs stick out of the nut on one side and some on the other side just do not look good. Generally, you will make the smooth side (stud and nut flush) on the side most open to view or on the side from which the next person would like to remove the nuts. Generally, they are complementary. If you have studs that are of uneven lengths or are too long, use your cutting torch.

Miscellaneous Fittings

Figure 14.32 shows a variety of standard and fabricated fittings.

To fabricate a crossover (Figure 14.32d), lay out D on lines 1 and 2. Mark points C on lines 3 and 4 at the end of the pipe. With a wraparound, draw cut lines C-A-C and C-B-C. The crossover is fabricated from four such pieces.

To fabricate a reducing lateral (Figure 14.32f), use the marking procedure shown in Figure 14.25. Use a straightedge and soapstone to prepare the branch. (See "Branches and Headers," this chapter.)

For the bull plug, preparing the end of the pipe (Figure 14.32h) is the same as for a branch (Figures 14.24 and 14.25). Piece 1 (Figure 14.32g and i), however, uses the inside diameter of the pipe, and the header uses the outside diameter.

A cap (Figure 14.32k) is welded in the same way as any butted pipe joint.

Pipe Fitting

If the cap is fabricated from a piece of pipe, it is an orange peel (Figure 14.32j) or a bull plug (Figure 14.32i).

Reading Isometric Pipe Sketches

Various isometric sketch notations are shown in Figure 14.33. On new construction, standard steel welding fittings will be used. All of the fittings shown in Figure 14.33 are available except the takeoff of ?°. In plotting the length of each piece of pipe, the travel of each fitting must be taken into account.

Figure 14.34 shows a 90° pipe turn. 90's are easily remembered: short-radius fittings have a linear travel the same as the pipe's diameter (I.D.), and long-radius fittings travel $1\frac{1}{2} \times$ pipe diameter. Deduct travel on both lengths of the pipe, before and after the turn. *Note:* All linear measurements are given center to center of the pipe:

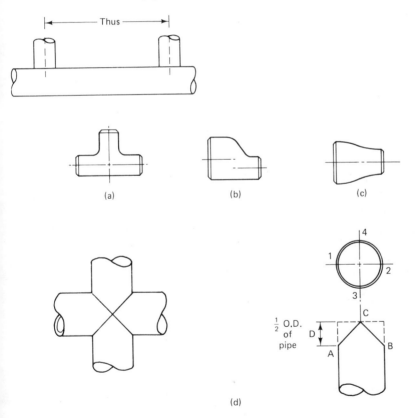

Figure 14.32 (a) Butt-welding T; (b) eccentric reducer; (c) concentric reducer; (d) crossover: standard and fabricated; (e) true Y; (f) reducing lateral; (g) fabricating a bull plug; (h) preparing end of pipe to be plugged; (i) completed bull plug; (j) orange peel; (k) cap; (l) blinds (for flanged joints); (m) boilermaker (also called a blind flange).

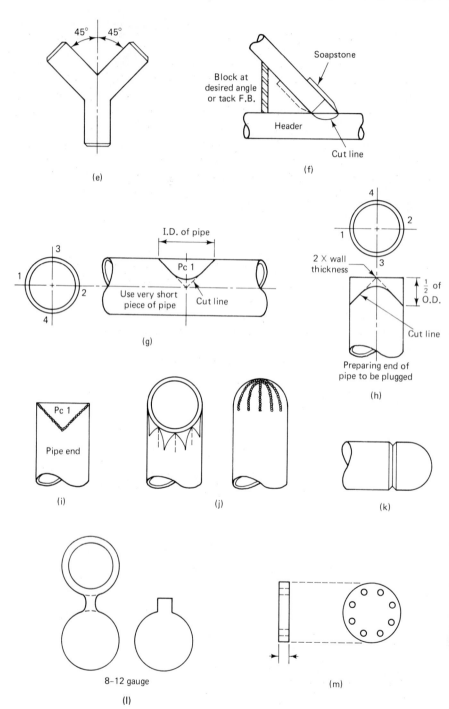

Figure 14.32 *(continued)*

Ancillary Skills

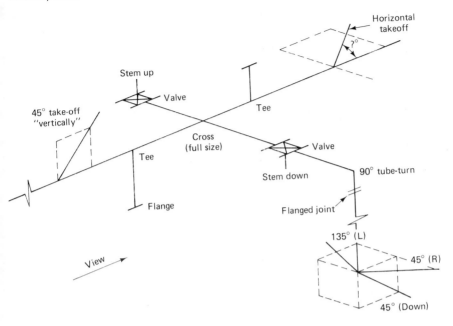

Figure 14.33 Isometric pipe sketch. Turns of less than 90° should be "boxed" or marked in degrees, or both.

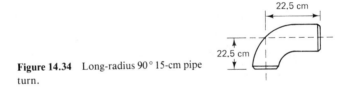

Figure 14.34 Long-radius 90° 15-cm pipe turn.

ANCILLARY SKILLS

Being Able to Fit in on the Job

1. Learn to use a wraparound properly so that you get a true line completely around the pipe. Watch for lumps (rust, dirt, old tack welds) that will throw the wraparound out of line.
2. Learn to cut and bevel at the same time. Be able to do it even if you prefer to radial-cut and then bevel. Some employers resent the extra time it takes for two cuts.
3. Learn how to center-punch a cut line and be able to leave all or half of the punch mark on the piece you want.
4. If pipe is lacquered and your soapstone burns off too far ahead of your kerf, center-punch the line. The same is true for paint, rust, and other materials that interfere with a good cut.

5. On old pipe, hammer around the cut line with a ball-peen hammer to dislodge any rust or residual material from the inner wall of the pipe that might interfere with the cut.
6. Learn to bevel outside in and inside out (the latter after a radial cut, of course).
7. When cutting a pipe in a horizontal run, start at the bottom (6 o'clock) and cut upward to about 12:30 o'clock; then restart at 6 o'clock and cut to 11:30 o'clock. Get above the pipe, make sure that it will fall safely,* and finish the cut. If you start at the top of the pipe, the iron oxide will go to the bottom of the pipe and will interfere with a good cut on the bottom.
8. When beveling the pipe after a radial cut, make sure that the end is square (check the same as for the flange face). Check several points around the pipe. If pipe is in place and cannot be rolled, start the bevel at the top and bevel downward on both sides as far as is comfortable and then bevel from inside out on the lower portion.
9. If you are cutting a pipe run and molten metal (actually, iron oxide) spews back at the torch, the pipe is rusty inside or has some substance adhering to the inside wall. Beat the pipe with a ball-peen hammer to shake the material loose from the wall. If this pounding does not do the job or if oil leaks out, make the best of a bad job and trim later.
10. Never lay a lighted torch down when working in an oil refinery. It is a bad idea anywhere, but in an oil refinery it can get you fired.
11. Carry only a striker that has a safety lock. Do not show up with any other type when asking for a job.

The First Day

1. Do not strike an arc or light a torch or cigarette, or even start a welding machine until the fire marshal has given clearance for *that day*. You must get a new welding permit every day even if the area is posted "Welding permitted in the area . . . from . . . to . . . " and you have 30 days to go. Conditions may change overnight. *Do not smoke* unless you have a smoking permit. You may think it is stupid that you are allowed to weld and burn and throw sparks from "hell to breakfast" but are not allowed to smoke.* That is the rule and to break it is to be fired. While waiting

---

*When working where you may cause someone to get hurt, the responsibility is yours. If the piece cut off is very long, and you have tied ropes around it, make sure that it will not swing into someone, or get someone to hold it.

for the fire marshal, get things ready: hook up your gauges, arrange the materials you will be using, and so on. If nothing else, clean up the area a little.

2. If after you have gotten your welding permit, you smell a new odor in the area, stop welding, shut off the machine, and notify the foreman, hail a passing fire marshal's car, or call the fire department.

3. Everyday tip: Do not skimp on your work clothes—you will be spending most of your day in work clothes; be comfortable. Clean clothes are warmer than dirty ones. Wear *good* shoes; the legs are the first to go. If there is a sitting-down weld to make, volunteer. Even lying in a ditch making a weld saves the legs.

Precautions

1. Treat all pipe runs as if they were pumping. Even if you are assured by everyone that the pipe is safe, observe certain precautions. Stand as far from the torch tip as you can reach. Blow a small hole in the pipe (be ready to retreat quickly), then enlarge the hole. Approach the hole with your nose and with some caution. If nothing is coming out, gas or liquid, proceed with the cut.

2. Cutting the wrong pipe is a very serious offense. Pipe to be cut should be marked by someone in authority and should be marked intermittently along its entire run. If the pipe run you are "cutting out" disappears under a walkway, or a portion of the pipe run is hidden from view, do not assume that the same pipe comes out the other side. Your pipe may turn, and another pipe run may take its place in the runway.

3. *Caution:* Never allow your fingers or hand to come between two flanges. If you have to stick anything in the joint in order to remove parts of the old gasket, to align a gasket, or to recover something, do it with a stick or a scraper. If there is no other way and you do have to get your hand in, put two nuts on a longer stud, run the two nuts to the center of the stud, insert the stud in the holes of the flanges and force the flanges apart by turning the nuts back toward the ends of the stud. For greatest safety, do this in two places and put a nut on each end of the stud that sticks out of the flange holes.

4. *Caution:* While we are on the subject: Take a short piece of standard pipe, of any diameter. Grasp it by the end by inserting your fingers inside the pipe and your thumb alongside the pipe. Feel the hard, cold, relatively

---

*Smoking "posts" are scattered about the plant; when you need a smoke, go to the smoking area, but not more than twice in the morning and twice in the afternoon.

sharp end of the pipe on your fingers. Now think of another pipe coming toward the end of your pipe because someone pushed it, or there was a strain in the line, or for any other reason. When you think about your severed fingers, you will never stick them in the end of a pipe again!

5. *Caution:* When making a rolling weld on a long line of pipe, and you are using a pipe turner, remember which way you twist the pipe. If you weld from one side and twist one way, on the next pass, weld from the other side and roll the pipe the opposite way. When you move to the next joint, remember which way you twisted the pipe. The better idea is to bring the pipe back to "neutral" as you leave one joint for the next. Failure to do this may put such a strain in the pipe that the welder who comes along later may be killed when he cuts the pipe.

APPENDIX: TRIGONOMETRY TABLE

See "Right Triangles," this chapter, for the ratios among the opposite and adjacent sides and the hypotenuse.

Note: Of little worth, perhaps, but interesting, the "co" in "cosine," "cotangent," and "cosecant" refers to the other angle making up the 90° of the right triangle. For example, assume a 30° angle. The tangent is 0.57735 and the cotangent is 1.73205.

$$90° - 30° = 60° \text{ (the other angle)}$$

The tangent of the 60° angle is 1.73205; the cotangent is 0.57735.

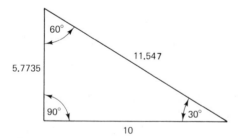

Trigonometric Functions

Deg.	Min.	Sine	Cosine	Tangent	Cotangent	Secant	Cosecant		
0	00	.00000	1.00000	.00000	Inf.	1.0000	Inf.	90	
1		.01745	.99985	.01745	57.290	1.0001	57.299	89	
1	30	.02618	.99966	.02618	38.188	1.0003	38.201	88	30
2		.03490	.99939	.03492	28.636	1.0006	28.654	88	
2	30	.04362	.99905	.04366	22.904	1.0009	22.925	87	30
3		.05234	.99863	.05241	19.081	1.0014	19.107	87	
3	30	.06105	.99813	.06116	16.350	1.0019	16.380	86	30
4		.06976	.99756	.06993	14.301	1.0024	14.335	86	
4	30	.07846	.99692	.07870	12.706	1.0031	12.745	85	30
5		.08715	.99619	.08749	11.430	1.0038	11.474	85	
5	30	.09584	.99540	.09629	10.385	1.0046	10.433	84	30
6		.10453	.99452	.10510	9.5144	1.0055	9.5668	84	
6	30	.11320	.99357	.11393	8.7769	1.0065	8.8337	83	30
7		.12187	.99255	.12278	8.1443	1.0075	8.2055	83	
7	30	.13053	.99144	.13165	7.5957	1.0086	7.6613	82	30
8		.13917	.99027	.14054	7.1154	1.0098	7.1853	82	
8	30	.14781	.98901	.14945	6.6911	1.0111	6.7655	81	30
9		.15643	.98769	.15838	6.3137	1.0125	6.3924	81	
9	30	.16505	.98628	.16734	5.9758	1.0139	6.0538	80	30
10		.17365	.98481	.17633	5.6713	1.0154	5.7588	80	
10	30	.18223	.98325	.18534	5.3955	1.0170	5.4874	79	30
11		.19081	.98163	.19438	5.1445	1.0817	5.2488	79	
11	30	.19937	.97072	.20345	4.9151	1.0205	5.0158	78	30
12		.20791	.97815	.21256	4.7046	1.0223	4.8097	78	
12	30	.21644	.97630	.22169	4.5170	1.0243	4.6201	77	30
13		.22495	.97437	.23087	4.3315	1.0263	4.4454	77	
13	30	.23344	.97237	.24008	4.1653	1.0284	4.2836	76	30
14		.24192	.97029	.24933	4.0108	1.0306	4.1336	76	
14	30	.25038	.96815	.25862	3.8667	1.0329	3.9939	75	30
15		.25882	.96592	.26795	3.7320	1.0353	3.8637	75	
15	30	.26724	.96363	.27732	3.6059	1.0377	3.7420	74	30

Deg.	Min.	Sine	Cosine	Tangent	Cotangent	Secant	Cosecant		
16		.27564	.96126	.28674	3.4874	1.0403	3.6270	74	
16	30	.28401	.95882	.29621	3.3759	1.0429	3.5209	73	30
17		.29237	.95630	.30573	3.2708	1.0457	3.4203	73	
17	30	.30070	.95372	.41530	3.1716	1.0485	3.3255	72	30
18		.30902	.95106	.32492	3.0777	1.0515	3.2361	72	
18	30	.31730	.94832	.33459	2.9887	1.0545	3.1515	71	30
19		.32557	.94552	.34433	2.9042	1.0576	3.0715	71	
19	30	.33381	.94264	.35412	2.8239	1.0608	2.9957	70	30
20		.34202	.93969	.36397	2.7475	1.0642	2.9238	70	
20	30	.35031	.93667	.37388	2.6746	1.0676	2.8554	69	30
21		.35837	.93358	.38386	2.6051	1.0711	2.7904	69	
21	30	.36650	.93042	.39391	2.5386	1.0748	2.7285	68	30
22		.37461	.92718	.40403	2.4751	1.0785	2.6695	68	
22	30	.38268	.92388	.41421	2.4142	1.0824	2.6131	67	30
23		.39073	.92050	.42447	2.3558	1.0864	2.5593	67	
23	30	.39875	.91706	.43481	2.2998	1.0904	2.5078	66	30
24		.40674	.91354	.44523	2.2460	1.0946	2.4586	66	
24	30	.41469	.90996	.45573	2.1943	1.0989	2.4114	65	30
25		.42262	.90631	.46631	2.1445	1.1034	2.3662	65	
25	30	.43051	.90258	.47697	2.0965	1.1079	2.3228	64	30
26		.43837	.89879	.48773	2.0503	1.1126	2.2812	64	
26	30	.44620	.89493	.49858	2.0057	1.1174	2.2411	63	30
27		.45399	.89101	.50952	1.9626	1.1223	2.2027	63	
27	30	.46175	.88701	.52057	1.9201	1.1274	2.1657	62	30
28		.46947	.88295	.53171	1.8807	1.1326	2.1300	62	
28	30	.47716	.87882	.54295	1.8418	1.1379	2.0957	61	30
29		.48481	.87462	.55431	1.8040	1.1433	2.0627	61	
29	30	.49242	.87035	.56577	1.7675	1.1489	2.0308	60	30
30		.50000	.86603	.57735	1.7320	1.1547	2.0000	60	
30	30	.50754	.86163	.58904	1.6977	1.1606	1.9703	59	30
31		.51504	.85717	.60068	1.6643	1.1666	1.9416	59	
31	30	.52250	.85264	.61280	1.6318	1.1728	1.9139	58	30

Deg.	Min.	Cosine	Sine	Cotangent	Tangent	Cosecant	Secant	Deg.	Min.
32		.52992	.84805	.62487	1.6003	1.1792	1.8871	58	30
32	30	.53730	.84339	.63707	1.5697	1.1857	1.8611	57	
33		.54464	.83867	.64941	1.5399	1.1924	1.8361	57	30
33	30	.55191	.83388	.66188	1.5108	1.1992	1.8118	56	
34		.55919	.82904	.67451	1.4826	1.2062	1.7883	56	30
34	30	.56641	.82413	.68728	1.4550	1.2134	1.7655	55	
35		.57358	.81915	.70021	1.4281	1.2208	1.7434	55	30
35	30	.58070	.81411	.71329	1.4019	1.2283	1.7220	54	
36		.58778	.80902	.72654	1.3764	1.2361	1.7013	54	30
36	30	.59482	.80386	.73996	1.3514	1.2442	1.6812	53	
37		.60181	.79863	.75355	1.3270	1.2521	1.6616	53	30
37	30	.60876	.79335	.76733	1.3032	1.2605	1.6427	52	
38		.61566	.78801	.78128	1.2799	1.2690	1.6243	52	30
38	30	.62251	.78261	.79543	1.2572	1.2778	1.6064	51	
39		.62932	.77715	.80978	1.2349	1.2867	1.5890	51	30
39	30	.63608	.77162	.82434	1.2131	1.2960	1.5721	50	
40		.64279	.76604	.83910	1.1917	1.3054	1.5557	50	30
40	30	.64945	.76041	.85408	1.1708	1.3151	1.5398	49	
41		.65606	.75471	.86929	1.1504	1.3250	1.5242	49	30
41	30	.66262	.74895	.88472	1.1303	1.3352	1.5092	48	
42		.66913	.74314	.90040	1.1106	1.3456	1.4945	48	30
42	30	.67559	.73728	.91633	1.0913	1.3563	1.4802	47	
43		.68200	.73135	.93251	1.0724	1.3673	1.4663	47	30
43	30	.68835	.72357	.94896	1.0538	1.3786	1.4527	46	
44		.69466	.71934	.96569	1.0355	1.3902	1.4395	46	30
44	30	.70091	.71325	.98270	1.0176	1.4020	1.4267	45	
45		.70711	.70711	1.00000	1.0000	1.4142	1.4142	45	
		Cosine	Sine	Cotangent	Tangent	Cosecant	Secant	Deg.	Min.

15

WEARFACING

The term *wearfacing* has been chosen over *hardfacing* because the degree of hardness desired varies with the type of wear involved. The hardest is not always the best.

The approach of this chapter will be that a knowledge of the types of wear to be expected, the types of overlaying materials available, and the wearfacing practices currently considered successful will enable the welder to select a surfacing material and the best method for its deposition, and enable him to weigh the cost of material and labor against the savings in downtime and replacement.

UNDERSTANDING WEARFACING

The first truly effective use of hardfacing probably was the "case" knife, which featured a surface harder than the core, thus providing a flexible blade with a surface that would hold a good cutting edge. (See "Case Hardening," Chapter 5.)

Since early experiments by blacksmiths with carbon in their forge and a handy water trough, wearfacing has progressed in importance to the point that many manufacturers of equipment subjected to localized wear will hardface or otherwise wearface those localized wearing areas prior to selling the equipment to the user.

It is not within the scope of this book to cover the vast field of wearfacing. Our coverage and the following examples are aimed at the heavy-duty mechanic who is also a relatively skilled welder. We have chosen, somewhat at random, a few of the hundreds of wearfacing applications presently used in the heavy industries. The welders in the lighter maintenance shops may adapt much of the information to wear problems in their field. For example, sprocket gears, cams, keyways, cable sheaves, and wobblers are subjected to similar wear regardless of their size.

The purposes of wearfacing are:

1. To restore a worn or eroded area to its original dimensions and wear resistance
2. To provide a surface that is superior to the original base metal in resisting the type of wear encountered
3. To provide the superior qualities of a nobler metal by overlaying a baser metal at a great savings in cost

Following are some guidelines for restoration.

1. Restoration of areas eroded by hostile media, such as acids, is usually a matter of electrode or filler metal selection and application. Except in cases where an entire surface is to be overlaid with a nobler metal, the electrode or filler metal generally matches or closely approximates the parent metal. Thus a 308 or 316 stainless steel will be restored with E-308 or E-316, respectively (see Chapter 9), and MONEL, INCONEL, INCOLOY, HASTELLOY, and others will be restored with like materials (see "Huntington Alloys," Chapter 10).
2. Restoration of worn or eroded areas of the lower-carbon steels may be made with the easily machinable, rapidly deposited E-6012, or their wearing areas may be enhanced with E-7018 and even further strengthened with the E-XX18 group plus the most suitable suffix, such as E-10018-D2.
3. Where simple restoration is required in the case of low-alloy steels, the electrode should approximate the alloy content of the parent metal (see Figure 6.3).

The importance of matching the wearfacing material to the in-service conditions is shown graphically by a partial listing of wearfacing materials manufactured by the Stoody Company (see Table D.1, Appendix D). The materials shown are for manual application. Welders in the earth-moving, dredging, mining, steelmaking, and petroleum industries and others will be particularly interested in the *Stoody Hardfacing Guidebook* (see Appendix E), which is available at nominal cost.

TYPES OF WEAR

The types of wear the welder may encounter, in brief, are:

1. *Corrosion:* Corrosion is also known as *oxidation* because it occurs by the

formation of oxides on the surface of metals. Some metals deteriorate by oxidation faster than others. Iron, for example, oxidizes rapidly (in the form of *rust*), whereas aluminum appears to oxidize very slowly. Aluminum actually reoxidizes the instant it is cleaned, but this oxidation resists further oxidation, whereas the rust of iron does not seem to retard further oxidation. Wear by corrosion is a form of erosion.

2. *Erosion:* Erosion is the wasting away of metals by prolonged contact with an unfriendly environment, such as acid, heat, and activating gases.
3. *Friction:* Friction is the rubbing of one surface against the other. Some metals tolerate friction better than others do. They are said to have a "low coefficient of friction." The extremes of the effect of friction are galling, seizing, and freezing.
4. *Abrasion:* Abrasion is the grinding action by a solid sliding, rubbing, or rolling against a surface.
5. *Impact:* Impact is the impinging or striking of one body against another. Wear is caused by a continued bombardment that gradually dislodges particles of a solid from their natural adherence to that solid.

When choosing filler metals that resist wear, the welder will read the words "subject to severe abrasion," "subject to severe impact," "subject to severe abrasion and moderate impact," or other variations of those words. What is the welder to do when he must make a compromise? As a general rule, the harder the surface is, the greater is its abrasion resistance; but the harder the material, the more it is liable to spall or chip away from impact.

One of the more common forms of wearing away by impact is the jet of water from a hose playing against a concrete wall; in time the wall will be seen to wear away. A jet of water of sufficient pressure can wear a hole through the wall in the same way that a driller punches a hole in concrete using a star drill and a hammer. The simplest form of abrasion may be seen from sand sliding down a metal chute: eventually the surface wears away, especially at the end. Sand blasting shows both wearing actions: the sand going through the nozzle wears the nozzle by abrasion, and the sand impinging on the object surface wears the surface away by impact.

An example of wear by friction is the front disk brake of an automobile: pads of one substance in sliding contact with the rotating disk of another substance. A welder who is familiar with surfacing might well ask: "Since pads are so inexpensive, why don't they come up with a metal for the rotors that won't wear away?" The answer may be practicality and/or economics. On the one hand, a certain amount of wearing away is necessary to prevent galling or seizing, and on the other, the hardness of some metals is an indication of their cost. Tungsten carbides, for example, are the hardest of the hardfacing compounds (9.5 on Mohs' scale), but that compound is also the most expensive, and the cost versus wear ratio may dictate the use of chromium carbides or vanadium carbides.

Types of Wear

In the case of our brakes, friction is a desirable destructor. The behavior of the mating metals must be controlled. We want the effect of the friction to vary according to the pressure on the brake pedal, and we try to develop a sense of touch that stops the car smoothly; and in emergencies, we do not want our braking surfaces to weld themselves together when we exert the maximum friction. Seizing by our oxyacetylene connections (brass on brass) is especially beneficial by enabling us to make connections *hand tight*. But the same binding action by disk brakes would be very annoying.

As a general rule, the harder the metal, the less its tendency to gall or seize. Another general rule is that like metals are more liable to gall than unlike metals—probably due to their natural molecular attraction. Some combinations of metals have less tendency to gall than others. Steelworking with brass or bronze does not gall. Cast irons wear well against many metals.

As mentioned previously, the hardest metal is not necessarily the best. Welders familiar with the words "Rockwell hardness test" may be tempted to choose the highest rating because of the general truism that "the harder the metal, the better its abrasion resistance." There are, however, exceptions to that general truth.

Tungsten carbides, for example, are compounded in varying screen sizes and as a general rule, the finer-screen-size tungsten carbides provide a harder structure, such as those carbides in a high-speed steel; but these small carbides do not resist the scratch of a sand crystal as well as will the larger carbides. The larger carbides offer an impact resistance to the sand crystal that reduces the scratching effect. Further, tungsten carbides are the heaviest of the carbides and in the welding puddle will tend to sink lower in the solidifying matrix and congregate on the lower portion of the total weld deposit. Thus the upper portion of the matrix must wear away before the harder underportion is exposed to wear. In some applications, then, it may be better to use a lighter-weight carbide, such as vanadium carbide, since the vanadium carbide will float higher in the weld puddle and solidify nearer the surface of the weld deposit. This factor certainly should be entertained in relatively heavy (thick) overlays, since vanadium carbides are much lower in price than tungsten carbides.

Abrasive wear must be countered by a surface that resists the scratching effect, that is, a force moving parallel to the surface. Wear by impact depends on properties in the metal below the surface: the property of shock absorbency, and the ability to cushion the force of the impact and distribute it over a broader area. Certain materials have this property to a greater degree than others, whether human-made or occurring in nature. Rubber, for example, may be the most shock resistant of all, as a hammer will show; but the ringing steel anvil of the blacksmith will bounce his hammer back up for the next strike and save him a lot of muscular energy. Thus the welder must decide:

1. Will he use a hard material to resist abrasion?
2. Will he use a tough material to resist impact?

3. Will he want a material that will wear smooth or rough, or maintain a good cutting edge?
4. Will he compromise?

The answer lies in the type of work the wearing surface must do. He cannot use the same surfacing material or buildup material for the frog or crossover of a railroad and the dipper teeth of an excavator. His choice is not made easy for him. He will find well over 300 wearfacing electrodes, rods, powders, and pastes from which to choose.

As a general rule (it seems we must always deal in generalities when welding), the metal that wears smoothest resists abrasion best; but the welder facing a well-drilling core head does not want the cutting surface to wear smooth. He wants it to wear so as always to present an irregularly serrated, rough or jagged cutting edge. The old-time whittler would turn up his nose at the smooth-wearing, nonoxidizing stainless steel knives so prized by the modern housewife. He says, "Nothin' cuts like a rusty knife" and "When I want a good whittlin' knife, I buy a good knife and bury it in the barnyard for about six months."

WEARFACING MATERIALS

The wearfacing alloys most used today may be, and have been, classified in several ways. We have elected to group them with respect to their relative hardness and have identified them as Class I through Class V in order to refer to them as we point out some of their respective applications.

Class I. The *stainless steels*, E-308, E-309, and E-310, are the most common. These electrodes and their barewire counterparts find their best use in rebuilding erosions caused by acids and other hostile media, and as cushions for harder facings. They are also used to tie down hard-facing layers that have peeled away from the base metal. Stainless steels are nearly equal to manganese steel in their work hardenability and will withstand more pounding without cracking. Their reluctance to pick up carbon is especially suited for joining semiaustenitic overlays to carbon steel. These steels are covered more fully in Chapters 9 and 10.

Class II. Austenitic manganese steel (also called Hadfield steel). These wearfacing materials are used primarily for buildup of worn areas. In some areas, the buildup is then hardfaced with abrasion-resistant material, such as Classes III, IV, and V. Their prime purpose is impact resistance. Their work hardenability and other characteristics are discussed in Chapter 12. Electrodes should contain up to 5% nickel in order to prevent migration of carbides to the grain boundaries. As deposited, these will be 17 to 22 Rockwell C and will work harden to 45 to 50 Rockwell C.

Class III. Semiaustenitic: iron base, less than 20% alloying constituents, such as 1 to 2% carbon, 5 to 12% chromium, and smaller amounts of other alloying elements. The deposit tends to remain austenitic as it solidifies, and there is not enough carbon and chromium for chromium carbides to form. The final structure of the deposit depends on the percentages of alloying constituents and the rate of cooling. Since some of the austenite decomposes to form martensite, these deposits are called *semi*austenitic. The two subclasses that result from controlled cooling can be described as follows:

a. The higher the percentage of austenite, the tougher and less hard the deposit as welded. They are, however, work hardenable. A typical deposit may be 20 to 30 Rockwell C as-welded but will become 40 to 50 Rockwell C after working or peening.
b. If the cooling rate is slower and gives the austenite time to transform into martensite (the first stage of the decomposition of austenite), the deposit will be hard as-deposited (55 to 65 Rockwell C).

The welder may control the hardness by controlling the rate of cooling. He can control the rate of cool-down by varying his welding procedure.

1. If he wants a predominantly austenitic deposit (tough, impact resistant, moderate abrasion resistance), he will use a small electrode and short beads, and start with the base metal cold and not let the base metal get so hot as to slow down the cooling rate of the deposit.
2. If he wants a higher percentage of martensite (hard, brittle; severe abrasion, mild impact) he will use larger electrodes, continuous welding, and preheat the base metal.
3. If he wants a deposit for moderate abrasion and moderate impact, he will vary his welding technique until he finds that mean result with his usual electrode applied to the usual weldments.

The semiaustenitic group are machinable only after a long anneal at high temperatures and only with tungsten-carbide tips.

Class IV. Chrome carbides, sometimes referred to as the "high-alloy" group. These materials may be ferrous (containing iron) or may be nonferrous. Those alloys in the ferrous subgroup, characteristically are nonmachinable, have excellent resistance to wear by abrasion and friction, and have a high hardness at elevated temperatures. In the nonferrous subgroup are alloys of chromium, carbon, cobalt, and tungsten. Although they cost more and are somewhat less abrasion resistant, they are machinable, more resistant to corrosion, and have good red hardness.

Class V. Diamond substitutes. These materials are called diamond substitutes because they approach the diamond on Mohs' scale of hardness (8.5

to 9.5). They are the most abrasion resistant of all wear facings. They may be carbides of tungsten, tantalum, titanium, or boron; or borides of chromium. In the rod form, the diamond-substitute particles are either encased in an alloy steel tube or are uniformly dispersed in a cast alloy rod. Diamond substitutes are also available in powder, paste, and insert form. The particles may be coarse or fine, depending on the type of wear to be encountered. The matrix may be a hard alloy or a softer carbon steel, depending on the worn surface (rough or smooth) the user wishes to maintain: rough, as in the case of the well driller's core heads; smooth, in the case of many earth-moving surfaces. Since deep penetration is undesirable,* these rods are generally best applied by a reducing ($2\times$) oxyacetylene flame, as in the sweating process. If stick welding is used, a sweating or diffusing application should be approximated as closely as possible, consistent with adequate bonding. A crescent weave or the Shuler weave should be used when stick welding.

Those alloys whose final properties depend on heat treatment (tool-and-die steels) are the subject of Chapter 7. It may serve a purpose, however, to state here that the welder should choose the overlay material whose recommended hardening process is the same as that of the parent metal; that is, choose an oil-hardening electrode or rod for an oil-hardening steel.

The automatic, semiautomatic, and submerged arc processes are used extensively as the size of the job and the cost of labor make the equipment desirable.

USE OF WEARFACING IN INDUSTRY

Steelmills

Coupling boxes. Build up work areas about 6 mm ($\frac{1}{4}$ in.) undersize with low-alloy carbon steel (about 20 to 25 Rockwell C) and finish with Class II or IIIa (as deposited, 16 to 20 Rockwell C; work hardening, to 40 to 45 Rockwell C). Preheating and slow cooling are recommended. Modern steel mill welders may be interested to know that in the 1930s we restored these boxes using 91 cm (36 in.) bare electrodes sheared 6.4 mm ($\frac{1}{4}$ in.) square from high-carbon steel plate. We laid beads 2.54 cm (1 in.) wide and 6.4 mm ($\frac{1}{4}$ in.) thick, using a modified crescent (Shuler) weave.

Roll and spindle wobblers. Restoration is done the same way as for couplers. Preheating and slow cooling are recommended. To boast a little again, in the 1930s we welded these in place using a wire rope and overhead crane to turn the roll. All we had then was a carbon-moly electrode that had a nasty habit of peeling away, a problem we solved by gouging out with a carbon arc

---

*It is easily understandable that an exceptionally high-priced overlay material should not be diluted unduly with an inferior base metal.

and tying the deposit down with a stainless steel electrode. If these wobblers were brought to the shop, we thermit-welded them.

Open-hearth charging pans (also called boxes). Build up worn charger ram slots slightly undersize with low-alloy carbon steel and finish (5-6 mm) with Class II or IIIa (work-hardening group).

Crane wheels. Rebuild undersize with low-alloy carbon steel and finish with Class II or IIIa. Preheat and slowly cool. Keep preheating during welding. Grind or machine to finish. (If factual analysis of parent metal is known, approximate that analysis as closely as possible.)

Earthmoving

Shovel teeth. Figure 15.1 illustrates hardface shovel teeth in new condition. The bead pattern is important.

If working dirt, clay, or sand, run beads transversely (Figure 15.1a). This pattern traps material between the beads and reduces wear on the base metal. We have the effect of dirt wearing on dirt, sand on sand, clay on clay.

When working in rock, run beads lengthwise (parallel to flow of material) (Figure 15.1b). This pattern permits sliding material along the ridges with very little wear on the base metal. Transverse beads are not suitable because of the impact of coarse material striking against the high spots.

If working a combination of materials, a mix of rock and other materials, apply facing in a waffle pattern (Figure 15.1c) as a compromise among the wearing effects.

Recommended facings are:

For severe abrasion, use Class IIIb.

For severe impact, use Class II.

For severe abrasion and moderate impact, use Class IIIa.

Repointing Teeth. Figure 15.2 illustrates the following procedure for repointing teeth.

1. Weld manganese repointers to manganese teeth with Class II (austenitic manganese steel).
2. Weld carbon steel repointers to carbon steel teeth with E-7018 or better.
3. Weld dissimilar steels with E-309 (E-310 is the second choice).
4. Face according to type of wear.

Caution: Do not hardface bottoms of shovel teeth. Bottoms should wear before top and sides. This eroding of the bottom creates a self-sharpening effect. Apply this self-sharpening principle to other wearing areas where deemed beneficial.

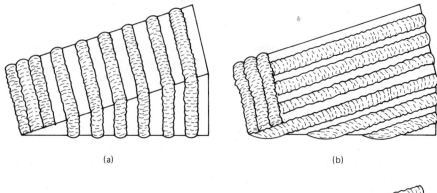

Figure 15.1 Shovel teeth: (a) transverse beads; (b) lengthwise beads; (c) waffle pattern.

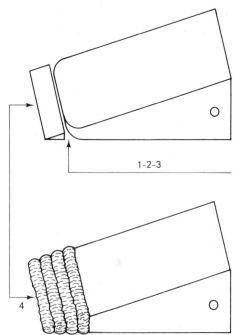

Figure 15.2 Repointing teeth.

Track rollers. Build up with low-alloy carbon steel, 20 to 25 Rockwell C, and face with the same or with Class II or Class IIIa. (Track rollers are generally carbon steel.)

Tractor idlers. *Carbon steel:* Jig up to weld downhand (flat position). Build up and face as for track rollers, above.

Shovel rollers. Rig up to turn for downhand welding. Use transverse or circumferential beads. Restore in the same way as for track rollers.

Shovel idlers. Rig up for downhand welding. Use a manual process. Build up undersize (6 mm) as above.

Shovel drive tumblers. *Carbon steel:* Build up and finish as for track rollers. *Manganese steel:* Build up and finish with Class II (as deposited, 20 Rockwell C; work hardening to 40 to 50 Rockwell C).

Shovel boom heels. *Carbon steel:* Build up with low-alloy carbon steel, 20 to 25 Rockwell C. Face with Class II or Class IIIa (work hardening). *Manganese steel:* Build up with Class II and face with Class II.

Shovel bucket-tooth adapters. *Carbon steel:* Build up with low-alloy carbon steel. *Manganese steel:* Build up with Class II. For moderate wear, use a waffle pattern with Class III, 45 to 56 Rockwell C. For severe abrasion, moderate impact, use Class III, 57 to 65 Rockwell C. Use a zigzag pattern on the sides.

Ripper teeth. Apply Class V (borides preferred) 5 cm (2 in.) back from the point. Face the remainder of the tooth and as high up the shank as desired with Class IIIb, 57 to 65 Rockwell C.

Dozer end bits. Restore worn corners with a low-alloy carbon steel. Face the lower edge solid about 5 cm (2 in.) and triangularly about halfway up the face. Apply a waffle pattern on a bias over the remainder of the face. For severe impact, use Class II. For severe abrasion, use Class IIIb. For abrasion and impact compromise, use Class IIIa.

Tractor grousers. Do not hardface new grousers. Anneal induction-hardened grousers before welding bar stock. Weld bar stock to pad with low-alloy carbon steel (not less than E-7018). Hardface cleat with Class IIIb, self-hardening 54 to 60 Rockwell C. Face only the outer thirds from the inside outward. Facing the entire length of cleat increases stress drastically.

Dragline buckets. *Carbon steel:* Hardface lips, top, and bottom, selecting the pattern to suit the wearing material. Use Class III to 50 Rockwell C for moderate wear. For severe abrasion, use Class III, 60 Rockwell C. *Manganese steel:* Buildup with Class II, face as above.

Dragline pins. *Carbon steel:* Hardface the ends of the pin with Class IIIb. Restore the worn shank with low-alloy carbon steel. Grind to size.

Dragline chains. Link-to-link wear, use Class II on manganese steel links and low-alloy carbon steel for carbon steel links. Face outside wear areas with Class IIIa or b.

Dragline clevis. Rebuild metal-to-metal wear on carbon steel with low-alloy carbon steel (E-7018 or better) and face with Class IIIa. On manganese steel, rebuild with Class II. On both, face the outside wear area with Class III 54 to 58 Rockwell C.

Tractor rails. Do not rebuild until worn to recommended minimum service limits. Build up the outside edges to the same level as the center. Use a 2-cm ($\frac{3}{4}$-in.) wash pass (crescent or Shuler). Do not overheat; weld alternate links (four should suffice). Use buildup only; do not hardface. Use low-alloy carbon steel, 22 to 28 Rockwell C.

Drive sprockets. Rig for flat-position welding. If badly worn, use E-7018 for fill and face with E-11018 or a slightly higher alloy (E-11018-D2). Use transverse beads. Use a template for approximate sizing, and grind to finish.

Dozer blades. Prebend or bolt the blade to the moldboard. Skip-weld stringers 4 to 5 cm wide and deposit on the front bottom. For moderate impact abrasion, use Class IIIb; for severe abrasion, use Class V (borides).

Shovel track pads. Rebuild worn areas to within 5 to 6 mm of original using low-alloy steel and face with Class IIIa. If manganese steel, build up with Class II and face with Class IIIa.

Cable sheaves. Jig up for downhand weld. Build up badly worn areas, and apply circumferential beads over the entire wearing surface. Use low-alloy carbon steel for buildup and finish. Preheat and postheat if necessary.

Shovel house rolls. These are usually made of high-carbon steel. Preheat to 260° C (500° F). Jig up for downhand welding. Rebuild with low-alloy steel and face with Class II. Machine or grind to size.

Top carrier rolls. *Cast iron:* Clean thoroughly. Preheat to 540 to 650° C (1000 to 1200° F). Keep the heat while welding. Use wide transverse beads. Weld alternately on opposite ends. Cool slowly (lime or asbestos). Use nickel-type electrode (see Chapter 8). *Carbon steel:* No preheat or postheat required. Use the same wide transverse beads and distribute stress. Build up with low-alloy carbon steel and face with Class II.

Ditcher drive segments. Rig for down-hand welding. Use a template for shaping. Use transverse beading. Grind the high spots. Use low-alloy carbon steel for buildup and face with Class II.

Ditcher rolls. *Carbon steel:* Rig so that the roller can turn. Rebuild the face and flange where worn. Use low-alloy steel. Grind the high spots. *Cast iron:* Preheat to 540 to 650° C (1000 to 1200° F). Maintain heat while welding. Use nickel-type electrode (see Chapter 8).

Ditcher drive sprockets. Rig for downhand welding. Use a template for shaping. Use transverse beads. Grind the high spots. Use low-alloy carbon steel.

Posthole augers. Use an oxyacetylene process (or Mapp gas) and bare rod of the Class V (WC or borides preferred). Apply alloy to the upper face of the flight periphery. Hardface the top side of the cutter teeth for self-sharpening action. Hardface the reaming edge and wings of the pilot bit.

Asphalt mixer paddles. If weldable, face the outer edge and leading edge solid. Apply stringer beads (spaced 4 to 1) on the leading face parallel to the abrasive flow. Rim the bolt holes with a stringer bead. Use Class IIIa or b.

Paving screw conveyors. *SMAW process:* Use Class IIIb, 56 Rockwell C or better. *Oxyacetylene process:* Use Class V (tungsten carbides preferred). In both cases, apply alloy to worn areas on the flight, faces, and edges.

Logging and Lumber

Hog rotors. Build up worn areas around knife slots, dress slots, and grind faces. Face the entire wearing surface with slightly weaved beads. Skip-weld. Use Class IIIa or b. Rebuilt rotors must be balanced. Preheat and cool slowly if necessary.

Anvil knives. Use an oxyacetylene process. Face the unbeveled side of the cutting edges with a single pass. Sharpen by grinding the beveled side. Use Class V (preferably borides or WC).

Knife keepers. Generally made of cast steel. Face the entire wearing surface with Class V (borides or WC), 56 Rockwell C or harder.

Saw carriage wheels. Position for easy turning and overbuild sufficiently for finishing (machining or grinding). Use low-alloy carbon steel.

Rotary-head chipper blades. Undercut the cutting edge 3 to 4 mm. Overlay with Class IV, 40 to 46 Rockwell C, and grind to a sharp edge.

Debarking hammers. Wearface the tips with Class V. Wearface the heels with Class II, 45 to 50 Rockwell C.

Debarking rotor knives. Use an oxyacetylene process. Hardface the points and cutting edges with Class V (preferably tungsten carbides or borides).

Debarker chain links. Restore the nipples on manganese links with Class II.

Bark conveyor trunnions. Position the trunnion for rolling and overbuild the running face with E-7018 or slightly higher (machinable) alloy. Finish by machining or grinding. With proper pre- and postheating, trunnions subject to extreme abrasion may be overlaid with harder facing, such as Class IV (Ni–Mo–Mn).

Log haul chairs. Hardface the nipples and other contact areas with Class V.

Conveyor chains. For chain links that wear on sprockets, use Class II. For link arms that wear on channel, use Class IIIb or Class V.

Chipper chutes. Use Class IIIa or b.

Chipper disks. Use Class IIIa or b.

Bed plates. Use Class IIIa or b.

Clutch fingers. Use Class IV.

Clutch jaws. Use Class III, self-hardening.

Log escalators. For drive sprockets and drums, build up teeth to template size with Class II alloy. Grind the high spots. Preheat and cool slowly if required. For a geared idler, build up the gearteeth to template size. Grind the high spots. Preheat and cool slowly if required. Use Class II alloy.

Chain drive tumblers. Flame-cut replacement blocks and weld them to the worn areas, using E-7018. Overlay the replacement blocks with Class IV, 45 Rockwell C or harder.

Conveyor chain links. Restore metal-to-metal contact areas with Class II. Rebuild the lugs with Class II. Face the outside wear area with Class III, 60 Rockwell C.

Mining

Clutch lugs. Rebuild to the original size. Peen to shape while the deposit is at red heat. Preheat and cool slowly when necessary. Use Class II.

Sprocket drums and traveling sprockets. Rebuild worn areas to size with E-7018 or better and grind to finish dimensions. Preheat and cool slowly if necessary.

Cutter chain lugs. Hardface the wear areas, using Class III, self-hardening.

Digging arms. Hardface new arms on carbon steel with Class V before putting in service. Re-hardface when necessary. On manganese steels, butter with E-308 or E-309 before facing with Class V.

Duck bills. Hardface all sides of point, and top, edge, and bottom of lip using Class III, 56 Rockwell C or harder.

Ball mill scoops. Apply stringer beads in a waffle pattern over a 45° segment, using Class III, 56 Rockwell C or harder.

Ball mill scoop lips. Overlay the leading edge and sides with Class V (borides or WC preferred). Overlay the remainder of the lip with Class III. Do not cover the entire side, only the small wearing area near the edge of the lip. Do not hardface the bottom.

Grizzlies. Hardface before putting in service. Overlay with Class IIIb over the entire wearing surface, using linear stringer beads. Run a single stringer bead on the sides from the end of the finger back about one-third.

Auger bits. Position for down-hand welding. Using Class V (WC or borides preferred). Overlay the tip 1.5 cm back from point. Do not overlay the bottom. Use the oxyacetylene process.

Undercutter bits. Same as auger bits.

Coal recovery augers. Using Class IIIb alloy, hardface the flight periphery. Do not overlay the flight face.

Core barrels (coal recovery). Using Class III, overlay a spiral around the barrel and flight edge of the center auger. Hardface the cutters mounted around the barrel rim, using Class V (borides or WC preferred).

Collar puller cutters. Build up the worn section with E-7018 or better. Hardface the entire tip with Class V (tungsten carbides—WC, or borides).

Skip hoists. Hardface the wearing areas inside the walls using a waffle pattern. Use Class III alloy, 57 to 61 Rockwell C.

Well Drilling

Core heads. Position core head with pointers up. Clean the surface. Maintain an even heat of about 260° C (500° F) with a torch. Build up worn areas with Class IV, 38 to 46 RC. Shape by hot wiping or hot knifing. Cool slowly in furnace or lime. For sharper teeth, grind off-hand. Grind for close-gauge tolerance on the outside diameter.

Automotive

Oxyacetylene or TIG process should be used for all of the automotive applications given here.

Engine valves. Overweld to accommodate machining or grinding. Experienced welders can use powder-spray torch (the hot-spray method) in one hand and turn the valve with the other hand. Undercut the valve seat about 4 mm before facing. Use Class IV, machinable.

Cams. Use Class IV, 56 Rockwell C.

Rocker arms. Use Class IV, 56 Rockwell C.

Tappets. Use Class IV, 56 Rockwell C.

Valve stem tips. Use Class IV, about 40 Rockwell C.

Valve rings. Use Class IV, about 40 Rockwell C.

Valve disks. Use Class IV, about 40 Rockwell C.

APPENDICES

A

GLOSSARY*

Abrasion. The scratching effect of a substance that wears away the surface of another substance, such as sand sliding down a metal chute.

Acetylene. A commercially available hydrocarbon gas (C_2H_2) derived from the chemical reaction of water and calcium carbide.

Age hardening. A hardening after heat treatment by a structural transformation that proceeds in a time–temperature ratio. See also *precipitation hardening*.

Air-hardening steel. Also called *self-hardening*. Steel with a deep-hardening characteristic in ambient air. They range from 1.00 to 2.25% carbon, with additives of chromium, molybdenum, and manganese for special benefits. See Chapter 7.

Allotropy (Gk. *allo,* other; *tropos,* turn). The ability of an element to exist in two or more forms: for example, diamond, carbon, and graphite.

Alloy. A metal composed of two or more elements; also the process of melding those elements. See also *meld*.

Alpha iron. Pure iron, body-centered cubic at room temperature, magnetic. See Chapter 5, particularly Figures 5.4 and 5.10.

Alternating current (ac *or* a.c.). An electric current that is alternately positive and negative. See Figure 1.3.

Ampere. The unit of electric current; the amount of such current that will flow in a circuit through 1 ohm of resistance under a potential (pressure) of 1 volt.

Annealing. The heating of a metal above its transformation range, soaking it at that heat for a predetermined time (1 hour for each 2.5 cm of thickness), and cooling it slowly for the purpose of softening it.

Arc blow. The erratic deviation of an arc from its normal course due to magnetic forces.

---

*Words in italic type are defined in the Glossary.

Glossary

Arc cutting. The cutting of metals by several processes that use the arc for melting the metal. See Chapter 13.

Argon (A). A chemical element. It is colorless, tasteless, odorless, and inert (nonreactive). As used in welding, it is a by-product of the extraction of oxygen from the atmosphere. It is relatively heavy, having atomic number 18.

Atmosphere. A unit of pressure equal to air pressure at sea level (14.696 psi, 29.92 inches of mercury, or 760 millimeters of mercury).

Austempering. The heat treatment of steel wherein both hardening and tempering are accomplished simultaneously by heating well above the transformation range, quick quenching to 290° (550° F) in a molten salt bath, and slow cooling to room temperature.

Austenite. Gamma iron and carbon in solid solution. See Figure 5.10.

Backhand welding. A torch-welding technique (gas or TIG) wherein the torch is angled into the puddle and opposite to the line of travel.

Backing ring. A ring used as a backup for a butted joint in pipe. (Do not confuse with a consumable insert wire.)

Backing weld. A weld made on the root side of a groove joint prior to filling the groove.

Backstep. A welding sequence wherein each weld increment is welded in the direction opposite to the direction of the total weld. See Figure 13.14.

Backweld. A weld deposited on the root side after a single groove weld has been made.

Bainite. Those structures occurring between the formation of martensite and pearlite; an aggregate of very fine needlelike ferrite and cementite; includes *troostite* and *sorbite*.

Bainite structure. See *Bainite*.

Bar. A unit of pressure equal to 1 million dynes per square centimeter (14.5 psi).

Base metal. A metal of lower value (lead, as opposed to a noble metal such as gold). Also often used in welding for the metal being welded, as opposed to the filler metal. "Parent metal" may be a less-confusing term.

Beading. Sometimes called *stringers* or *stringer beads*. The deposition of filler metal without oscillation of the electrode or torch. The result is a narrow deposit. A technique often used for overlays and the cover layer of a horizontal groove weld, such as the horizontal butt weld of a vertically positioned pipe run.

Bell-hole welding. The pipe welder's term for laying a line by welding the pipe sections in position. See also *stovepipe welding*.

Beta iron. Also known as nonmagnetic alpha iron. The body-centered iron that exists between temperatures 767° and 900° C (1414° and 1652° F). The change from beta iron to alpha iron is thought to be within the atom.

Beta martensite. Martensite that is not magnetic.

Bevel. A relatively full chamfer in edge preparation. See Figure 13.27.

Bevel angle. The angle of cut in a single- or double-V edge preparation. See the cutback diagram, Figure 13.27.

Birdnesting. The backcoiling of an automatic wire feed between the cable and drive roll. See "Wire Feed," Chapter 4.

Block sequence. Also called blockweld. A welding sequence wherein separated predetermined sections of a longitudinal weld are fully (sometimes partially) filled before subsequent sections are welded. See also *cascade sequence*.

Blowhole. A gas pocket formed in a weld because the metal solidified before gases could escape.

Blowpipe. The gas-welding torch.

Body-centered cubic lattice (bcc). A grouping of nine atoms to form a cube with one of the atoms at the center of the cubic form. See Figure 5.1.

Boilermaker. The pipe welder's colloquialism for a blind flange used to blank the end of a pipe.

Boxing. The continuance of a weld bead around a corner; a required procedure in ship welding, even on a 16-gauge coaming.

Brass. A group of copper-based alloys whose principal alloying constituent is zinc. See Chapter 10.

Braze welding. A metal-joining process wherein a nonferrous filler metal of a lower melting point than the parent metal is deposited with a beading-up technique as differentiated from the thin-flow technique of brazing.

Brinell hardness number. A measure of a metal's hardness by impacting a hardened steel ball against the metal under a known load. The hardness number is the quotient of the known load divided by the area of the impression:

$$\text{BHN} = \frac{\text{applied load (in kg)}}{\frac{\pi \times D}{2} \times D - (\sqrt{D^2 d^2})} = \text{kg/mm}^2$$

where D is the diameter of the ball and d is the diameter of the impression.

Brittleness. The tendency of a metal to fracture without deformation; for example, the fracture of white cast iron or glass.

Bronze. A group of copper-based alloys whose principal alloying constituent is an element other than zinc, such as aluminum bronze, manganese bronze, and so on.

Bronze welding. A misnomer for *braze welding*.

Buildup. A weld deposited to restore a weldment to its original dimensions. It is usually overbuilt and ground or machined to finish. Also applies to restoration of a worn area prior to a specialized wearfacing.

Buildup sequence. The order in which beads or passes are deposited in relatively deep groove welds or multipass fillet welds. Sequence sometimes assigns each bead a number and size.

Burner. A skilled tradesman, coined from the fact that steel is cut by burning it. The term is in a moribund state but clings to life as tenaciously as *straight* and *reverse polarities*.

Buttering. The overlaying of one metal's surface with a dissimilar but compatible metal which is also compatible with yet another metal that is to be joined to the buttered metal. Similar to *cladding*.

Butt joint. A juncture of two sections of material brought into alignment in the same plane.

Butt weld. A weld made in a butted joint. See also *butt joint*.

Calescent. The temperature at which steel loses its magnetism; the temperature line where beta iron changes to alpha iron; the temperature at the low end of the critical range.

Glossary

Capillary action. The product or result of *capillary attraction*.
Capillary attraction. The tendency of a liquid to flow between two closely abutted surfaces and to flow out on a solid surface. See Chapter 2, especially Figure 2.1.
Cap pass. The pipe welder's term for the final or covering weld layer that slightly overwelds or reinforces a groove weld.
Carbide. A compounding of carbon with another element. See also *-ide*; *iron carbide*; *tungsten carbide*.
Carbide precipitation. The separation of carbon out of solution to form carbides. See the discussion of chrome carbides of the 3xx series, Chapter 9.
Carbon arc. An arc produced by using a carbon electrode in a welding circuit for cutting and welding. The carbon electrode is nonconsumable (to a degree), as is the TIG electrode, and may be used similarly with or without shielding.
Carbon dioxide (CO_2). A gas extracted from the waste gases given off by the burning of natural gas, coke, and fuel oil. Used extensively in MIG welding of steel due to its production of deep sound welds and its relatively low cost. See Chapter 4.
Carburizing flame. Any flame that can introduce carbon into the metal being heated. As applied to the oxyacetylene torch, it means an excess of acetylene. It has good and bad connotations. See *sweating*; *wearfacing*; "Welding Stainless Steel," in Chapter 9; and the Chapter 10 discussions of welding nickel and copper.
Cascade sequence. A sequence wherein a longitudinal weld is made by depositing overlapping layers, each layer extending a few centimeters beyond its predecessor. Used in deep-groove welds. See also *block sequence*.
Case hardening. A hardening process that hardens the surface area of steel to a desired depth but leaves a ductile core; a form of differential hardening. See Chapter 5.
Cast iron. An iron–carbon alloy in the carbon range 1.7 to 4.5% that is formed by casting in a mold. Silicon is usually an important alloying constituent. See also *malleable iron*; *nodular iron*; *white iron*; and Chapter 8.
Cast steel. Molten steel poured into a mold so as to be solidified in a predetermined shape.
Caulk weld. Also called *seal weld*. Any weld made for the simple purpose of making a joint liquid-tight.
Cementite. See *iron carbide*.
Chain intermittent. A welding pattern on a T-joint wherein only a predetermined percentage of a joint is welded: for example, "weld 3, skip 6," with each weld exactly opposite the other. (Sometimes erroneously called *skip welding*.) See Figure 13.29.
Chamfering. Any edge preparation that deviates from the square edge. Beveling is a form of chamfering.
Chrome irons. A somewhat generalized term for the martensitic and ferritic stainless steels of the 4xx series which contain little or no nickel and are magnetic. See Chapter 9, in particular, Figure 9.3.
Cladding. The overlaying of a metal surface with a dissimilar metal, usually a nobler metal or a metal that will resist a particular destructive force.
Cleavage line. See *slip line*. Slip lines become cleavage lines when a metal ruptures.
CO_2 welding. GMAW using carbon dioxide for shielding.
Cold rolling. The steelmaker's term for the reduction of a metal's cross-sectional dimen-

sion by a rolling compression at a temperature below the top of the transformation range.

Cold welding. The fusion of metals by hammering or other impacting while the metal is at room temperature. Readily accomplished with hand tools, such as hammer and chisel or centerpunch, or compressed-air gun and suitable metalworking tool.

Cold working. Any deformation of shape or structure by mechanical means while the metal is in or below its critical temperature, such as *cold rolling*.

Constant potential. See *constant voltage*.

Constant voltage. The characteristic of a welding machine that keeps the voltage at a constant value regardless of fluctuations in current, as in GMAW and FCAW.

Cover pass. The final layer of a groove weld.

Crater. The depression in a weld pool caused by the penetration of the arc.

Creep. A slow, continued deformation of a metal under stress as opposed to a sudden deformation; often used also to describe unstable oxyacetylene regulators.

Critical range. See *transformation range*.

Cryogen. A refrigerant.

Cryogenics (Gk. *kryos*, icy cold). The branch of physics dealing with the production and effects of extreme cold.

Cyaniding. Case hardening by heating in a bath of molten cyanide salt, then quick quenching.

Decarburization. The loss of surface carbon by steel when heated in an oxidizing atmosphere.

Deoxidizing. The removal of oxygen from a molten metal by chemical reactions between elements introduced into the weld metal by way of the electrode coating or core.

Differential hardening. Surface or localized hardening. See "Case Hardening," Chapter 5.

Dip. See *nozzle dip*.

Double ending. The pipe welder's term for joining two lengths of pipe by making a rolling weld; a means of reducing the number of in-position welds.

Down-hand welding. The welder's colloquialism for welding in the flat position. (Do not confuse with *vertical-down welding*.)

Downhill welding. The welding of any joint inclined from the vertical to the horizontal, regardless of the degree of the incline. Also the pipe welder's term for welding from 12 o'clock down to 6 o'clock when making an in-position weld.

Drag. The distance between the entry and exit of the oxygen jet stream at the top and bottom of the kerf. The *draglines* show the path of the jet stream and are indicative of many factors in the cut. See Figure 13.58.

Dragline. See *drag*.

Drawing. See *tempering*.

Ductility. The ability of a metal to withstand permanent deformation without rupturing.

Duty cycle. A percentage rating assigned to a welding machine that indicates the limits of its use at maximum loading. For example, a welding machine rated at 300 amperes and an 80% duty cycle means that the machine may operate 8 minutes in 10 at a 300-ampere load.

Dyne. Unit of force that accelerates 1 gram of free mass 1 centimeter per second per second.
Elasticity. The ability of a metal to return to its original dimensions when stress causing a deformation is removed. See Figure 5.3.
Elastic limit. The maximum stress a material can withstand without permanent deformation. See Figure 5.3.
Elongation. The measure of a metal's ability to stretch in a straight-line dimension, expressed as a percentage of the original gauge. See also *elastic limit*.
Endurance limit. The stress that a metal will endure indefinitely uder repetitive loadings. See also *fatigue failure*.
Erg. The cgs unit of work or energy equaling the force of 1 dyne acting through 1 centimeter.
Eutectic. The elemental constituency of an alloy at the intersection of two descending liquidi. See Figures 5.7, 5.9, and 5.10.
Eutectic alloy. An alloy wherein the compositional ratio of two elements occurs at the intersection of their respective liquidi. See Figures 5.7 and 5.9.
Eutectoid steel. A steel composed of pure pearlite. So-called because the iron–carbide eutectic is at 4.3% carbon, yet pearlite occurs and behaves as a eutectic at 0.85% carbon at 700° C (1292° F). See Figure 5.10.
Eyeballing. A pipewelder's colloquialism for alignment by sight, using other structures as reference lines or his own natural sense of the horizontal and vertical planes. Shipwrights find it useful, since spirit levels do not work well when the ship is waterborne.
Face-centered cubic lattice (fcc). A grouping of 14 atoms to form a cube with one atom centered in each facet of the cube. See Figure 5.1c.
Fatigue failure. The fracture or rupture of a metal due to stress from repeated loadings.
Fatigue limit. The stress that a metal will support indefinitely. See also *endurance limit*.
Ferrite. A solid solution in which alpha iron is the solvent. Pure iron (alpha) will hold less than 0.015% carbon in solution but will hold relatively large amounts of nickel and others.
Filler pass. Those weld layers that fill the void in a groove weld between the *hot pass* and the cap or cover pass.
Fingernailing. The welder's colloquialism for the uneven melting of the coating of a stick electrode due to the eccentricity of the core.
Firecracker weld. A SMAW technique wherein the arc is struck manually and the electrode is laid in the groove so as to continue welding voluntarily; similar to gravity welding.
Firing line welder. The pipe welder who makes the hot pass in the laying of a line wherein the welding of the joints is specialized.
Fisheye. A tiny nucleic circular void caused by entrapped slag or gas (usually hydrogen), so called because of its appearance under the microscope.
Flame shrinking. A method of removing a warp or buckle in a deck plate, bulkhead, or similar structure by means of heating and cooling. See Chapter 13.
Flame straightening. See *flame shrinking*.

Flashback. The torch flame recedes back into the mixing chamber of the torch.

Flop weld. The welding of a pipe joint by welding two quadrants and flopping the assembly 180° to weld the other two quadrants. Used when the pipe cannot be rolled because of attached fittings or takeoffs.

Flux. Any material or gas that is used to dissolve or prevent the formation of oxides prior to and during welding, and to aid in breaking down the surface tension of the metal.

Flux-core arc welding (FCAW). Arc welding with a continuous automatically fed tubular-type electrode whose shielding is provided by a flux material within the core of the electrode. Additional alloying ingredients may also be included in the core material. Gas shielding may also be added.

Foldback. A reversal of arc travel that folds back the weld puddle on the deposited bead for the purpose of eliminating crater cracks. See also *side swing*.

Forge welding. Pressure welding by means of hot or cold working by rolling, hammering, or smithing.

Forward swing. A method of eliminating crater cracks by a gradual thinning-out of the weld puddle as the arc is lifted from the joint with a forward sweep instead of an abrupt perpendicular lift. See also *Sideswing*.

Free carbon. Sometimes called uncombined carbon. Carbon that is not in solution or compounded. This carbon may exist as free atoms at grain boundaries, or as flakes or nodules of graphite. See Chapter 8.

Free ferrite. Alpha iron with very little carbon in solution (less than 0.015%) found in *hypoeutectoid steel*.

Free machining steel. Steel containing sulfur or silicon which has been introduced to make the steel more easily machinable. See also *high-sulfur steel*.

Freezing. Solidification of a hot liquid metal. See Chapter 5.

Fusion weld. A weld made by melting the two surfaces to be joined and allowing those surfaces to flow together and solidify as an amalgamated mass. Fusion welds may be made with or without filler metal.

Gall. To wear away and become pocked and blistered because of friction, such as fretting, chafing, or rubbing. Extreme galling may lead to seizing.

Galvanized steel. Also called galvanized iron (G.I.). A steel that has been coated with zinc. SMAW is the best method for fusion welding. Use E-6010 or E-6011, whip ahead to burn off the galvanized coating, and work the puddle to allow vaporized zinc to escape. When welding with E-7018 or better for high-quality welds, grind off the zinc coating.

Gamma iron. Face-centered cubic, nonmagnetic iron (see Figures 5.7 and 5.10); the solvent in austenite.

Gas metallic-arc welding (GMAW). The more formal term for MIG welding. A method of fusion welding using a consumable electrode automatically fed at a preset rate and using gas for shielding. See Chapter 4.

Gas pocket. See *blowhole*.

Globular transfer. The transfer of metal across the arc in globules. See "Spray Transfer" and "Globular Transfer," Chapter 4.

Glossary

Grain. A crystalline, haphazard formation of atomic groups around a nucleus, so called because their varieties of shapes as are grains of sand.

Gram calorie. Also called calorie. The amount of heat required at a pressure of 1 atmosphere to raise the temperature of 1 gram of water 1 degree Celsius.

Grapes. The welder's colloquialism for irregular and excessive melt-through of a root pass.

Graphite. An allotrope of carbon. Carbon that has separated out of solution (decomposition of cementite) and exists as structural discontinuities in a metal in the form of flakes or nodules. See "Gray Cast Iron" and "Nodular Iron," Chapter 8.

Graphitic carbon. See *graphite*.

Graphitic tool steel. Tool steel featuring a relatively high percentage of free graphite.

Graphitization. The separation of carbon out of solution in cementite. See also *graphite*.

Gray iron. A member of the cast iron family. It is so called because of its appearance when fractured. See "Gray Cast Iron," Chapter 8; and Figure 8.1.

Hadfield steel. Also called austenitic manganese steel. See Chapter 12.

Hardfacing. The overlay or cladding of a metal with a harder metal. See also *wearfacing* and Chapter 15.

HASTELLOY. A registered trademark of the Cabot Corporation for a series of specialized alloys. See Chapter 10, in particular, Figure 10.3.

Heat treatment. The heating and cooling of metals for the purpose of effecting structural changes. See Chapters 5 and 7.

Heliarc. A registered trade name. See also *TIG*.

Helium (He). A very light, inert gas, atomic number 4; a by-product of the gas and petroleum industries.

Hexagonal close-packed space lattice (hcp). A grouping of 17 atoms into a six-sided polygon of six angles and six sides, with one atom centered in each hexagonal facet and three atoms triangularly spaced equidistant from center to facet at the midpoint of the polygon. See Figure 5.1d.

High-speed steel. Steel for tools that are designed for high-speed removal of metal which necessitates a good red-hardness characteristic. They deep-harden in oil (some in air). See Chapter 7.

High-sulfur steel. Also called *free-machining steel*. Steel containing relatively high amounts of sulfur in order to facilitate machining.

Hot pass. A pipe welder's term for the second pass in a groove weld that is calculated to burn deep into the root pass and clean out all slag inclusions, and leave a relatively clean, smooth weld face for the filler pass or cover pass.

Hot quench. The cooling of metals in a bath of molten metal or molten salt. See also *austempering*.

Hot shortness. The tendency of a metal to become brittle at elevated temperatures.

Hot working. Any method of deforming, shaping, or refining of a metal while that metal is at a temperature above the upper transformation-range temperature. See Figure 5.13.

Hot-working tool-and-die steel. Steel used in tools and dies that become relatively hot in service.

Huntington alloys. A group of specialized alloys, including but not limited to, MONEL, INCONEL, and INCOLOY. See Chapter 10.

Hypereutectic. Containing the minor alloying constituent in an amount in excess of that contained in the eutectic mixture.

Hypereutectoid steel. A steel composed of pearlite and free cementite. See Figure 5.10.

Hypoeutectic. Containing the minor alloying constituent in an amount less than that contained in the eutectic mixture.

Hypoeutectoid steel. A steel composed of pearlite and free ferrite. See Figure 5.10.

-ide. A suffix denoting the compounding of the root element with another element. For example, the word *carbide* implies a compounding of carbon with another element; nitride, the combining of nitrogen with another element.

Idiot rod. A contemptuous colloquialism of the experienced welder when referring to electrodes whose coating may be kept in contact with the weldment.

INCOLOY. A registered trademark of Huntington Alloys, Inc.; see Chapter 10.

INCONEL. See *Incoloy*.

Inert gas. A gas that will not combine with other elements.

Ingot iron. Iron that is very low in carbon, almost a pure iron; usually strengthened by cold rolling.

Innershield. A registered trademark of the Lincoln Electric Company. See also *flux-core arc welding*.

Insert. Any preplaced alloy that will be melted in the welding joint, such as preplaced silver rings for brazing, consumable rings for TIG welding of root passes in pipe, and special inserts for the welding of incompatible metals.

Intergranular corrosion. Corrosion occurring at the grain boundaries. See also *carbide precipitation* and Chapter 9.

Intermittent weld. See *chain intermittent* and *staggered intermittent*.

Iron carbide. Also called *cementite*. A compound of iron and carbon (Fe_3C). See Chapter 5.

Joule. The mks unit of work or energy (10^7 ergs, or about 0.7373 foot pound).

Kerf. The space resulting from the removal of metal by a cutting process. See "Oxyacetylene Cutting," Chapter 13.

Keyholing. A welding technique wherein the digging action of the arc enlarges the root opening of a groove weld and produces a weld metal deposit on the back side of the joint. See Figure 1.6.

Kip (kilo + pound). A unit of weight. One kip equals 1000 pounds.

Land. The *root face* of a beveled or gouged groove joint.

Lead burning. An almost archaic term for lead welding.

Liquidus (*plural, liquidi*). The lowest temperature a liquid can reach before solidifying. See the graph of descending liquidi in Figure 5.10.

Load. The force or stress applied to a material during its in-service applications or for testing.

Magnetic pinch effect. The radial force exerted by a current flowing in a conductor which tends to reduce the diameter of that conductor. As applied to arc welding, the pinch effect squeezes a portion of the melting electrode off and away from the electrode. (This radial force is proportional to the square of the current.)

Glossary

Malleability. The ability of a metal to be permanently deformed without rupturing.

Malleabilizing. An annealing procedure for transforming white cast iron to malleable iron by slow annealing at high temperatures in order to free and refine much of the combined carbon.

Malleable iron. A member of the cast iron family which is essentially a white iron that has been subjected to an extensive heat-treating process called malleabilizing. See also *malleabilizing*.

MAPP gas. MAPP is a registered trade name. It is liquefied acetylene whose oxy-flame is about midway between oxy-propane and oxyacetylene. Its neutral flame heat is 2926° C (5300° F).

Martempering. Hardening and tempering simultaneously by heating steel above the transformation range and quenching in a hot-salt bath at a temperature just above the M-point and then cooling in air to room temperature. Similar to *austempering* but quick-quenched to a lower temperature. See Figure 7.2

Martensite. The first stage of the decomposition of austenite. An unstable structure obtained by quick-quenching carbon steels to avoid further decomposition of the austenite. See Figure 7.2; "Martempering," Chapter 7; and Chapter 5.

Meld. The molten state of an alloy. The word derives from a blend of melt and weld.

Melder. A VIP on the furnace floor; he is responsible for the composition of the heat and the time of heating.

Micrometer (micron). One thousandth of a millimeter (0.000001 meter).

MIG welding. The welder's colloquial acronym for metal inert gas. See *gas metallic-arc welding*.

Miscibility. The ability of elements to alloy or dissolve with each other.

Mohs' scale of hardness. A measure of a material's hardness on a scale of 1 to 10. 1, talc; 2, gypsum; 3, calcite; 4, fluorspar; 5, agatite; 6, feldspar; 7, quartz; 8, topaz; 9, sapphire; and 10, diamond. For example, cementite is 8.5 on Mohs' scale; tungsten carbide is 9.5.

MONEL. A registered trade name of Huntington Alloys, Inc. A nickel–copper alloy of 67% nickel–28% copper. See Chapter 10.

Mottled iron. A member of the cast-iron family whose composition is pearlite, free graphite, and free cementite.

M-point. The temperature at which austenite begins to transform to martensite.

Muntz metal. A copper-based alloy very high in zinc; a member of the leaded-brass family. See Chapter 10.

Neutral flame. The oxy-fuel flame that totally consumes both gases. See neutral, carburizing, and oxidizing flames shown in Figure 2.6.

Newton (N). A unit of force that will accelerate a free mass of 1 kilogram 1 meter per second per second.

Nickel silver. A term applied to copper–nickel–zinc alloys. See Chapter 10.

Nodular iron. Also called ductile iron. A member of the cast-iron family in which the free graphite exists as nodules or spheroids. See "Nodular Iron," Chapter 8.

Nonferrous metals. A generalized term referring to metals containing less than 50% iron.

Normalizing. Heating above the transformation range and cooling in air (room temperature). *Annealing* leaves steel a little too gummy for machining; normalizing is just about right.

Nozzle dip. A commercially available antispatter compound that inhibits the adherence of weld spatter on the MIG gun nozzle.

Ohm. The unit of electrical resistance. The amount of resistance that will restrict the flow of current to 1 ampere at a pressure of 1 volt.

Open-circuit voltage. Also called no-load voltage. The voltage measured at the output terminals of a welding machine when the machine is running but not welding.

Oxide. The combining of oxygen with another element, such as iron oxide.

Oxidizing flame. An oxy-fuel flame that does not consume all of the oxygen. Characteristically, it introduces oxygen into the weld puddle and is disastrous in most cases, but useful in several applications, such as welding high-zinc brasses.

Oxy-arc cutting. The cutting of metals as made by the chemical reaction of oxygen with superheated molten metal in the cut.

Pascal (Pa). The unit of pressure exerted by a liquid or gas in all directions that is equal to the force of one newton per square meter. This name has been accepted by the International Organization for Standardization.

Pearlite. An iron–carbon alloy of eutectoid structure, containing 0.85% carbon and existing as alternate layers (or plates) of ferrite and cementite.

Peeling. The separation of a weld deposit from the parent metal at or just below the fusion line. Sometimes encountered in hardfacing. See Chapter 15.

Peening. The mechanical working of a metal by repeated impact, such as hammer blows, used as a method of stress relief.

Phosphor bronze. A copper–tin alloy containing phosphorus as a deoxidizer of tin. See Chapter 10.

Pig iron. Relatively pure iron as extracted from the ore and cast into small blocks for easy handling. The name is derived from the molds, which feed from a casting duct that resemble suckling pigs.

Pinch effect. A radial force exerted by a flow of current in a conductor that tends

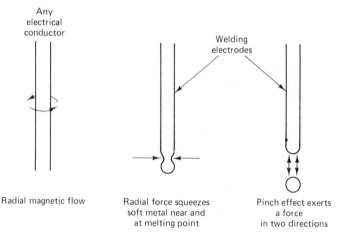

to reduce the cross-sectional area of that conductor, the magnitude of the pinch (squeeze) being proportional to the square of the current. As it applies to welding,

Glossary

the pinch effect must overcome surface tension and other factors in order to separate globules of molten metal from the consumable electrode for transfer across the arc.

Plasticity. See *ductility*.

Precipitation. The separation of an alloying element out of solution. See the cementite solubility curve in Figure 5.10 and the solubility curve *JLM* in Figure 5.9.

Precipitation hardening. Hardening by heat treatment that precipitates hardening agents out of solution, such as $CuAl_2$ in some aluminums. See also *age hardening*.

Quench crack. Crack occurring during the quick quenching of a hardening procedure.

Radial crack. Crack originating in the fusion zone and continuing on into the parent metal at right angles to the axis of the weld. See Chapter 7.

Red hardness. The ability of a metal to retain its hardness at very high temperatures.

Red shortness. Similar to hot shortness but at a specified red temperature range.

Reducing flame. See *carburizing flame*.

Refractory metal. A metal that resists many of the harmful effects of high temperatures. See Chapter 10.

Reverse polarity. Welding electrode positive (dc +) and weldment negative (dc −); a somewhat archaic, moribund term that clings tenaciously to respectability.

Rockwell hardness test. A measure of a metal's hardness. The Rockwell C scale measures the harder metals by means of the depth of penetration by a steel cone point. The B scale uses a steel ball and is used to measure the softer metals.

Root face. Also called *land*. The shoulder or unremoved portion at the root of a beveled or gouged groove when the angle face does not extend the full thickness of the section.

Root pass. The initial bead of a groove weld, plate or pipe.

Roping. See *stubbing*.

Rosebud. Welder's colloquialism for an oxy-fuel multiorifice heating nozzle.

Scarfing. The removal of unwanted metal, such as routing out defective welds and rough machining, although rough machining has all but been replaced by the term "flame machining" when referring to the oxy-fuel method.

Scleroscope (Shore scleroscope). A pointed cylinder in a glass tube measures the hardness of a metal by dropping 25.4 cm (10 in.) and registering its bounce on a scale. The height of the bounce determines the hardness.

Seizing. The cohesion of one surface to a relatively moving surface caused by excessive friction.

Self-hardening steel. See *air-hardening steel* and Chapter 7.

Semiautomatic welding. A welding process using automatic wirefeed and manual control of the weld deposition, as in MIG welding.

Shielded metal-arc welding (SMAW). Arc welding with a flux-coated consumable electrode. See also *stick welding*.

Short-circuiting arc welding. One of the across-the-arc transfers used in MIG welding. The welder's colloquial term is "short arc." See the discussion of GMAW in Chapter 4.

Shuler weave. A modified crescent weave wherein the crescents are deposited in the sequence 1, 3, 2, 5, 4, 7, 6 and so on, with crescent 3, for instance, swinging back between crescents 1 and 2. (Alternate lines are broken only as an aid in perspective.

If the weave is properly applied, the deposit will be smooth, with indiscernible travel lines.

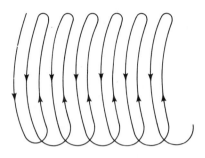

Side swing. A welding technique wherein the welder swings the arc puddle to one side of the deposited bead before breaking the arc for the purpose of eliminating crater cracks. This technique makes restarts less troublesome than does the *foldback* method.

Silver brazing. The modern term for "silver soldering," a term which stubbornly clings to the trades. See "Brazing," Chapter 2.

Skip welding. Also called *wandering welds*. Skip welding is used for broader distribution of heat input into the parent metal and also to reduce distortion. (Do not confuse with *intermittent welding*.) See Chapter 13.

Slip line. The plane of the interstices of paralled alignments of atom groups along which a metal may be stretched or deformed. See also *cleavage line* and Figures 5.2 and 5.3.

Slip-on flange. See Figure 14.17a.

Soaking. Heating a metal at a specific temperature for a specified period (usually given as a time–thickness ratio) in order to distribute the heat uniformly throughout the metal and allow the desired structural changes to take place. A steelmaker's term for the heating of an ingot in the soaking pits prior to blooming or other hot working.

Solid solution. A solidified alloy in which the constituents are dissolved in each other; for example, the iron and carbon in austenite.

Solidus (*plural,* **solidi**). The temperature at which a pure metal freezes and at which an alloy has completely solidified. See Figures 5.7 and 5.10.

Sorbite. The decomposition of austenite is complete but in a finely divided state. The stage of transformation just prior to the formation of pearlite. See also *bainite*.

Space lattice. The geometrical configuration of an atom group that is peculiar to each pure metal. See Figures 5.2 and 5.3.

Spall. To break away in chips, scales, or particles as a result of shock from repetitive percussive impact. For example, excessive peening can embrittle metal to the point that it flakes away.

Spheroidizing. A heat treatment to obtain spheroidic (nodular) free graphite in cast iron or spheroidic cementite in steel.

Spool. A short length of pipe flanged at both ends. Also, the reel that contains continuous welding wire.

Spray transfer. Metal transferred across the arc in very minute drops as compared to *globular transfer*.

Glossary

Squirt-through. A colloquialism for the penetration of an automatically fed electrode through the root opening of a groove weld without consummation in the weld puddle. See also *whiskers*.

Staggered intermittent. Used in welding T and lap joints when continuous welding is not necessary in order to reduce distortion and reduce costs. See Figure 13.26.

Stainless steel. A term applied to very high alloy steels characterized by their excellent corrosion resistance in hostile environments. See Chapter 9, especially Figure 9.1.

Stick-out. The length of protrusion of the electrode in TIG and semiautomatic welding.

Stick welding. The welder's colloquial term for SMAW, to distinguish it from GMAW.

Stitch welding. See *intermittent welding*.

Stovepipe welding. The pipe welder's term for running a pipeline by adding single pipe sections and welding the joints in position.

Straight polarity. Welding electrode negative (dc−) and weldment positive (dc+). See also *reverse polarity*.

Strain. The unstable condition produced in a metal by welding, hot working, cold working, and others. In welding it is caused by uneven expansion and contraction during the heating and cooling cycles. See Chapter 13.

Stress. The force that produces strain, deformation, or rupture.

Stress crack. Cracking of the weld or parent metal caused by residual stresses induced into the metal by the act of welding.

Stress relief. A heat treatment for the purpose of stabilizing a weldment; that is, heating uniformly after welding to remove residual stresses. Stress relief temperatures are well below the transformation range, hence should not be called *annealing* or *normalizing*.

Stubbing. Also called *roping*. The action of an electrode failing to burn off in the arc, producing a lumpy convex bead contour. See "Wirefeed," Chapter 4.

Surfacing. A catchall term for *buttering*; *cladding*; *hardfacing*; *overlaying*; plating; *wearfacing*.

Sweating. A method of depositing a filler metal with the oxyacetylene torch wherein an excess acetylene introduces carbon into the surface of the steel, thereby reducing the melting point well below the normal melting point.

Tempering. Also called drawing. A heat treatment of hardened and martensitic steels. These temperatures range from 100 to 650° C (212 to 1200° F).

Tensile strength. A metal's resistance to a linear stress that tends to rupture the metal by stretching. See Chapter 13.

Thermit. A specially prepared mixture of finely divided aluminum and the oxides of a metal whose chemical interaction will, when ignited, transfer the oxygen of the oxides to the aluminum, thus producing aluminum oxide and a superheated pure metal. See "Thermit-Welding Cast Iron," Chapter 8.

Thermit welding. Also called foundry weld. A weld made by casting a superheated liquid metal into a mold shaped about the weld area of the section to be welded or of sections to be restored.

TIG. TIG is the welder's acronym for tungsten inert gas; although GTAW, gas tungsten-arc welding, is more complete.

Transformation range. Also called *critical range*. The temperature range encompassing the structural changes a metal undergoes during a normal cool-down to room temperatures. See Chapter 5.

Troostite. A very fine aggregate of ferrite and cementite found in steels cooled at a slower rate than the quick quench of martensite; appears as a very fine pearlite. See also *bainite*.

Tube-Turn. A registered trade name which is so established in the trade that many pipe welders refer to their prefabricated pipe turns by that name. Pipe turns are also called by the degree of the turn, such as 45's, 90's.

Tungsten inert-gas. The full name of the acronym TIG. Same as GTAW.

Two-holing. Pipe welder's term for the alignment of the holes of a flange with other fittings and the run of the pipe.

Undercut. The eroding of the parent metal at the toe of a weld caused by faulty welding technique. See "Understanding the Weld Puddle," Chapter 1.

Upset. Any deformation of a metal that results in a perpendicular displacement of that metal's mass, such as upset joint, upset welding, or upset expansion.

Vertical-down welding. The welding of a vertical joint starting at the top and proceeding toward the bottom. Compare with *downhill welding*.

Vertical-up welding. The welding of a vertical joint starting at the bottom and proceeding to the top.

Wagon tracks. Voids that occur in multipass welding due to inadequate purging and fusion along the toes (edges) of parallel beads. Excessively convex beads and failure to clean the edges of beads are contributing factors. The name derives from the appearance of these voids under the microscope.

Water-hardening steel. Steel from .050 to 1.40% carbon that are water-quenched to obtain their hardness. See Chapter 7.

Watt. The unit of electrical power. The product of voltage times current equals the power consumed in the welding circuit. See Chapter 1.

Wearfacing. A relatively new term, gaining in popularity. The cladding, buttering, or overlaying of an inferior metal with a weld deposit designed to resist specific types of wear, such as abrasion, corrosion, impact, and friction.

Weldment. Any material structure of any size or shape that will be, is being, or has been welded.

Weld-neck flange. A pipe flange formed with a neck the same size as its mating pipe section so as to be welded as a butted groove joint. See Figure 14.17b.

Wetting out. A term used to describe the tendency of a liquid to flow out over a solid surface. See also *capillary attraction*.

Whipping. An inward-upward movement of the electrode in vertical-up welding for the purpose of managing the weld puddle distribution. Whipping may be employed in all positions for specific purposes; for example, burning off galvanized coatings (or paint, etc.) ahead of the weld puddle.

Whiskers. The unconsumed portions of automatic feed electrodes in MIG welding that are caused by *squirt-through*.

White iron. A member of the cast-iron family. See also *cast iron*.

Work hardening. The hardening of a metal by cold working. See Chapter 12.

B
SYMBOLS AND ABBREVIATIONS

A	Area
AAC	Air carbon-arc cutting
Ac	Actinium (at. no. 89)
Ag	Silver (at. no. 47)
AH	Air hardening
AISI	American Iron and Steel Institute
aka	Also known as
Al	Aluminum (at. no. 13)
Am	Americium (at. no. 95)
AOC	Oxygen-arc cutting
Ar	Argon (at. no. 18)
As	Arsenic (at. no. 33)
ASME	American Society of Mechanical Engineers
ASTM	American Society for Testing Metals
At	Astatine (at. no. 85)
Au	Gold (at. no. 72)
AWS	American Welding Society
B	Boron (at. no. 5)
Ba	Barium (at. no. 56)
bcc	Body-centered cubic
Be	Beryllium (at. no. 4)
Bi	Bismuth (at. no. 83)
Bk	Berkelium (at. no. 97)
Br	Bromine (at. no. 35)
Btu	British thermal unit

C	Celsius/centigrade; carbon (at. no. 6); circumference
Ca	Calcium (at. no. 20)
CAC	Carbon-arc cutting
CAW	Carbon-arc welding
Cb	Columbium (see niobium, Nb)
Cd	Cadmium (at. no. 48)
Ce	Cerium (at. no. 58)
Cf	Californium (at. no. 98)
cfh	Cubic feet per hour
cfm	Cubic feet per minute
cgs	Centimeter–gram–second
CI	Cast iron
Cl	Chlorine (at. no. 17)
cm	Centimeter
Cm	Curium (at. no. 96)
Co	Cobalt (at. no. 27)
Co_2	Carbon dioxide
Cr	Chromium (at. no. 24)
Cu	Copper (at. no. 29; from *cuprum*)
CW	Cold welding
D	Diameter (in formulas); Roman 500
DB	Dip brazing
dc	Direct current
DCEN	Direct-current electrode negative
DCEP	Direct-current electrode positive
DCRP	Direct-current reverse polarity
DCSP	Direct-current straight polarity
deg	Degree
DFB	Diffusion brazing
DFW	Diffusion welding
diam	Diameter
dm	Decimeter
Dy	Dysprosium (at. no. 66)
EBC	Electron beam cutting
EBW	Electron beam welding
Er	Erbium (at. no. 68)
Es	Einsteinium (at. no. 99)
Eu	Europium (at. no. 63)
EW	Electroslag welding
EXW	Explosion welding
F	Fahrenheit; fluorine (at. no. 9)
FB	Furnace brazing
FCAW	Flux-cored arc welding
FCAW-EG	Flux-cored arc welding, electrogas
fcc	Face-centered cubic

Symbols and Abbreviations

F & D	Faced and drilled
Fe	Iron (at. no. 26; from *ferrum*)
F to F	Face to face
Flg	Flange
Flgd	Flanged
Fm	Fermium (at. no. 100)
FOC	Chemical flux cutting
Fr	Francium (at. no. 87)
FRW	Friction welding
FW	Flash welding
g	Gram
G	Gauge, gage
Ga	Gallium (at. no. 31)
Galv	Galvanized
Gd	Gadolinium (at. no. 64)
Ge	Germanium (at. no. 32)
GMAW	Gas metallic-arc welding
GMAW-EG	Gas metallic-arc welding, electrogas
GMAW-P	Gas metallic-arc welding, pulsed arc
GMAW-S	Gas metallic-arc welding, short-circuit arc
GSSW	Gas shielded-stud welding
GTAC	Gas tungsten-arc cutting
GTAW	Gas tungsten-arc welding
GTAW-P	Gas tungsten-arc welding, pulsed arc
H	Hydrogen (at. no. 1)
hcp	Hexagonal close packed
He	Helium (at. no. 2)
hex.	Hexagonal, six-sided
Hf	Hafnium (at. no. 72)
Hg	Mercury (at. no. 80; from *hydrargyrum*)
Ho	Holmium (at. no. 67)
HPW	Hot pressure welding
HSS	High-speed steel
I	Iodine (at. no. 53)
IB	Induction brazing
IBBM	Iron body bronze (brass) mounted
ID	Inside diameter (pipe); inside dimension
In	Indium (at. no. 49)
IPS	Iron pipe size (see NPS, preferred)
Ir	Iridium (at. no. 77)
IS	Induction soldering
IW	Induction welding
k	1000; kip
K	Kelvin; potassium (at. no. 19; from kalium)
kg	Kilogram

kilo	1000
kl	Kiloliter
km	Kilometer
kPa	Kilopascal
Kr	Krypton (at. no. 36)
ksi	1000 pounds per square inch (kips/square inch)
La	Lanthanum (at. no. 57)
LBC	Laser beam cutting
LBW	Laser beam welding
Li	Lithium (at. no. 3)
LOC	Oxygen lance cutting
lpm	Liters per minute
Lu	Lutetium (at. no. 71)
Lw	Lawrencium (at. no. 103)
MAC	Metal arc cutting
Md	Mendelevium (at. no. 101)
mdd	Milligrams per square decimeter per day
mega	One million of
mg	Milligram
Mg	Magnesium (at. no. 12)
micro	One millionth of
mks	Meter–kilogram–second
mm	Millimeter
Mn	Manganese (at. no. 25)
Mo	Molybdenum (at. no. 42)
MPa	Megapascal
N	Nitrogen (at. no. 7); Newton
Na	Sodium (at. no. 11)
Nb	Niobium (at. no. 41)
Nd	Neodymium (at. no. 60)
Ne	Neon (at. no. 10)
Ni	Nickel (at. no. 28)
No	Nobelium (at. no. 102)
Np	Neptunium (at. no. 93)
NPS	Nominal pipe size
O	Oxygen (at. no. 8)
OAW	Oxyacetylene welding
OC	Oxygen cutting
O.D.	Outside diameter (pipe); outside dimension
OFC	Oxy-fuel-gas cutting
OFC-A	Oxygen fuel-gas cutting, acetylene
OFC-H	Oxygen fuel-gas cutting, hydrogen
OFC-N	Oxygen fuel-gas cutting, natural gas
OFC-P	Oxygen fuel-gas cutting, propane

Symbols and Abbreviations

OHW	Oxy-hydrogen welding
Os	Osmium (at. no. 76)
OS	Oven soldering
OS & Y	Outside screw and yoke
P	Phosphorus (at. no. 15); pressure
Pa	Protactinium (at. no. 91); Pascal
PAC	Plasma arc cutting
PAW	Plasma arc welding
Pb	Lead (at. no. 82; from *plumbum*)
Pd	Palladium (at. no. 46)
PEW	Percussion welding
PGW	Pressure gas welding
Pm	Promethium (at. no. 61)
Po	Polonium (at. no. 84)
POC	Metal powder cutting
Pr	Praseodymium (at. no. 59)
psi	Pounds per square inch
Pt	Platinum (at. no. 70)
Pu	Plutonium (at. no. 94)
QED	Quod erat demonstrandum: that which was to be proven (mathematical)
qv	Quod vides: which see, refer to
Ra	Radium (at. no. 88)
Rb	Rubidium (at. no. 37)
RB	Resistance brazing
Re	Rhenium (at. no. 75)
Rh	Rhodium (at. no. 45)
Rn	Radon (at. no. 86)
RPW	Projection welding
RS	Resistance soldering
RSEW	Resistance seam welding
RSW	Resistance spot welding
Ru	Ruthenium (at. no. 44)
s	Stere
S	Sulfur (at. no. 16)
SAW	Submerged arc welding
Sb	Antimony (at. no. 51; from *stibium*);
Sc	Scandium (at. no. 62)
Se	Selenium (at. no. 34)
Si	Silicon (at. no. 14)
Sm	Samarium (at. no. 62)
SMAW	Shielded metal arc welding
Sn	Tin (at. no. 50; from *stannum*)
Sr	Strontium (at. no. 38)

Ta	Tantalum (at. no. 73)
Tb	Terbium (at. no. 65)
TB	Torch brazing
Tc	Technetium (at. no. 43)
Te	Tellurium (at. no. 52)
Temp.	Temperature
Th	Thorium (at. no. 90)
Ti	Titanium (at. no. 22)
Tl	Thallium (at. no. 81)
ts	Tensile strength
TW	Thermit welding
U	Uranium (at. no. 92)
USW	Ultrasonic welding
UW	Upset welding
V	Vanadium (at. no. 23)
Vol.	Volume
W	Tungsten (at. no. 74); from "wolfram")
WC	Tungsten carbide
Wt.	Weight
Xe	Xenon (at. no. 54)
XS	Extra strong
XXS	Double extra strong
Y	Yttrium (at. no. 39)
Yb	Ytterbium (at. no. 70)
Zn	Zinc (at. no. 30)
Zr	Zirconium (at. no. 40)

C

UNIT CONVERSION FORMULAS

Basic metric units:

liter
meter
gram
bar
pascal

Prefixes:

Prefix	Multiple	Symbol
mega	10^6	M
kilo	10^3	k
hecto	10^2	h
deka	10	da
deci	10^{-1}	d
centi	10^{-2}	c
milli	10^{-3}	m
micro	10^{-6}	μ

Celsius-Fahrenheit

°C × 1.8 + 32 = °F
°F − 32 × 0.55555 = °C

Linear

2.54 × in. = cm
25.40 × in. = mm
0.0393 × mm = in.
0.621 × km = miles
1.609 × miles = km

Fraction of Inch

top number ÷ bottom number × 25.4 = mm
inch decimal × 25.4 = mm

Flow Rate

0.4719 × cu ft/hr (cfh) = liters/min (1pm)
2.119 × 1pm = cfh
3.785 × gal/min = 1pm
0.264 × 1pm = gal/min

Travel Speed

0.4233 × in./min = mm/sec
2.362 × mm/sec = in./min

Area

645.2 × sq in. = mm^2
0.00155 × mm^2 = sq in.
6.452 × sq in. = cm^2
0.1550 × cm^2 = sq in.

Volume

0.061 × cm^3 = cu in.
0.00061 × mm^3 = cu in.
16.4 × cu in. = cm^3
1640.00 × cu in. = mm^3
0.02832 × cu ft = m^3 (stere)
35.3147 × m^3 = cu ft

Capacity

1.0 × m^3 (stere) = kl

Unit Conversion Formulas

1000.0	×	m^3	=	liters
1000.0	×	cm^3	=	ml
231.0	×	gal	=	cu in.
0.13368	×	gal	=	cu ft
3.785412	×	gal	=	liters
61.0328	×	liters	=	cu in.
0.03531	×	liters	=	cu ft
0.26417	×	liters	=	gal
7.481	×	cu ft	=	gal

Stress-Load-Pressure

0.000703	×	psi	=	kg/mm^2
0.07030	×	psi	=	kg/cm^2
14.2234	×	kg/cm^2	=	psi
1422.34	×	kg/mm^2	=	psi
0.0006895	×	psi	=	hectobar
0.06895	×	psi	=	bar
6.895	×	psi	=	kilopascal (kPa)
6.89476	×	kips/sq in. (ksi)	=	MPa
6894.76	×	psi	=	newtons/square meter (N/m^2)
0.000145	×	N/m^2	=	psi
6894.76	×	psi	=	Pa
0.000145	×	Pa	=	psi
0.145	×	kPa	=	psi
14.5033	×	bar	=	psi
9.807	×	kg	=	N
4.448	×	lb	=	N
0.2248	×	N	=	lb
1009.000	×	N	=	kg
2.205	×	kg	=	lb
0.4536	×	lb	=	kg

Vacuum

14.696	×	atmosphere (atm)	=	psi
29.92	×	atm	=	in. of mercury
760.00	×	atm	=	mm of mercury
760,000.00	×	atm	=	micrometer (μm)

0.06895	× psi	= atm
2.036	× psi	= in. of mercury
51.5	× psi	= mm of mercury
51,500.00	× psi	= µm
0.0334	× in. of mercury	= atm
0.491	× in. of mercury	= psi
25.4	× in. of mercury	= mm of mercury
25,400.0	× in. of mercury	= µm
0.00132	× mm of mercury	= atm
0.0195	× mm of mercury	= psi
0.0394	× mm of mercury	= in. of mercury
1,000.0	× mm of mercury	= µm
1.32×10^{-6}	× µm	= atm
1.95×10^{-5}	× µm	= psi
3.94×10^{-5}	× µm	= in. of mercury
0.001	× µm	= mm of mercury
3,386.39	× in. of mercury	= N/m^2
0.0002953	× N/m^2	= in. of mercury

Corrosion

372,000.0	× g/sq in./hr	= mg/dm^2/day (mdd)
0.0274	× g/m^2/yr	= mdd
0.0003277	× mg/dm^2	= oz/sq ft
0.00000269	× mdd	= g/sq in./hr
36.5	× mdd	= g/m^2/yr
0.00748	× mdd	= lb/sq ft/yr
3,052.00	× oz/sq ft	= mg/dm^2
133.80	× lb/sq ft/yr	= mdd

Thermal Conduction

5.191	× Btu/sq ft/sec/°F/in.	= joule/cm^2/sec/°C/cm
1.2404	× Btu/sq ft/sec/°F/in.	= g cal/cm^2/sec/°C/cm
4.185	× g cal/cm^2/sec/°C/cm	= joule/cm^2/sec/°C/cm
0.8062	× g cal/cm^2/sec/°C/cm	= Btu/sq ft/sec/°F/in.
2,902.00	× g cal/cm^2/sec/°C/cm	= Btu/sq ft/hr/°F/in.
0.2389	× joules/cm^2/sec/°C/cm	= g gal/cm^2/sec/°C/cm
0.1926	× joules/cm^2/sec/°C/cm	= Btu/sq ft/sec/°F/in.
693.4	× joules/cm^2/sec/°C/cm	= Btu/sq ft/hr/°F/in.

Unit Conversion Formulas

$0.1442 \times$ Btu/sq ft/hr/in./°F $=$ W/m-°K
$6.9335 \times$ W/m-°K $=$ Btu/sq ft/hr/in./°F

Electrical Resistivity

$0.16624 \times$ ohms/circ. mil/ft $=$ microhms/cm^2/cm
$6.0153 \times$ microhms/cm^2/cm $=$ ohms/circ. mil/ft

D | TABLES

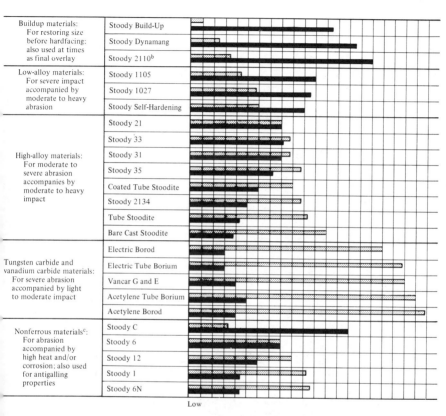

[a] Stoody, Dynamang, Stoodite, Electric Borod, Electric Tube Borium, and Vancar are registered trademarks.

[b] Stoodite, Borod, and Vancar are registered trademarks of Stoody Co., Industry, Calif.

[c] The mechanical properties and wear resistance of the nonferrous alloys shown here are superior to iron-based alloys at elevated temperatures.

Source: Reproduced with permission of Stoody Company, Industry, California.

TABLE D.1 Relative Abrasion Resistance and Impact Strength[a]

TABLE D.2 Nominal Chemical Composition[a] (%) of Huntington Alloys

Designation	Ni	C	Mn	Fe	S	Si	Cu	Cr	Al	Ti	Others
Nickel Alloys											
Nickel 200	99.5[b]	0.08	0.2	0.2	0.005	0.2	0.1	—	—	—	—
Nickel 201	99.5[b]	0.01	0.2	0.2	0.005	0.2	0.1	—	—	—	—
Nickel 205	99.5[b]	0.08	0.2	0.1	0.004	0.08	0.08	—	—	0.03	Mg 0.05
Nickel 211	95.0[b]	0.1	4.8	0.4	0.008	0.08	0.1	—	—	—	—
Nickel 220	99.5[b]	0.04	0.1	0.05	0.004	0.03	0.05	—	—	0.03	Mg 0.05
Nickel 230	99.5[b]	0.05	0.08	0.05	0.004	0.02	0.05	—	—	0.003	Mg 0.06
Nickel 270	99.98	0.01	<0.001	0.003	<0.001	<0.001	<0.001	<0.001	—	<0.001	Co<0.001
Duranickel alloy 301	96.5[b]	0.2	0.2	0.3	0.005	0.5	0.1	—	4.4	0.6	Mg<0.001
Permanickel alloy 300	98.5[b]	0.2	0.2	0.3	0.005	0.2	0.1	—	—	0.4	Mg 0.4
Monel Nickel–Copper Alloys											
Monel alloy 400	66.5[b]	0.2	1.0	1.2	0.01	0.2	31.5	—	—	—	—
Monel alloy 404	54.5[b]	0.08	0.05	0.2	0.01	0.05	44.0	—	0.03	—	—
Monel alloy R-405	66.5[b]	0.2	1.0	1.2	0.04	0.2	31.5	—	—	—	—
Monel alloy K-500	66.5[b]	0.1	0.8	1.0	0.005	0.2	29.5	—	2.7	0.6	—
Monel alloy 502	66.5[b]	0.05	0.8	1.0	0.005	0.2	28.0	—	3.0	0.2	—
Monel Copper–Nickel Alloy											
Monel alloy 401	42.5[b]	0.05	1.6	0.4	0.008	0.1	Bal.	—	—	—	—
Inconel Nickel–Chromium Alloys											
Inconel alloy 600	76.0[b]	0.08	0.5	8.0	0.008	0.2	0.2	15.5	—	—	—
Inconel alloy 601	60.5	0.05	0.5	14.1	0.007	0.2	0.5	23.0	1.4	—	—
Inconel alloy 625	61.0[b]	0.05	0.2	2.5	0.008	0.2	—	21.5	0.2	0.2	Mo 9.0 Cb + Ta 3.6
Inconel alloy 702	79.5[b]	0.05	0.5	1.0	0.005	0.4	0.2	15.5	3.2	0.6	—
Inconel alloy 706	41.5	0.03	0.2	40.0	0.008	0.2	0.2	16.0	0.2	1.8	Cb + Ta 2.9
Inconel alloy 718	52.5	0.04	0.2	18.5	0.008	0.2	0.2	19.0	0.5	0.9	Mo 3.0 Cb + Ta 5.1
Inconel alloy 721	71.0[b]	0.04	2.2	6.5	0.005	0.08	0.1	16.0	—	3.0	—
Inconel alloy 722	75.0[b]	0.04	0.5	7.0	0.005	0.4	0.2	15.5	0.7	2.4	—
Inconel alloy X-750	73.0[b]	0.04	0.5	7.0	0.005	0.2	0.2	15.5	0.7	2.5	Cb + Ta 1.0
Inconel alloy 751	72.5[b]	0.05	0.5	7.0	0.005	0.2	0.2	15.5	1.2	2.3	Cb + Ta 1.0
Incoloy Nickel–Iron–Chromium Alloys											
Incoloy alloy 800	32.5	0.05	0.8	46.0	0.008	0.5	0.4	21.0	0.4	0.4	—
Incoloy alloy 801	32.0	0.05	0.8	44.5	0.008	0.5	0.2	20.5	—	1.1	—
Incoloy alloy 802	32.5	0.4	0.8	46.0	0.008	0.4	0.4	21.0	—	—	—
Incoloy alloy 804	41.0	0.05	0.8	25.4	0.008	0.4	0.2	29.5	0.3	0.6	—
Incoloy alloy 825	42.0	0.03	0.5	30.0	0.02	0.2	2.2	21.5	0.1	0.9	Mo 3.0
Ni-Span-C alloy 902	42.2	0.03	0.4	48.5	0.02	0.5	0.05	5.3	0.6	2.6	—

[a] Not for specification purposes.
[b] Includes cobalt.

TABLE D.3 Reference Guide for Huntington Welding Products

Product[a]	Recommended for:	Nominal Chemical Composition (%)	MIL-E-22200/4[b]—Type	AWS A5.6—Class
Copper-Nickel Electrode (Metal-Arc Welding)				
Monel Welding Electrode 187	Wrought or cast 70/30, 80/20 and 90/10 copper-nickel alloys to themselves or each other.	32.0 Ni, 2.0 Mn, 65.0 Cu, 0.2 Si	MIL-CuNi (70/30)	ECuNi
Nickel and Nickel-Iron Electrodes for Welding Cast Irons (Metal-Arc Welding)				AWS A5.15-Class
Ni-Rod Welding Electrode	Welding various grades of cast iron, particularly thin sections or when maximum machinability is required.	95.0 Ni, 1.0 C, 3.0 Fe, 0.7 Si, 0.2 Mn	—	ENi-Cl
Ni-Rod 55 Welding Electrode	Welding gray iron, ductile iron, Ni-Resist corrosion-resisting iron, and welding cast irons to various wrought alloys. Particularly useful for heavy sections and high-phosphorus irons.	53.0 Ni, 1.2 C, 45.0 Fe, 0.7 Si, 0.3 Mn	—	ENiFe-Cl
Ni-Rod FC 55 Cored Wire	Welding gray iron, ductile iron, malleable iron, and Ni-Resist corrosion-resistant iron to themselves, to mild steel, or to high-nickel alloys using either automatic or semiautomatic welding processes.	50.0 Ni, 44.0 Fe, 1.0 C, 0.6 Si, 4.2 Mn	—	—

Product[a]	Recommended for:	Nominal Chemical Composition (%)	MIL-E-22200/3[b]—Type	AWS A5.11—Class
Nickel and Nickel Alloy Welding Electrodes (Metal-Arc Welding)				
Inconel Welding Electrode 132	Inconel nickel-chromium alloy 600 to itself or to stainless or carbon steel and for overlaying on steel. Also for joining dissimilar alloys.	73.0 Ni, 8.5 Fe, 15.0 Cr, 2.1 Cb, 1.8 Mn, 0.2 Si	—	ENiCrFe-1
Incoloy Welding Electrode 135	Incoloy (nickel-iron-chromium) alloy 825 and similar alloys to themselves.	36.0 Ni, 2.0 Mn, 26.0 Fe, 1.8 Cu, 29.0 Cr, 3.75 Mo, 0.4 Si	—	—
Nickel Welding Electrode 141	Nickel 200 and 201 to themselves or to each other; clad side of nickel 200 or 201 clad steel; overlaying on steel; and welding nickel 200 or 201 to steel.	96.0 Ni, 0.6 Si, 0.2 Al, 2.5 Ti, 0.3 Mn	MIL-4N11	ENi-1
Inconel Welding Electrode 182[c]	Inconel alloys 600 and 601 to themselves or to carbon or stainless steels	67.0 Ni, 7.8 Mn, 7.5 Fe, 14.0 Cr, 1.8 Cb, 0.5 Si, 0.4 Ti	MIL-8N12	ENiCrFe-3
Monel Welding Electrode 190[c]	Monel alloys 400 and 404 to themselves, to each other, to carbon steel or for overlaying on steel.	65.0 Ni, 3.1 Mn, 30.5 Cu, 0.8 Si, 0.6 Ti, 0.2 Al	MIL-9N10	ENiCu-7
Inco-Weld A Welding Electrode	Dissimilar alloys such as austenitic and ferritic steels to each other or to high-nickel alloys. Also, Inconel alloys 600 and 601 and Incoloy alloy 800 to themselves and for welding 9% nickel steel.	70.0 Ni, 2.0 Mn, 9.0 Fe, 15.0 Cr, 1.8 Cb, 1.5 Mo, 0.3 Si	—	ENiCrFe-2

TABLE D.3 Reference Guide for Huntington Welding Products (Continued)

Product[a]	Recommended for:	Nominal Chemical Composition (%)	Common Specifications	
Inco-Weld B Welding Electrode	Electrodes of the ENiCrFe-4 classification are ac electrodes with higher yield and tensile strengths than the ENiCrFe-2 classification and are used for welding 9% nickel steel. ENiCrFe-4 electrodes are similar to ENiCrFe-2 electrodes except they are designed for use where ac current is desirable to combat magnetic arc blow.	70.0 Ni, 2.0 Mn, 9.0 Fe, 15.0 Cr, 2.5 Cb, 2.0 Mo, 0.3 Si	—	ENiCrFe-4
Inco-Weld C Welding Electrode	Broad range of materials, including many difficult-to-weld compositions. Stainless steels, mild- and medium-carbon steels, spring steels, low-alloy steels, and dirty steels.	General-purpose electrode— composition considered proprietary	—	—
Inconel Welding Electrode 112[c]	Inconel alloys 601 and 625 and Incoloy alloys 800 and 801 to themselves, and for joining many dissimilar combinations of nickel-based or nickel-containing alloys.	61.0 Ni, 21.5 Cr, 9.0 Mo, 4.0 Fe, 3.6 Cb and Ta, 0.3 Mn, 0.4 Si	MIL-1N12	ENiCrMo-3
Inconel Welding Electrode 113	Welding of cryogenic alloy steels e.g. 5% nickel steels, 9% nickel steels, etc. This electrode is recommended for cryogenic applications to −320° F (−196° C).	62.5 Ni, 21.5 Cr, 9.0 Mo, 2.5 Fe, 2.8 Cb and Ta, 1.0 Mn, 0.4 Si	—	—
Inconel Welding Electrode 117	Inconel alloy 617 to itself and for many dissimilar wrought metals, such as Inconel alloys 600 and 601, Incoloy alloys 800H and 802, and cast alloys such as HK-40, HP, and HP-45 modified. Especially suited to high-temperature applications.	52.0 Ni, 23.5 Cr, 12.0 Co, 9.0 Mo, 0.2 Al, 0.1 C, 0.6 Mn, 1.7 Fe, 0.008 S, 0.5 Si, 0.2 Cu[d]	—	—
Nickel and Nickel Alloy Bare Welding Rods and Electrodes			MIL-E-21562[b]—Type	AWS A5.14—Class
Monel Filler Metal 60[e]	TIG or MIG welding Monel alloys 400 and 404 to themselves or each other. MIG overlay on steel after first layer of Nickel Filler Metal 61. Use with Incoflux 5 Submerged Arc Flux for joining alloy 400 to itself or for overlaying on steel without nickel barrier layer.	65.0 Ni, 2.0 Mn, 1.0 Si, 27.0 Cu, 2.2 Ti	MIL-EN60 MIL-RN60	ERNiCu-7
Nickel Filler Metal 61[e]	TIG or MIG welding nickel 200 or 201 to themselves or each other, or overlaying on steel. Use with Incoflux 6 Submerged Arc Flux for joining nickel or overlaying on steel.	96.0 Ni, 3.0 Ti, 0.4 Si, 0.3 Mn	MIL-EN61 MIL-RN61	ERNi-1

Filler Metal	Welding Use	Composition	Spec	AWS A5.7—Class
Inconel Filler Metal 62	TIG or MIG welding Inconel alloy 600.	74.0 Ni, 7.5 Fe, 16.0 Cr, 2.2 Cb, 0.1 Mn, 0.1 Si	MIL-EN62 MIL-RN62	ERNiCrFe-5
Incoloy Filler Metal 65	TIG or MIG welding Incoloy alloy 825	42.0 Ni, 30.0 Fe, 1.7 Cu, 21.0 Cr, 1.0 Ti, 3.0 Mo, 0.7 Mn, 0.3 Si	MIL-RN65	ERNiFeCr-1
Inconel Filler Metal 82[e]	TIG and MIG welding Inconel alloys 600 and 601 and Incoloy alloy 800 to themselves or to stainless or carbon steels; for overlaying on steel. Use with Incoflux 4 Submerged Arc Flux for joining alloys 600 and 800 and 9% nickel steel to themselves, or for overlaying on steel.	72.0 Ni, 3.0 Mn, 1.0 Fe, 20.0 Cr, 0.6 Ti, 2.5 Cb, 0.2 Si	MIL-EN82 MIL-RN82	ERNiCr-3
Inconel Filler Metal 92	TIG and MIG welding dissimilar alloys such as austenitic and ferritic steels to each other and to high-nickel alloys. Overlaying on steel and welding 9% nickel steel. Weld deposits will respond to age-hardening procedures.	71.0 Ni, 2.3 Mn, 6.6 Fe, 16.4 Cr, 3.2 Ti, 0.1 Si	MIL-EN6A MIL-RN6A	ERNiCrFe-6
Inconel Filler Metal 601	TIG welding of Inconel alloy 601	60.5 Ni, 14.1 Fe, 23.0 Cr, 1.4 Al, 0.5 Mn, 0.2 Si	—	—
Inconel Filler Metal 617	Inconel alloy 617 to itself and for many dissimilar wrought metals such as Inconel alloys 600 and 601, Incoloy alloys 800H and 802, and cast, high-temperature alloys such as HK-40, HP, and HP-45 Modified.	52.0 Ni, 22.0 Cr, 12.5 Co, 9.0 Mo, 1.2 Al, 1.5 Fe, 0.5 Mn, 0.5 Si, 0.3 Ti	—	—
Inconel Filler Metal 625[e]	TIG or MIG welding Inconel alloys 601 and 625 and for high-strength welds in 9% nickel steel; also for joining many dissimilar combinations of nickel-based or nickel-containing alloys.	61.0 Ni, 2.5 Fe, 21.5 Cr, 3.6 Cb, 9.0 Mo, 0.2 Mn, 0.2 Si, 0.2 Al, 0.2 Ti	MIL-EN625 MIL-RN625	ERNiCrMo-3
Inconel Filler Metal 718	TIG welding Inconel alloys 718, 706, and X-750. Weld will age harden.	52.5 Ni, 18.5 Fe, 18.6 Cr, 0.4 Al, 0.9 Ti, 5.0 Cb, 3.1 Mo, 0.2 Mn, 0.3 Si	—	—
Copper-Nickel Alloy Bare Welding Rods and Electrodes			MIL-E-21562—Type	AWS A5.7—Class
Monel Filler Metal 67[e]	TIG and MIG welding 70/30, 80/20 and 90/10 copper-nickel alloys. MIG overlay on steel after first layer of Nickel Filler Metal 61. Use with Incoflux 5 for joining copper-nickel alloys.	31.0 Ni, 0.75 Mn, 67.5 Cu, 0.3 Ti, 0.1 Si	MIL-EN67 MIL-RN67	ERCuNi

[a] Monel, Inconel, Incoloy, Incoflux, Inco-Weld, and Ni-Rod are registered trademarks.
[b] Latest revision.
[c] Meets stringent radiographic requirements in all positions.
[d] Incoflux 4, Incoflux 5, and Incoflux 6 Submerged Arc Fluxes are available for welding with these filler metals.
[e] For optimum weldability, electrodes of $\frac{3}{32}$ in. (2.4 mm) and $\frac{1}{8}$ in. (3.2 mm) diameter contain 1.4% manganese and 0.5% columbium plus tantalum.

Source: Courtesy of Huntington Alloys, Inc.

TABLE D.4 Specification Index for Huntington Alloys

Specification	Huntington Alloy No.	Specification	Huntington Alloy No.	Specification	Huntington Alloy No.	Specification	Huntington Alloy No.	Specification	Huntington Alloy No.
ASTM A 461	X-750	ASTM B 444	625	ASME SB-407	800	SAE AMS 5596	718	MIL-N-7786	X-750
B 127	400	B 446	625	SB-408	800	5597	718	MIL-N-8550	X-750
B 160	200	F 9	205	SB-409	800	5598	X-750	MIL-N-17506	K-500
	201	F 239	220	SB-423	825	5599	625	MIL-N-22986	600
B-161	200		230	SB-424	825	5662	718	MIL-N-22987	600
	201		270	SB-425	825	5663	718	MIL-N-23228	600
B 162	200	F 290	211	SB-443	625	5664	718	MIL-N-23229	600
	201		300	SB-444	625	5665	600	MIL-N-24106	400
B 163	200	ASME SB 127	400	SB-446	625	5666	625	MIL-N-24114	X-750
	201	SB 160	200	SAE AMS 4544	400	5667	X-750	MIL-N-46025	205
	400		201	4574	400	5668	X-750	MIL-S-21977	X-750
	600	SB-161	200	4674	R-405	5670	X-750	MIL-S-23192	X-750
	800		201	4675	400	5671	X-750	MIL-T-842	400
	825	SB-162	200	4676	K-500	5687	600	MIL-T-1368	400
B 164	400		201	5221	902	5698	X-750	MIL-T-7840	600
	R-405	SB-163	200	5223	902	5699	X-750	MIL-T-22945	600
B 165	400		201	5225	902	5742	801	MIL-T-23227	600
B 166	600		400	5540	600	5832	718	MIL-T-23520	400
B 167	600		600	5541	722	5837	625	MIL-W-4471	K-500
B-168	600		800	5542	X-750	7232	600	NAVSHIPS 250-1500-1	600
B 407	800		825	5550	702	7233	400	QQ-N-281	400
B 408	800	SB-164	400	5552	801	7234	R-405	QQ-N-286	R-405
B 409	800		R-405	5553	201	MIL-F-23999	K-500	QQ-W-390	K-500
B 423	825	SB-165	400	5580	600	MIL-N-894	400		502
B 424	825	SB-166	600	5582	X-750		R-405		600
B 425	825	SB-167	600	5589	718	MIL-N-6710	600		
B 443	625	SB-168	600	5590	718	MIL-N-6840	600		

Source: Courtesy of Huntington Alloys, Inc.

Alloy[b]	Ni	Co	Cr	Mo	W	Fe	Si	Mn	C	Others
HASTELLOY alloy B-2	Balance	1.00[c]	1.00[c]	26.00–30.00	—	2.00[c]	0.10[c]	1.00[c]	0.02[c]	P–0.40[c] S–0.030[c]
HASTELLOY alloy C-276	Balance	2.50[c]	14.50–16.50	15.00–17.00	3.00–4.50	4.00–7.00	0.08[c]	1.00[c]	0.02[c]	V–0.35[c] P–0.040[c] S–0.030[c]
HASTELLOY alloy C-4	Balance	2.00[c]	14.00–18.00	14.00–17.00	—	3.00[c]	0.08[c]	1.00[c]	0.015[c]	Ti–0.70[c] P–0.04[c] S–0.03[c]
HASTELLOY alloy G	Balance	2.50[c]	21.00–23.50	5.50–7.50	1.00[c]	18.00–21.00	1.00[c]	1.00–2.00	0.05[c]	Cu–1.50–2.50 Cb + Ta–1.75–2.50 P–0.04[c] S–0.03[c]
HASTELLOY alloy X	Balance	0.50–2.50	20.50–23.00	8.00–10.00	0.20–1.00	17.00–20.00	1.00[c]	1.00[c]	0.05–0.15	—
Haynes alloy No. 20 Mod	25.00–27.00	—	21.00–23.00	4.00–6.00	—	Balance	1.00[c]	2.50[c]	0.05[c]	Ti–4 × C min.
Haynes alloy No. 25	9.00–11.00	Balance	19.00–21.00	—	14.00–16.00	3.00[c]	1.00[c]	1.00–2.00	0.05–0.15	P–0.030[c] S–0.030[c]
Haynes alloy No. 625	Balance	1.00[c]	20.00–23.00	8.00–10.00	—	5.00[c]	0.50[c]	0.50[c]	0.10[c]	Cb + Ta–3.15–4.15 Al–0.40[c] Ti–0.40[c] P–0.015[c] S–0.015[c]
Multimet alloy	19.00–21.00	18.50–21.00	20.00–22.50	2.50–3.50	2.00–3.00	Balance	1.00[c]	1.00–2.00	0.08–0.16	Cb + Ta–0.75–1.25 N–0.10–0.20

[a]The undiluted deposited chemical compositions of some of these alloys may vary in carbon, silicon, and manganese content beyond the limits shown.
[b]HASTELLOY, Haynes, and Multimet are registered trademarks.
[c]Maximum.

Source: Courtesy of High Technology Materials Division, Cabot Corporation.

TABLE D.6 Wrought Stainless and Heat-Resisting Steels

AISI Type	Chemical Composition (%)[a]						Density (lb/cu. in.)	Thermal Expansion (in./in./°F × 10⁻⁶), 32-600°F	Thermal Conductivity (Btu/sq. ft./hr./in./°F), 212°F	Electrical Resistivity (ohm/circ mil/ft), 70°F	Modulus of Elasticity (10⁴ psi), Tension	Modulus of Elasticity (10⁴ psi), Torsion
	C	Mn	Si	Cr	Ni	Others[b]						
201	0.15	5.50- 7.50	1.00	16.00-18.00	3.50- 5.50	N-0.25 P-0.060	—	9.7	—	—	28.6	—
202	0.15	7.50-10.00	1.00	17.00-19.00	4.00- 6.00	N-0.25 P-0.060	—	10.2	—	—	—	—
301	0.15	2.00	1.00	16.00-18.00	6.00- 8.00	—	0.29	9.5	113	432	28	12.5
302	0.15	2.00	1.00	17.00-19.00	8.00-10.00	—	0.29	9.9	113	432	28	12.5
302B	0.15	2.00	2.00-3.00	17.00-19.00	8.00-10.00	—	0.29	10.0	110	432	28	—
303	0.15	2.00	1.00	17.00-19.00	8.00-10.00	S-0.15 min. P-0.20	0.29	9.9	113	432	28	—
303Se	0.15	2.00	1.00	17.00-19.00	8.00-10.00	Se-0.15 min. S-0.060	0.29	9.9	113	432	28	—
304	0.08	2.00	1.00	18.00-20.00	8.00-12.00	Mo-0.60 opt. P-0.20	0.29	9.9	113	432	28	12.5
304L	0.03	2.00	1.00	18.00-20.00	8.00-12.00	—	—	—	—	—	—	—
305	0.12	2.00	1.00	17.00-19.00	10.00-13.00	—	0.29	9.9	113	432	28	12.5
308	0.08	2.00	1.00	19.00-21.00	10.00-12.00	—	0.29	9.9	106	432	28	—
309	0.20	2.00	1.00	22.00-24.00	12.00-15.00	—	0.29	9.3	108	468	29	—
309S	0.08	2.00	1.00	22.00-24.00	12.00-15.00	—	0.29	9.3	108	468	29	—
310	0.25	2.00	1.50	24.00-26.00	19.00-22.00	—	0.29	9.0	96	468	29	—
310S	0.08	2.00	1.50	24.00-26.00	19.00-22.00	—	0.29	9.0	96	468	29	—
314	0.25	2.00	1.50-3.00	23.00-26.00	19.00-22.00	—	0.28	8.4	121	462	29	—
316	0.08	2.00	1.00	16.00-18.00	10.00-14.00	Mo-2.00-3.00	0.29	9.0	113	444	28	—
316L	0.03	2.00	1.00	16.00-18.30	10.00-14.00	Mo-2.00-3.00	—	—	—	—	—	—
317	0.08	2.00	1.00	18.00-20.00	11.00-15.00	Mo-3.00-4.00	0.29	9.0	113	444	28	—
321	0.08	2.00	1.00	17.00-19.00	9.00-12.00	Ti-5 × C min.	0.29	9.5	112	432	28	—
347	0.08	2.00	1.00	17.00-19.00	9.00-13.00	Cb + Ta – 10 × C min.	0.29	9.5	112	438	28	—
348	0.08	2.00	1.00	17.00-19.00	9.00-13.00	Cb + Ta – 10 × C min[f]	0.29	9.5	112	438	28	—
384	0.08	2.00	1.00	15.00-17.00	17.00-19.00	—	0.29	10.0	113	476	28	—
385	0.08	2.00	1.00	11.50-13.50	14.00-16.00	—	0.29	10.4[h]	115	447	28	—

403	0.15	1.00	0.50	11.50–13.00	—	P–0.040	0.28	6.3	173	342	29	—
405	0.08	1.00	1.00	11.50–14.50	—	Al–0.10–0.30 P–0.040	0.28	6.4	187	360	29	—
410	0.15	1.00	1.00	11.50–13.50	—	P–0.040	0.28	6.3	173	342	29	—
414	0.15	1.00	1.00	11.50–13.50	1.25– 2.50	P–0.040	0.28	6.1	173	420	29	—
416	0.15	1.25	1.00	12.00–14.00	—	S–0.15 min. P–0.060	0.28	6.1	173	342	29	—
416Se	0.15	1.25	1.00	12.00–14.00	—	Se–0.15 min. S–0.060 Mo–0.60 opt. P–0.060	0.28	6.1	173	342	29	—
420	0.15	1.00	1.00	12.00–14.00	—	P–0.040	0.28	6.0	173	330	29	—
420F	0.15 min.	1.25	1.00	12.00–14.00	—	P–0.060 Mo–0.60 S–0.15 min.	0.28	5.7[h]	174	331	29	11.7
429	0.12	1.00	1.00	14.00–16.00	—	P–0.040	0.28	5.7[h]	178	355	29	—
430	0.12	1.00	1.00	14.00–18.00	—	P–0.040	0.28	6.1	181	360	29	—
430F	0.12	1.25	1.00	14.00–18.00	—	S–0.15 min. P–0.060	0.28	6.1	181	360	29	—
430FSe	0.12	1.25	1.00	14.00–18.00	—	Se–0.15 min. S–0.060 Mo–0.60 opt. P–0.060	0.28	6.1	181	360	29	—
431	0.20	1.00	1.00	15.00–17.00	1.25– 2.50	P–0.040	0.28	6.7	140	432	29	—
434	0.12	1.00	1.00	16.00–18.00	—	P–0.040 Mo–0.75–1.25	0.28	—	182	361	29	—
436	0.12	1.00	1.00	16.00–18.00	—	P–0.040 Mo–0.75–1.25 Cb + Ta – 5 × C min., 0.70 max.	0.28	5.2[h]	166	361	29	—
440A	0.60–0.75	1.00	1.00	16.00–18.00	—	Mo–0.75 P–0.040	0.28	5.6[h]	168	360	29	—
440B	0.75–0.95	1.00	1.00	16.00–18.00	—	Mo–0.75 P–0.040	0.28	5.6[h]	168	360	29	—
440C	0.95–1.20	1.00	1.00	16.00–18.00	—	Mo–0.75 P–0.040	0.28	5.6[h]	168	360	29	—
442	0.20	1.00	1.00	18.00–23.00	—	P–0.040	0.28	5.6[i]	150	384	29	—
446	0.20	1.50	1.00	23.00–27.00	—	N–0.25 P–0.040	0.27	6.0	145	402	29	—
501	0.10 min.	1.00	1.00	4.00– 6.00	—	Mo–0.40–0.65 P–0.040	0.28	6.8	254	240	29	—
502	0.10	1.00	1.00	4.00– 6.00	—	Mo–0.40–0.65 P–0.040	0.28	6.8	254	240	29	—

[a] Maximum unless shown otherwise.
[b] S–0.030 unless otherwise shown. P–0.045 unless otherwise shown.
[c] Sheet or strip.
[d] Minimum.
[e] Bar.
[f] Ta–0.10 max., Co–0.20 max.
[g] H—hardening temperature, °F.
   T—tempering temperature, °F.
[h] 32°–212°F.
[i] 68°–212°F.

Source: *Stainless and Heat-Resisting Steels Manual*, April 1963, courtesy of the Committee of Stainless Steel Producers, American Iron and Steel Institute.

TABLE D.6 (continued)

AISI Type	Condition	Tensile Strength (1000 psi)	Yield Strength (0.2% Offset) (1000 psi)	Elongation (%)	Reduction of Area (%)	Hardness, Rockwell	Condition	Tensile Strength (1000 psi)	Yield Strength (0.2% Offset) (1000 psi)	Elongation (%)	Reduction of Area (%)	Hardness, Rockwell
201	Annealed[c]	115	55	55	—	90B	Half-hard[c]	150[d]	110[d]	15[d]	—	32C
202	Annealed[c]	105	55	55	—	90B	—	—	—	—	—	—
301	Annealed[c]	110	40	60	—	85B	Half-hard[c]	150[d]	110[d]	18[d]	—	32C
302	Annealed[c]	90	40	50	—	85B	Quarter-hard[c]	125[d]	75[d]	12[d]	—	25C
302B	Annealed[c]	95	40	55	—	85B	—	—	—	—	—	—
303	Annealed[e]	90	35	50	55	84B	—	—	—	—	—	—
303Se	Annealed[e]	90	35	50	55	84B	—	—	—	—	—	—
304	Annealed[e]	85	35	60	70	80B	—	—	—	—	—	—
304L	Annealed[c]	75	28	55	—	79B	—	—	—	—	—	—
305	Annealed[c]	85	38	50	—	80B	—	—	—	—	—	—
308	Annealed[c]	85	35	50	—	80B	—	—	—	—	—	—
309	Annealed[c]	90	45	45	—	85B	—	—	—	—	—	—
309S	Annealed[c]	90	45	45	—	85B	—	—	—	—	—	—
310	Annealed[c]	95	45	45	—	85B	—	—	—	—	—	—
310S	Annealed[c]	95	45	45	—	85B	—	—	—	—	—	—
314	Annealed[c]	100	50	40	—	85B	—	—	—	—	—	—
316	Annealed[c]	84	42	50	—	79B	—	—	—	—	—	—
316L	Annealed[c]	81	42	50	—	79B	—	—	—	—	—	—
317	Annealed[c]	90	40	45	—	85B	—	—	—	—	—	—

321	Annealed[c]	90	35	45	—	80B	—	—	—	—	—	—
347	Annealed[c]	95	40	45	—	85B	—	—	—	—	—	—
348	Annealed[c]	95	40	45	—	85B	—	—	—	—	—	—
384	Annealed[e]	75	35	55	72	70B	—	—	—	—	—	—
385	Annealed[e]	72	30	55	78	66B	—	—	—	—	—	41C
403	Annealed[e]	75	40	35	70	82B	—	—	—	—	—	—
405	Annealed[e]	70	40	30	60	80B	—	—	—	—	—	—
410	Annealed[e]	75	40	35	70	82B	H-1800 T-400[g]	190	145	15	55	41C
414	Annealed[e]	115	90	20	60	99B	H-1800 T-400[g]	190	145	15	55	41C
416	Annealed[e]	75	40	30	60	82B	H-1800 T-400[g]	200	150	15	55	43C
416Se	Annealed[e]	75	40	30	60	82B	H-1800 T-400[g]	190	145	12	45	41C
420	Annealed[e]	95	50	25	55	92B	H-1800 T-400[g]	190	145	12	45	41C
420F	Annealed[e]	95	55	22	50	97B	H-1900 T-600[g]	230	195	8	25	50C
429	Annealed[e]	71	45	30	65	82B	Cold-drawn	110	100	14	40	98B
430	Annealed[e]	75	45	30	65	82B	—	—	—	—	—	—
430F	Annealed[e]	80	55	25	60	86B	—	—	—	—	—	—
430FSe	Annealed[e]	80	55	25	55	86B	H-1900 T-400[g]	205	155	15	55	43C
431	Annealed[e]	125	95	20	55	24C	—	—	—	—	—	—
434	Annealed[c]	77	53	23	—	83B	—	—	—	—	—	—
436	Annealed[c]	77	53	23	—	83B	—	—	—	—	—	—
440A	Annealed[e]	105	60	20	45	95B	H-1900 T-600[g]	260	240	5	20	51C
440B	Annealed[e]	107	62	18	35	96B	H-1900 T-600[g]	280	270	3	15	55C
440C	Annealed[e]	110	65	14	25	97B	H-1900 T-600[g]	285	275	2	10	57C
442	Annealed[e]	80	45	20	40	90B	—	—	—	—	—	—
446	Annealed[e]	80	50	25	45	86B	—	—	—	—	—	—
501	Annealed[e]	70	30	28	65	84B	H-1650 T-1000[g]	175	135	15	50	41C
502	Annealed[e]	65	25	30	75	80B	—	—	—	—	—	—

TABLE D.7 Wrought Coppers and Copper Alloys[a]

CDA Number	Previous Commonly Accepted Trade Name	Nominal Chemical Composition (%)					Melting Range (°F)
		Cu	Zn	Sn	Pb	Other Named Elements	
C10200	Oxygen Free	99.95	—	—	—	99.95 min. Cu + Ag	1981[d]
C11000	Electrolytic Tough Pitch	99.90	—	—	—	99.90 min. Cu + Ag[e] 0,0.04	1949–1981
C11300	Silver-Bearing Tough Pitch	99.90	—	—	—	99.90 min. Cu + Ag[e] 0,0.04	1980[d]
C12200	Phosphorus Deoxidized (HRP)	99.90	—	—	—	99.90 min. Cu + Ag P 0.02	1981[d]
C21000	Gilding, 95%	95	5	—	—	—	1920–1950
C22000	Commercial Bronze, 90%	90	10	—	—	—	1870–1910
C22600	Jewelry Bronze, 87.5%	87.5	12.5	—	—	—	1840–1895
C23000	Red Brass, 85%	85	15	—	—	—	1810–1880
C24000	Low Brass, 80%	80	20	—	—	—	1770–1830
C26000	Cartridge Brass, 70%	70	30	—	—	—	1680–1750
C26800	Yellow Brass	65	35	—	—	(268—Sheet) (270—Rod and Wire)	1660–1710
C28000	Muntz Metal	60	40	—	—	—	1650–1660
C31400	Leaded Commercial Bronze	89	9.25	—	1.75	—	1850–1900
C33000	Low-Leaded Brass Tube	66	33.5	—	0.5	—	1660–1720
C33200	High-Leaded Brass Tube	66	32.4	—	1.6	—	1650–1710
C33500	Low-Leaded Brass	65	34.5	—	0.5	—	1650–1700
C34000	Medium-Leaded Brass	65	34	—	1.0	—	1630–1700
C34200	High-Leaded Brass	65	33	—	2.0	353—Suitable for hot-working	1630–1670
C35600	Extra High-Leaded Brass	63	34.5	—	2.5	—	1630–1660
C36000	Free-Cutting Brass	61.5	35.5	—	3.0	—	1630–1650
C36500	Leaded Muntz Metal	60	39.4	—	0.6	(366—.06As)(367—.06Sb)(368—.06P)	1630–1650

Code	Name	Cu	Zn	Sn	Pb	Other	Melting Range
C37000	Free-Cutting Muntz Metal	60	39	—	1.0	—	1630-1650
C37700	Forging Brass	59	39	—	2.0	—	1620-1640
C38500	Architectural Bronze	57	40	—	3.0	—	1610-1630
C44300	Inhibited Admiralty	71	28	1	—	(443—.06As)(444—.06Sb)(445—.06P)	1650-1720
C46400	Naval Brass	60	39.25	0.75	—	(465—.06As)(446—.06Sb)(467—.06P)	1630-1650
C48500	Leaded Naval Brass	60	37.5	0.75	1.75	—	1630-1650
C50200	Phosphor Bronze, 1.25% E	98.75	—	1.25	—	P-trace	1900-1970
C51000	Phosphor Bronze, 5% A	95	—	5	—	0.20 P	1750-1920
C52100	Phosphor Bronze, 8% C	92	—	8	—	0.20 P	1620-1880
C52400	Phosphor Bronze, 10% D	90	—	10	—	0.20 P	1550-1830
C54400	Free-Cutting Phosphor Bronze	88	4	4	4	0.25 P	1700-1830
C61400	Aluminum Bronze, D	91	—	—	—	7 Al, 2 Fe	1905-1915
C65100	Low-Silicon Bronze, B	98.5	—	—	—	1.5 Si	1890-1940
C65500	High-Silicon Bronze, A	97	—	—	—	3 Si	1780-1880
C67500	Manganese Bronze, A	58.5	39	1	—	0.1 Mn, 1.4 Fe	1590-1630
C68700	Aluminum Brass	77.5	20.5	—	—	2 Al	1710-1780
C70600	Copper Nickel, 10%	88.7	—	—	—	10 Ni, 1.3 Fe	2010-2100
C71500	Copper Nickel, 30%	70	—	—	—	30 Ni	2140-2260
C74500	Nickel Silver, 65-10	65	25	—	—	10 Ni	1870[d]
C75200	Nickel Silver, 65-18	65	17	—	—	18 Ni	1960-2030
C75400	Nickel Silver, 65-15	65	20	—	—	15 Ni	1970[d]
C75700	Nickel Silver, 65-12	65	23	—	—	12 Ni	1900[d]
C77000	Nickel Silver, 55-18	55	27	—	—	18 Ni	1930[d]

[a] For detailed analyses of subcategories within a series, write to Copper Development Association, Inc., 405 Lexington Ave., New York, NY 10174.

[b] 0.5% extension under load.

[c] In 2 inches.

[d] Melting point—liquidus.

[e] 113, 8 oz/ton Ag; 114, 10 oz/ton Ag; 116, 25 oz/ton Ag.

TABLE D.7 (continued)

CDA Number	Density (lb/cu in.), 68°F	Thermal Expansion (in./in./°F × 10⁻⁶), 68–572°F	Thermal Conductivity (Btu/sq ft/hr/in./°F), 68°F	Electrical Resistivity (ohm circ mil ft), 68°F	Modulus of Elasticity (10⁶ psi), Tension	Modulus of Elasticity (10⁶ psi), Torsion	Form and Size	Annealed Temper Grain Size (mm)	Tensile Strength (psi)	Yield Strength[b] (psi)	Elong[c] (%)	Hardness Rockwell	Shear Strength (psi)
C10200	0.323	9.8	2712	10.3	17	6.4	Flat—0.040 in.	0.050	32,000	10,000	45	40F	22,000
C11000	0.322	9.8	2712	10.3	17	6.4	Flat—0.040 in.	0.050	32,000	10,000	45	40F	22,000
C11300	0.322	9.8	2688	10.4	17	6.4	Flat—0.040 in.	0.025	34,000	11,000	45	45F	23,000
C11200	0.323	9.8	2352	12.2	17	6.4	Flat—0.040 in.	0.050	32,000	10,000	45	40F	22,000
C21000	0.320	10.0	1620	18.5	17	6.4	Flat—0.040 in.	0.050	34,000	10,000	45	46F	—
C22000	0.318	10.2	1308	23.6	17	6.4	Flat—0.040 in.	0.050	37,000	10,000	45	53F	28,000
C22600	0.317	10.3	1200	25.9	17	6.4	Flat—0.040 in.	0.050	39,000	11,000	46	55F	29,000
C23000	0.316	10.4	1104	28.0	17	6.4	Flat—0.040 in.	0.050	40,000	12,000	47	59F	31,000
C24000	0.313	10.6	972	32.4	16	6.0	Flat—0.040 in.	0.050	44,000	14,000	50	61F	32,000
C26000	0.308	11.1	840	37.0	16	6.0	Flat—0.040 in.	0.050	47,000	15,000	62	64F	—
C26800	0.306	11.3	804	38.4	15	5.6	Flat—0.040 in.	Soft anneal	47,000	15,000	62	64F	—
C28000	0.303	11.6	852	37.0	15	5.6	Bar—0.250 in.	Half hard	54,000	21,000	45	80F	40,000
C31400	0.319	10.2	1248	24.7	17	6.4	Tube—1×.065 in.	0.050	55,000	50,000	12	61B	31,000
C33000	0.307	11.2	804	39.9	15	5.6	Tube—1×.065 in.	0.050	47,000	15,000	60	64F	—
C33200	0.308	11.3	804	39.9	15	5.6	Flat—0.040 in.	0.025	52,000	20,000	50	75F	—
C33500	0.306	11.3	804	39.9	15	5.6	Flat—0.040 in.	0.050	47,000	15,000	62	64F	—
C34000	0.306	11.3	804	39.9	15	5.6	Flat—0.040 in.	0.050	47,000	15,000	60	64F	—

Alloy							Form	Temper					
C34200	0.306	11.3	804	39.9	15	5.6	Flat—0.040 in.	0.035	49,000	17,000	52	68F	34,000
C35600	0.307	11.4	804	39.9	14	5.3	Flat—0.040 in.	0.035	49,000	17,000	50	68F	—
C36000	0.307	11.4	804	39.9	14	5.3	Plate—0.250 in.	Quarter hard	56,000	45,000	20	62B	33,000
C36500	0.304	11.6	852	37.0	15	5.6	Plate—1.0 in.	Hot rolled	54,000	20,000	45	80F	40,000
C37000	0.304	11.6	828	38.4	15	5.6	Tube—1.5 × .125 in	Light anneal	54,000	20,000	40	80F	—
C37700	0.305	11.5	828	38.4	15	5.6	Rod—1.0 in.	As-extruded	52,000	20,000	45	78F	—
C38500	0.306	11.6	852	37.0	14	5.3	Rod-1.0 in.	As-extruded	60,000	20,000	30	65B	35,000
C44300	0.308	11.2	768	41.5	16	6.0	Tube—1 × .065 in.	0.025	53,000	22,000	65	75F	—
C46400	0.304	11.8	804	39.9	15	5.6	Flat—0.040 in.	Light anneal	62,000	30,000	40	60B	41,000
C48500	0.305	11.8	804	39.9	15	5.6	Rod—1.0 in.	Soft anneal	57,000	25,000	40	55B	36,000
C50200	0.321	9.9	1440	21.6	17	6.4	Flat—0.040 in.	0.025	40,000	14,000	48	60F	—
C51000	0.320	9.9	480	69.1	16	6.0	Flat—0.040 in.	0.050	47,000	19,000	64	73F	—
C52100	0.318	10.1	432	79.8	16	6.0	Flat—0.040 in.	0.050	55,000	—	70	75F	—
C52400	0.317	10.2	348	94.3	15	6.0	Flat—0.040 in.	0.035	66,000	28,000	68	55B	—
C54400	0.321	9.6	600	54.6	15	5.6	Flat—0.125 in.	0.035	44,000	19,000	50	65F	—
C61400	0.285	9.0	468	74.1	17	6.4	Rod—1.0 in.	Soft anneal	82,000	45,000	40	84B	45,000
C65100	0.316	9.9	396	86.4	17	6.4	Flat—0.040 in.	0.035	40,000	15,000	50	55F	—
C65500	0.308	10.0	252	148	15	5.6	Flat—0.040 in.	0.070	56,000	21,000	63	76F	42,000
C67500	0.302	11.8	732	43.2	15	5.6	Rod—1.0 in.	Soft anneal	65,000	30,000	33	65B	42,000
C68700	0.301	10.3	696	45.1	16	6.0	Tube—1.0 × .065 in	0.025	60,000	27,000	55	77F	—
C70600	0.323	9.5	312	115	18	6.8	Tube—1.0 × .065 in	0.025	44,000	16,000	42	65F	—
C71500	0.323	9.0	204	225	22	8.3	Flat—1.0 in.	As hot-rolled	55,000	20,000	45	35B	—
C74500	0.314	9.1	312	115	17.5	6.6	Flat—0.040 in.	0.050	51,000	19,000	46	28B	—
C75200	0.316	9.0	228	173	18	6.8	Flat—0.040 in.	0.035	58,000	25,000	40	40B	—
C75400	0.314	9.0	252	148	18	6.8	Flat—0.040 in.	0.050	55,000	19,000	42	31B	—
C75700	0.314	9.0	276	180	18	6.8	Flat—0.040 in.	0.050	54,000	19,000	45	30B	—
C77000	0.314	9.3	204	189	18	6.8	Flat—0.040 in.	0.035	60,000	27,000	40	55B	—

Source: *Copper Development Association Standards Handbook for Copper and Copper Alloy Wrought Mill Products*, 5th ed., August 1964, courtesy of Copper Development Association, Inc.

TABLE D.8 High-Alloy Castings

ACI Designation	Chemical Composition (%)[a]							
	C	Mn	Si	P	S	Cr	Ni	Other
CA-15	0.15	1.00	1.50	0.04	0.04	11.5–14.0	1.0	Mo 0.5 max.[b]
CA-40	0.20–0.40	1.00	1.50	0.04	0.04	11.5–14.0	1.0	Mo 0.5 max.[b]
CB-30	0.30	1.00	1.50	0.04	0.04	18.0–22.0	2.0	—
CB-7Cu	0.07	1.00	1.00	0.04	0.04	15.5–17.0	3.6–4.6	Cu 2.3–3.3
CC-50	0.50	1.00	1.50	0.04	0.04	26.0–30.0	4.0	—
CD-4MCu	0.040	1.00	1.00	0.04	0.04	25.0–27.0	4.75–6.00	Mo 1.75–2.25, Cu 2.75–3.25
CE-30	0.30	1.50	2.00	0.04	0.04	26.0–30.0	8.0–11.0	—
CF-3	0.03	1.50	2.00	0.04	0.04	17.0–21.0	8.0–12.0	—
CF-8	0.08	1.50	2.00	0.04	0.04	18.0–21.0	8.0–11.0	—
CF-20	0.20	1.50	2.00	0.04	0.04	18.0–21.0	8.0–11.0	—
CF-3M	0.03	1.50	1.50	0.04	0.04	17.0–21.0	9.0–13.0	Mo. 2.0–3.0
CF-8M	0.08	1.50	2.00	0.04	0.04	18.0–21.0	9.0–12.0	Mo. 2.0–3.0
CF-8C	0.08	1.50	2.00	0.04	0.04	18.0–21.0	9.0–12.0	Cb 8 × C min., 1.0 max.

Alloy	C	Mn	Si	P	S	Cr	Ni	Other
CF-16F	0.16	1.50	2.00	0.17	0.04	18.0-21.0	9.0-12.0	Mo 1.5 max., Se 0.20-0.35
CG-8M	0.08	1.50	1.50	0.04	0.04	18.0-21.0	9.0-13.0	Mo 3.0-4.0
CH-20	0.20	1.50	2.00	0.04	0.04	22.0-26.0	12.0-15.0	—
CK-20	0.20	1.50	2.00	0.04	0.04	23.0-27.0	19.0-22.0	—
CN-7M	0.07	1.50	c	0.04	0.04	18.0-22.0	21.0-31.0	Mo, Cu^c
CY-40	0.40	1.50	3.00	0.015	0.015	14.0-17.0	Bal	Fe 11.0 max.
CZ-100	1.00	1.50	2.00	0.015	0.015	—	95 min.	Fe 1.50 max.
M-35	0.35	1.50	2.00	0.015	0.015	—	Bal	Cu 26.0-33.0, Fe 3.50 max.
HA	0.20	0.35-0.65	1.00	0.04	0.04	8.0-10.0	—	Mo 0.90-1.20
HC	0.50	1.00	2.00	0.04	0.04	26.0-30.0	4.0	Mo 0.5 max.^{b}
HD	0.50	1.50	2.00	0.04	0.04	26.0-30.0	4.0-7.0	Mo 0.5 max.^{b}
HE	0.20-0.50	2.00	2.00	0.04	0.04	26.0-30.0	8.0-11.0	Mo 0.5 max.^{b}
HF	0.20-0.40	2.00	2.00	0.04	0.04	19.0-23.0	9.0-12.0	Mo 0.5 max.^{b}
HH	0.20-0.50	2.00	2.00	0.04	0.04	24.0-28.0	11.0-14.0	Mo 0.5 max.^{b}, N 0.2 max.
HI	0.20-0.50	2.00	2.00	0.04	0.04	26.0-30.0	14.0-18.0	Mo 0.5 max.^{b}
HK	0.20-0.60	2.00	2.00	0.04	0.04	24.0-28.0	18.0-22.0	Mo 0.5 max.^{b}
HL	0.20-0.60	2.00	2.00	0.04	0.04	28.0-32.0	18.0-22.0	Mo 0.5 max.^{b}
HN	0.20-0.50	2.00	2.00	0.04	0.04	19.0-23.0	23.0-27.0	Mo 0.5 max.^{b}
HT	0.35-0.75	2.00	2.50	0.04	0.04	13.0-17.0	33.0-37.0	Mo 0.5 max.^{b}
HU	0.35-0.75	2.00	2.50	0.04	0.04	17.0-21.0	37.0-41.0	Mo 0.5 max.^{b}
HW	0.35-0.75	2.00	2.50	0.04	0.04	10.0-14.0	58.0-62.0	Mo 0.5 max.^{b}
HX	0.35-0.75	2.00	2.50	0.04	0.04	15.0-19.0	64.0-68.0	Mo. 0.5 max.^{b}

^{a}Maximum unless shown as range, balance Fe.

^{b}Molybdenum not intentionally added.

^{c}There are several proprietary alloy compositions falling within the stated chromium and nickel ranges, and containing varying amounts of silicon, molybdenum, and copper.

TABLE D.8 (continued)

ACI Designation	Density (lb/cu in.)	Thermal Expansion (in/in./°F × 10⁻⁴) 70–1000°F	Thermal Conductivity (Btu/sq ft/hr/in./°F), 212°F	Electrical Resistivity (ohm/circ mil/ft), 70°F	Modulus of Elasticity, (10⁶psi) Tension	Tensile Strength (psi) As-Cast	Tensile Strength (psi) Aged	Yield Strength (psi) As-Cast	Yield Strength (psi) Aged	Elongation (%) As-Cast	Elongation (%) Aged	Hardness, BHN As-Cast	Hardness, BHN Aged	Thermal Treatment
CA-15	0.275	6.4	174	469	29	—	200,000	—	150,000	—	7	—	390	1800°F, A.C. + 600°F
CA-40	0.275	6.4	174	457	29	—	220,000	—	165,000	—	1	—	470	1800°F, A.C. + 600°F
CB-30	0.272	6.5	154	457	29	—	95,000	—	60,000	—	15	—	195	1450°F, F.C. 1000°F, A.C.
CB-7Cu	—	—	—	—	—	—	—	—	—	—	—	—	—	—
CC-50	0.272	6.4	151	463	29	97,000	—	65,000	—	18	—	210	—	1900°F, A.C.
CD-4MCu	—	—	—	—	—	—	—	—	—	—	—	—	—	—
CE-30	0.277	9.6	—	511	25	95,000	—	60,000	—	15	—	170	—	—
CF-3	—	—	—	—	—	—	—	—	—	—	—	—	—	—
CF-8	0.280	—	110	457	28	77,000	—	37,000	—	55	—	140	—	1950°–2050°F, W.Q.
CF-20	0.280	10.4	110	469	28	77,000	—	36,000	—	50	—	163	—	2000 + °F, W.Q.
CF-3M	—	—	—	—	—	—	—	—	—	—	—	—	—	—
CF-8M	0.280	9.7	113	493	28	80,000	—	42,000	—	50	—	163	—	1950°–2100°F, W.Q.
CF-8C	0.280	10.3	112	427	28	77,000	—	38,000	—	39	—	149	—	1950°–2050°F, W.Q.
CF-16F	0.280	9.9	113	433	28	77,000	—	40,000	—	52	—	150	—	2000 + °F, W.Q.
CG-8M	—	—	—	—	—	—	—	—	—	—	—	—	—	—

Alloy													Heat treatment
CH-20	0.279	9.6	98	505	28	88,000	—	50,000	38	—	190	—	2000 + °F, W.Q.
CK-20	0.280	9.2	98	541	29	76,000	—	38,000	37	—	144	—	2100°F, W.Q.
CN-7M	0.289	9.7	145	541	24	69,000	—	31,500	48	—	130	—	1950°–2050°F, W.Q.
CY-40	—	—	—	—	—	—	—	—	—	—	—	—	
CZ-100	—	—	—	—	—	—	—	—	—	—	—	—	
M-35	—	—	—	—	—	—	—	—	—	—	—	—	
HA	0.279	7.1	182	421	29	95,000[d]	107,000	65,000[d]	23[d]	21	180[d]	220	1825°F + 1250°F
HC	0.272	6.3	151	463	29	110,000	115,000	75,000	19	18	223	—	1400°F/24 hr, F.C.
HD	0.274	7.7	151	487	27	85,000	—	48,000	16	—	190	270	1400°F/24 hr, F.C.
HE	0.277	9.6	120[e]	511	25	95,000	90,000	45,000	20	10	200	190	1400°F/24 hr, F.C.
HF	0.280	9.9	108	481	28	85,000	100,000	45,000	35	25	165	200[f]	1400°F/24 hr, A.C.
HH	0.279	9.4	98	451–511	27	85,000[f]	92,000[f]	40,000[f]	15[f]	8[f]	180[f]	200	1400°F/24 hr, A.C.
HI	0.279	9.9	98	—	27	80,000	90,000	45,000	12	6	170	190	1400°F/24 hr, A.C.
HK	0.280	9.2	98	541	29	75,000	85,000	50,000	17	10	170	—	1400°F/24 hr, A.C.
HL	0.279	9.2	98	565	27	82,000	—	52,000	19	—	192	—	
HN	0.283	—	—	—	29	68,000	—	38,000	17	—	160	—	
HT	0.286	8.5	92	602	27	70,000	75,000	40,000	10	5	180	200	1400°F/24 hr, A.C.
HU	0.290	8.8	—	632	27	70,000	73,000	40,000	9	5	170	190	1800°F/48 hr, A.C.
HW	0.294	7.8	92	674	25	68,000	84,000	36,000	4	4	185	205	1800°F/48 hr, F.C.
HX	0.294	7.8	—	—	25	65,000	73,000	36,000	9	9	176	185	1800°F/48 hr, A.C.

[d] Annealed.
[e] 70°–1500°F.
[f] Type II, austenitic.

Source: Courtesy of Alloy Casting Institute.

TABLE D.9 Properties of Some Elements

Element	Melting Point (°F)	Boiling Point, (°F)	Specific Heat (70°F) (Btu/lb/°F)	Specific Gravity	Modulus of Elasticity, Tension (10⁶ psi)	Closest Approach of Atoms (angstroms)	Crystal Structure
Aluminum	1220	4442	0.215	2.70	9	2.862	Face-centered cubic
Antimony	1167	2516	0.049	6.62	11.3	2.904	Rhombohedral
Barium	1337	2084	0.068	3.5	—	4.348	Body-centered cubic[a]
Beryllium	2332	5020	0.45	1.85	40-44	—	Close-packed hexagonal
Bismuth	520	2840	0.029	9.80	4.6	3.111	Rhombohedral
Boron	3690	—	0.309	2.34	—	—	—
Cadmium	610	1409	0.055	8.65	8	—	Close-packed hexagonal
Calcium	1540	2625	0.149	1.55	3.2–3.8	—	Face-centered cubic[a]
Carbon (graphite)	6740[b]	8730	0.165	2.25	0.7	1.42	Hexagonal[a]
Cerium	1479	6280	0.045	6.77	6	—	Face-centered cubic[a]
Cesium	83.6	1273	0.048	1.90	—	—	Body-centered cubic
Chromium	3407	4829	0.11	7.19	36	2.498	Body-centered cubic[a]
Cobalt	2723	5250	0.099	8.85	30	2.4967	Close-packed hexagonal[a]
Columbium	4474	8901	0.065	8.57	—	2.859	Body-centered cubic
Copper	1981	4703	0.092	8.96	16	2.556	Face-centered cubic
Gallium	85.6	4059	0.079	5.91	—	2.437	Orthorhombic
Germanium	1720	5125	0.073	5.32	—	2.449	Diamond cubic
Gold	1945	5380	0.031	19.32	11.6	2.882	Face-centered cubic
Hafnium	4032	9750	0.035	13.09	—	—	Close-packed hexagonal[a]
Indium	313	3632	0.057	7.31	1.57	3.25	Face-centered tetragonal
Iridium	4449	9570	0.031	22.5	76	2.714	Face-centered cubic
Iron	2798	5430	0.11	7.87	28-29	2.482	Body-centered cubic[a]
Lead	621	3137	0.031	11.36	2	3.499	Face-centered cubic

Lithium	357	2426	0.79	0.53	—	3.039	Body-centered cubic
Magnesium	1202	2025	0.245	1.74	6.35	3.196	Close-packed hexagonal
Manganese	2273	3900	0.115	7.43	23	—	Cubic (complex)[a]
Mercury	37	675	0.033	13.55	—	3.005	Rhombohedral
Molybdenum	4730	10,040	0.066	10.22	47	2.725	Body-centered cubic
Nickel	2647	4950	0.105	8.90	30	2.491	Face-centered cubic
Osmium	4500(?)	9950	0.031	22.57	81	—	Close-packed hexagonal
Palladium	2826	7200	0.058	12.02	16.3	2.750	Face-centered cubic
Platinum	3217	8185	0.031	21.45	21.3	2.775	Face-centered cubic
Potassium	147	1400	0.177	0.86	—	4.624	Body-centered cubic
Rhenium	5755	10,650	0.033	21.04	66.7	2.74	Close-packed hexagonal
Rhodium	3571	8130	0.059	12.44	42.5	2.689	Face-centered cubic[a]
Rubidium	102	1270	0.080	1.53	—	4.88	Body-centered cubic
Ruthenium	4530	8850	0.057	12.2	60	—	Close-packed hexagonal[a]
Silicon	2570	4860	0.162	2.33	16.4	2.351	Diamond cubic
Silver	1761	4010	0.056	10.49	11	2.888	Face-centered cubic
Sodium	208	1638	0.295	0.97	—	3.714	Body-centered cubic
Strontium	1414	2520	0.176	2.60	—	4.31	Face-centered cubic[a]
Tantalum	5425	9800	0.034	16.6	27	2.859	Body-centered cubic
Thallium	577	2655	0.031	11.85	—	3.408	Close-packed hexagonal[a]
Thorium	3182	7000	0.034	11.66	—	3.60	Face-centered cubic
Tin	450	4120	0.054	7.30	6-6.5	—	Tetragonal[a]
Titanium	3035	5900	0.124	4.51	16.8	—	Close-packed hexagonal[a]
Tungsten	6170	10,706	0.033	19.3	50	2.734	Body-centered cubic[a]
Vanadium	3450	6150	0.119	6.1	18-20	2.632	Body-centered cubic
Zinc	787	1663	0.092	7.13	—	2.6648	Close-packed hexagonal
Zirconium	3366	6470	0.067	6.49	13.7	3.17	Close-packed hexagonal[a]

[a]Ordinary form; other structures known or possible.
[b]Sublimes

TABLE D.10 Approximate Relationships between Hardness Values, Nickel, and High-Nickel Alloy[a]

Diamond Pyramid Hardness Number, DPH (Indenter-1, 5, 10, 30 kgf Load)	Brinell Hardness Number, BHN (10 mm Standard Ball, 3000 kgf Load)	Rockwell Hardness Number									Rockwell Superficial Hardness Number						Knoop Hardness Number[b] KHN (Knoop Indenter 500 and 1000 gf Load)
		A Scale (Diamond Penetrator 60 kgf Load)	B Scale (1.588 mm Ball, 100 kgf Load, $\frac{1}{16}$ in.)	C Scale (Diamond Penetrator 150 kgf Load)	D Scale (Diamond Penetrator 100 kgf Load)	E Scale (3.175 mm Ball, 100 kgf Load, $\frac{1}{8}$ in.)	F Scale (1.588 mm Ball, 60 kgf Load, $\frac{1}{16}$ in.)	G Scale (1.588 mm Ball, 150 kgf Load, $\frac{1}{16}$ in.)	K Scale (3.175 mm Ball, 150 kgf Load, $\frac{1}{8}$ in.)	15-N Scale (Superficial Diamond Penetrator, 15 kgf Load)	30-N Scale (Superficial Diamond Penetrator, 30 kgf Load)	45-N Scale (Superficial Diamond Penetrator, 45 kgf Load)	15-T Scale (1.588 mm Ball, 15 kgf Load, $\frac{1}{16}$ in.)	30-T Scale (1.588 mm Ball, 30 kgf Load, $\frac{1}{16}$ in.)	45-T Scale (1.588 mm Ball, 45 kgf Load, $\frac{1}{16}$ in.)	
513	479	75.5	—	50.0	63.0	—	—	—	—	85.5	68.0	54.5	—	—	—	—
481	450	74.5	—	48.0	61.5	—	—	—	—	84.5	66.5	52.5	—	—	—	—
452	425	73.5	—	46.0	60.0	—	—	—	—	83.5	64.5	50.0	—	—	—	—
427	403	72.5	—	44.0	58.5	—	—	—	—	82.5	63.0	47.5	—	—	—	—
404	382	71.5	—	42.0	57.0	—	—	—	—	81.5	61.0	45.5	—	—	—	436
382	363	70.5	—	40.0	55.5	—	—	—	—	80.5	59.5	43.0	—	—	—	413
362	346	69.5	—	38.0	54.0	—	—	—	—	79.5	58.0	41.0	—	—	—	392
344	329	68.5	—	36.0	52.5	—	—	—	—	78.5	56.0	38.5	—	—	—	372
326	313	67.5	—	34.0	50.5	—	—	—	—	77.5	54.5	36.0	—	—	—	352
309	298	66.5	**106**	32.0	49.5	—	**116.5**	94.0	—	76.5	52.5	34.0	94.5	85.5	77.0	325
285	275	64.5	**104**	28.5	46.5	—	**115.5**	91.0	—	75.0	49.5	30.0	94.0	84.5	75.0	304
266	258	63.0	**102**	25.5	44.5	—	**114.5**	87.5	—	73.5	47.0	26.5	93.0	83.0	73.0	283
248	241	61.5	100	22.5	42.0	—	**113.0**	84.5	—	72.0	44.5	23.0	92.5	81.5	71.0	267
234	228	60.5	98	20.0	40.0	—	**112.0**	81.5	—	70.5	42.0	20.0	92.0	80.5	69.0	251
220	215	59.0	96	**17.0**	38.0	**108.5**	**111.0**	78.5	100.0	69.0	39.5	17.0	91.0	79.0	67.0	239
209	204	57.5	94	**14.5**	36.0	107.0	**110.0**	75.5	98.0	68.0	37.5	14.0	90.5	77.5	65.0	226
198	194	56.5	92	**12.0**	34.0	**106.0**	**108.5**	72.0	96.5	66.5	35.5	11.0	89.5	76.0	63.0	215
188	184	55.0	90	**9.0**	32.0	104.5	**107.5**	69.0	94.5	65.0	32.5	7.5	89.0	75.0	61.0	204
179	176	53.5	88	**6.5**	30.0	**107.0**	**106.5**	65.5	93.0	64.0	30.5	5.0	88.0	73.5	59.5	195
171	168	52.5	86	**4.0**	28.0	**106.0**	**105.0**	62.5	91.0	62.5	28.5	2.0	87.5	72.0	57.5	187
164	161	51.5	84	**2.0**	26.5	**104.5**	**104.0**	59.5	89.0	61.5	26.5	−0.5	87.0	70.5	55.5	

322

157	155	50.0	82	—	24.5	**103.0**	**103.0**	56.5	87.5	—	86.0	69.5	53.5	179
151	149	49.0	80	—	22.5	**102.0**	**101.5**	53.0	85.5	—	85.5	68.0	51.5	173
145	144	47.5	78	—	21.0	**100.5**	**100.5**	50.0	83.5	—	84.5	66.5	49.5	166
140	139	46.5	76	—	**19.0**	99.5	99.5	47.0	82.0	—	84.0	65.5	47.5	160
135	134	45.5	74	—	**17.5**	98.0	98.5	43.5	80.0	—	83.0	64.0	45.5	154
130	129	44.0	72	—	**16.0**	97.0	97.0	40.5	78.0	—	82.5	62.5	43.5	149
126	125	43.0	70	—	**14.5**	95.5	96.0	37.5	76.5	—	82.0	61.0	41.5	144
122	121	42.0	68	—	**13.0**	94.5	95.0	34.5	74.5	—	81.0	60.0	39.5	140
119	118	41.0	66	—	**11.5**	93.0	93.5	31.0	72.5	—	80.5	58.5	37.5	136
115	114	40.0	64	—	**10.0**	91.5	92.5	—	71.0	—	79.5	57.0	35.5	—
112	111	39.0	62	—	**8.0**	90.5	91.5	—	69.0	—	79.0	56.0	33.5	—
108	108	—	60	—	—	89.0	90.0	—	67.5	—	78.5	54.5	31.5	—
106	106	—	58	—	—	88.0	89.0	—	65.5	—	77.5	53.0	29.5	—
103	103	—	56	—	—	86.5	88.0	—	63.5	—	77.0	51.5	27.5	—
100	100	—	54	—	—	85.5	87.0	—	62.0	—	76.0	50.5	25.5	—
98	98	—	52	—	—	84.0	85.5	—	60.0	—	75.5	49.0	23.5	—
95	95	—	50	—	—	83.0	84.5	—	58.0	—	74.5	47.5	21.5	—
93	93	—	48	—	—	81.5	83.5	—	56.5	—	74.0	46.5	19.5	—
91	91	—	46	—	—	80.5	82.0	—	54.5	—	73.5	45.0	17.0	—
89	89	—	44	—	—	79.0	81.0	—	52.5	—	72.5	43.5	14.5	—
87	87	—	42	—	—	78.0	80.0	—	51.0	—	72.0	42.0	12.5	—
85	85	—	40	—	—	76.5	79.0	—	49.0	—	71.0	41.0	10.0	—
83	83	—	38	—	—	75.0	77.5	—	47.0	—	70.5	39.5	7.5	—
81	81	—	36	—	—	74.0	76.5	—	45.5	—	70.0	38.0	5.5	—
79	79	—	34	—	—	72.5	75.5	—	43.5	—	69.0	36.5	3.0	—
78	78	—	32	—	—	71.5	74.0	—	42.0	—	68.5	35.5	1.0	—
77	77	—	30	—	—	70.0	73.0	—	40.0	—	67.5	34.0	**−1.5**	—

[a] Hardness Conversion Chart for Nickel and High-Nickel Alloys. A.S.T.M., E140-65. The use of hardness scales for hardness values shown in boldface type are not recommended by the manufacturers of hardness testing machines since they are beyond the ranges recommended for accuracy. Such values are shown for comparative purposes, only, where comparisons may be desired and the recommended machine and scale are not available.

[b] For Knoop hardness determinations the specimen must be polished, etched, and repolished until a final light etch shows a clearly defined microstructure free from disturbed metal. Care must be exercised to ensure that the top and bottom of the mounted specimen are parallel. In no case shall the departure from symmetry in the longitudinal direction of the indentation be greater than 5 filar microscope units.

TABLE D.11 Approximate Comparison of Gauges

	Inches						Millimeters	
Gauge No.	American or Brown & Sharpe's	Birmingham or Stubs'	Washburn & Moen's	Imperial S.W.G.	London or Old English	United States Standard	United States Standard	Stubs'
7/0	—							
6/0	0.5800	—		0.500	—	0.5000	12.700	—
5/0	0.5165	—	0.4900	0.464	—	0.4687	11.906	—
		—	0.4615	0.432	—	0.4375	11.113	—
4/0	0.4600	0.454	0.4305	0.400	0.454	0.4062	10.319	11.532
3/0	0.4096	0.425	0.3938	0.372	0.425	0.3750	9.525	10.795
2/0	0.3648	0.380	0.3625	0.348	0.380	0.3437	8.731	9.652
1/0	0.3249	0.340	0.3310	0.324	0.340	0.3125	7.938	8.636
1	0.2893	0.300	0.3065	0.300	0.300	0.2812	7.144	7.620
2	0.2576	0.284	0.2830	0.276	0.284	0.2656	6.747	7.214
3	0.2294	0.259	0.2625	0.252	0.259	0.2500	6.350	6.579
4	0.2043	0.238	0.2437	0.232	0.238	0.2343	5.953	6.045
5	0.1819	0.220	0.2253	0.212	0.220	0.2187	5.556	5.588
6	0.1620	0.203	0.2070	0.192	0.203	0.2031	5.159	5.156
			0.1920					

7	0.1443	0.180	0.1770	0.176	0.1875	4.763	4.572
8	0.1285	0.165	0.1620	0.160	0.1718	4.366	4.191
9	0.11440	0.148	0.1483	0.144	0.1562	3.969	3.759
10	0.10190	0.134	0.1350	0.128	0.1406	3.572	3.404
11	0.09074	0.120	0.1205	0.116	0.1250	3.175	3.048
12	0.08081	0.109	0.1055	0.104	0.10930	2.778	2.769
13	0.07196	0.095	0.0915	0.092	0.09375	2.381	2.413
14	0.06408	0.083	0.0800	0.080	0.07812	1.984	2.108
15	0.05707	0.072	0.0720	0.072	0.07031	1.786	1.829
16	0.05082	0.065	0.0625	0.064	0.06250	1.588	1.651
17	0.04526	0.058	0.0540	0.056	0.05625	1.429	1.473
18	0.04030	0.049	0.0475	0.048	0.05000	1.270	1.245
19	0.03589	0.042	0.0410	0.040	0.04375	1.111	1.067
20	0.03196	0.035	0.0348	0.036	0.03750	0.953	0.889
21	0.02846	0.032	0.0317	0.0315	0.03437	0.873	0.813
22	0.02535	0.028	0.0286	0.0295	0.03125	0.784	0.711
23	0.02257	0.025	0.0258	0.0270	0.02812	0.714	0.635
24	0.02010	0.022	0.0230	0.0250	0.02500	0.635	0.559
25	0.01790	0.020	0.0204	0.0230	0.02187	0.556	0.508
26	0.01594	0.018	0.0181	0.0205	0.01875	0.476	0.457

TABLE D.12 Dimensions of Standard Pipe Sizes

Nominal Pipe Size (in.)	Outside Diameter (in.)	Nominal Wall Thickness for:				
		Schedule 5	Schedule 10	Schedule 40	Schedule 80	Schedule 160
$\frac{1}{8}$	0.405	—	0.049	0.068	0.095	—
$\frac{1}{4}$	0.540	—	0.065	0.088	0.119	—
$\frac{3}{8}$	0.675	—	0.065	0.091	0.126	—
$\frac{1}{2}$	0.840	0.065	0.083	0.109	0.147	0.187
$\frac{3}{4}$	1.050	0.065	0.083	0.113	0.154	0.218
1	1.315	0.065	0.109	0.133	0.179	0.250
$1\frac{1}{4}$	1.660	0.065	0.109	0.140	0.191	0.250
$1\frac{1}{2}$	1.900	0.065	0.109	0.145	0.200	0.281
2	2.375	0.065	0.109	0.154	0.218	0.343
$2\frac{1}{2}$	2.875	0.083	0.120	0.203	0.276	0.375
3	3.500	0.083	0.120	0.216	0.300	0.438
$3\frac{1}{2}$	4.000	0.083	0.120	0.226	0.318	—
4	4.500	0.083	0.120	0.237	0.337	0.531
5	5.563	0.109	0.134	0.258	0.375	0.625
6	6.625	0.109	0.134	0.280	0.432	0.718
8	8.625	0.109	0.148	0.322	0.500	0.906
10	10.750	0.134	0.165	0.365	0.593	—
12	12.750	0.165	0.180	0.406	0.687	—

E

BIBLIOGRAPHY

There are hundreds of good books on the subject of welding and metallurgy. The author has selected those he has used in his work and in his teaching.

1. *Metallurgy*, Carl G. Johnson, American Technical Society, Chicago, IL, 1956.
2. *Metals and How to Weld Them*, T. B. Jefferson, James F. Lincoln Arc Welding Foundation, Cleveland, OH, 1963.
3. *Modern Welding Technology*, Howard B. Cary, Prentice-Hall, Inc., Englewood Cliffs, NJ, 1979.
4. *Pipefitter's and Pipewelder's Handbook*, Thos. W. Frankland, P.O. Box 297, Chicago, IL 60617, 1972. (A small, pocket-sized book that every pipewelder/fitter should have in his toolbox.)
5. *Stoody Hardfacing Guidebook*, Stoody Company, P.O. Box 1901, Industry, CA 91749, 1966.
6. *The Weld Tech Series in Welding*, Chrysler Learning Inc., Prentice-Hall, Inc., Englewood Cliffs, NJ, 1983.
7. *Welding Aluminum*, American Welding Society, Miami, FL, 1972.
8. *Welding Encyclopedia*, T. B. Jefferson, Monticello Books, Morton Grove, IL, 1974.
9. *Welding Handbook*, American Welding Society (a work of many volumes).
10. *Welding Practices and Procedures*, Richard Carr and Robert O'Con, Prentice-Hall, Inc., Englewood Cliffs, NJ, 1983.
11. *Welding: Principles and Practices*, Raymond J. Sacks, Bennett Publishing Company, Peoria, IL, 1976.

INDEX

Abbreviations, 289
Abrasion, 260, 274
Age hardening, 162, 274
Air-arc, 209, 212
Air-hardening steels, 97, 102, 106, 274
Allotropic, 53, 66, 68, 274
Alloys, classes of, 58
Alpha iron, 57, 66, 274
Alumina, 78
Aluminum, 159-74
  heat treatment, 161, 162
  preheating, 164
  understanding, of, 159
Aluminum alloys, 160
Aluminum bronze, 151
Aluminum oxides, 159
Aluminum welding, 163-73
  brazing, 172, 173
  filler rods, 168
  gas shielding, 165
  to other metals, 158
  oxy-acetylene welding, 164, 165
  results in welding, 173
  soldering, 171, 172
  stick welding, 170, 171
  TIG welding, 165, 170
  welding defects, 174
Amorphous cement theory, 56
Annealing, 56, 73, 274
  (*see also* Heat treatment)
Antimony, 60, 63, 320

Arc welding, stick, 1-20
  resistance in the arc, 4, 5
  shielding, 6
  transfer, 5
  understanding of, 1
  (*see also* Welding techniques)
Argon, 38, 40, 41, 49, 275
Austempering, 100, 275
Austenite, 66, 67, 69, 275
Austenitic manganese steels, 175-78
  characteristics that affect welding, 176
  understanding of, 175
  wearfacing applications, 262
  welding procedures, 177, 178
  work hardening, 175, 176
Axes of solidification, 54

Backfire, 33
Backhand welding, 23, 275
Backing ring, 275
Backstep welding, 275
Bainite, 73, 275
Bar, (atmosphere), 275
  (*see also* English-Metric conversions)
Base metal, 275
Beading, 275
Bell-hole welding, 275
Beryllium, 154, 320
Beryllium-copper alloys, 151
Beta iron, 57, 275
Beta martensite, 275

Beveling:
  angle of, 198, 275
  cutback formula, 198
Birdnesting, 20, 275
Black-heart malleable iron, 113
Block sequence, 275
Blowhole, 276
Blowpipe, 276
Body-centered cubic lattice, 53, 276
Boilermaker, 250, 276
Boxing, 276
Brass, 152, 276
Brazing, 22, 23, 26, 27
  aluminum, 172, 173
  brasses and bronzes, 151–54
  copper, 150
  nickel-based alloys, 145, 148
  stainless steel, 134
  (*see also* each metal separately)
Brinnell hardness number, 276
Brittleness, 276
Bronze, 151, 153, 276
Bronze welding (*see* Brazing)
Build-up, 276
Buried arc, 48
Burner, 276
Buttering, 276
Butt joint, 276 (*see* Joint design)

Calescent, 276
Capillary action, 8, 276
Capillary attraction, 8, 277
Cap pass, 277
Carbide, 277
  chromium carbide, 263, 264
  precipitation, 277
  tungsten carbide, 264
  vanadium carbide, 264
Carbon-arc cutting, (CAC), 208–10
Carbon dioxude, 277
  as shielding, 49
Carbon equivalent, 87, 90
Carbon steel, (*see* High-, Low-, and Medium-carbon steels)
Carburizing, 30, 75, 277
Cascade sequence, 277
Case hardening, 74, 277
Castings, high-alloy, 316
Cast iron, 108–27, 277
  black-heart malleable, 113
  braze welding, (OAW) 118
  braze welding, (stick) 119
  equilibrium diagram, 67, 115
  function of alloying elements, 110, 111
  fusion welding, 117, 118
  gray iron, 108, 110
  heat treatment, 116
  malleable iron, 108, 109, 113, 120

  nodular iron, 113, 114, 121
  pearlitic gray iron, 110
  soldering, 124
  tempering of fusion zone, 122
  thermit welding, 120, 124, 125
  transformations, 67, 114, 115
  welding, 114, 116, 117–27
  welding to other metals, 158
  weld joint preparation, 116, 121, 122
  white iron, 108, 112, 120
Cast steel, 277
Caulk weld, 277
Cementite, (*see* Iron-carbide)
Chain intermittent, 187, 277
Chamfering, 209, 277
Chrome irons, 131, 277
Circuit, the welding, 4
  troubleshooting of, 13
Cladding, 277
Classes of alloys, 58
Cleavage lines, 54
Cold rolling, 277
Cold welding, 278
Cold working, 278
Columbium, 290, 320
Compound, 58
Constant potential, (*see* Constant voltage)
Constant voltage, 45, 278
Conversion formulas, 295–99
Copper and copper-based alloys:
  aluminum-copper, 151
  ambracs, 154
  chemical composition and characteristics, 312–15
  copper-silicon, 153
  copper-zinc, 152
  copper-zinc-tin, 153
  cupro-nickel, 154
  deoxidized copper, 150
  manganese bronze, 154
  nickel-silver, 154
  tough-pitch copper, 150
Corrosion (*see* Conversion formulas)
Cover pass (*see* Cap-pass)
Cracking, weld:
  crater cracks, 83
  radial cracks, 84
  root cracks, 84
  underbead cracks, 84
Crater, 278
Creep, 278
Critical range (*see* Transformation range)
Cyrogen, 278
Cryogenics, 278
Cryogenic steel, welding electrodes for, 304
Current, 4, 12
Cyaniding, 75, 278
Cyber-Tig, 38
Cycle, Duty (*see* Duty cycle)

Index

Decarburization, 278
Deformation, 55, 183, 185-88
Delta iron, 57, 67
Deoxidizing, 278
Diamond substitute, 263, 264
Differential hardening, 74, 112, 278
Dip (*see* Nozzle dip)
Dissimilar metals, welding of, 156
Distortion:
   angular displacement, 183, 185, 186
   axial displacement, 183, 186
   prevention of, 181, 185-88
Double-ending, 228, 278
Downhill welding, 278
Drag, 278
Drag line (*see* Drag)
Drawing (*see* Tempering)
Ductility, 278
Duty cycle, 278
Dyne, 279

Elasticity, 55, 279
Elastic limit, 55, 279
Electrodes, welding:
   classification of, 17, 18
   (*see* each metal separately)
   shielding, 14, 15
Elements, the properties of, 320, 321
Elongation, 279
Emulsion, 58
Endurance limit, 279
English-Metric conversions, 295
Eutectic, 279
   class II alloys, 61, 279
   defined and explained, 61, 279
   iron-carbon eutectic, 46, 67
   lead-antimony eutectic, 61, 62
   lead-tin eutectic, 63, 64
Eutectic alloy, 61, 62
Eutectoid, 66, 67
Eutectoid steel, 67, 279
Eyeballing, 279

Face-centered cubic lattice (fcc), 53, 279
Fatigue failure, 279
Fatigue limit, 279
Ferrite, 68, 69, 93, 279
Filler pass, 279
Fingernailing, 279
Firing-line welder, 279
Fisheye, 279
Flame hardening, 75
Flames, Oxy-acetylene:
   carburizing, neutral, oxidizing, 30, 31
   washing (*see* Scarfing)
Flame shrinking, 211, 212, 279

Flame straightening, 211, 212
Flashback, 34, 280
Flop weld, 280
Flux, 280
   cast iron, 118, 119
   characteristics and purposes, 25
   copper-silicon, 153
   nickel-based alloys, 142, 145
   stainless steel, 133
Foldback, 280
Forehand welding, 23
Forge welding, 280
Formulas (*see* Appendix C):
   conversion, 295-99
   pipefitting, 220-28
Forward swing, 280
Free carbon, 280
Free ferrite, 68, 69, 280
Free-machining steel, 280
Freezing, 280
Fusion welding, 280

Galling, 280
Galvanized steel, 280
Gamma iron, 57, 68, 69, 280
Gas, shielding:
   argon, 49
   argon-CO_2, 49
   argon-helium, 49
   argon-oxygen, 49
   carbon-dioxide (CO_2), 49
Gas-flow rate, 41, 43
Gas metallic-arc welding:
   arc length, 50
   arc transfer (*see* Transfer, arc)
   arc voltage, 46, 47
   backcoiling of wirefeed, 50
   birdnesting, 50
   current, 45, 46, 47
   nozzle angle, 50
   spatter, 50
   squirt-through, 49
   stick-out, 47, 50
   stubbing (roping), 50
   techniques, 49
   travel speed, 50
   understanding the arc, 45
   wirefeed, 45, 50
Gas pocket (*see* Blowhole)
Gas Tungsten-Arc Welding (GTAW):
   current selection, 39, 43
   electrode tip, 39
   electrode type, 42
   filler metals (*see* specific metal)
   gas flow rate, 41
   lpm/cfh conversion formula (Appendix C)
   shielding gases, 39. 40

Gas Tungsten-Arc Welding (cont.)
  stick-out, length of, 38
  (*see also* Aluminum welding)
Globular transfer, 47, 280
Gold, welding, 156
Grain, 38, 281
Gram (*see* Metric system)
Graphite, 93, 281
Graphitic carbon (*see* Graphite)
Graphitic tool steel, 103, 281
Gray iron (*see* Cast iron)

Hadfield steel (*see* Austenitic manganese steels)
Hardfacing (*see* Wearfacing)
Hardness, 79, 80
  controlling hardness, 82
  cross reference of hardness numbers, 322, 323
HASTELLOY alloys:
  chemical composition, 147
  welding electrodes, 147
  welding processes, 147
  welding techniques, 139–42
Heat treatment:
  aluminum, 161, 162
  cast iron, 114, 116, 117, 120, 121, 127
  free-machining steel, 65
  nickel-based alloys, 139
  stainless steel, 136, 137
  steel, 73–75, 88, 89
  tool-and-die steel (*see* Tool-and-die steels)
Heliarc, 37
Helium, 49, 281, 320
Heli-weld, 38
Hexagonal space lattice (hcp), 53, 281
High-carbon steel, 83
High-frequency stabilization, 39
High-speed steel (HSS), 102, 107, 281
High-sulfur steel, 281
Hot pass, 281
Hot quench, 100, 281
Hot shortness, 281
Hot working, 281
Hot-working dies, 103, 281
Huntington alloys, 282
  electrode reference guide, 145, 146, 303–7
  filler wire, 146, 303–7
  specifications index, 303–7
  welding processes, 146
  welding techniques, 139–42
Hydrogen, 40, 78, 320
Hyper-eutectic, 282
Hyper-eutectoid, 282
Hypo-eutectic, 282
Hypo-eutectoid, 282

Impact, 260
INCOLOY, 144, 303–17
INCONEL, 143, 303–17
Inert gas, 38, 282
Ingot iron, 282
Innershield, 282
Insert, 282
Intergranular corrosion, 282
Intermittent welding:
  chain intermittent, 191
  staggered intermittent, 191
Iron (*see* cast iron)
Iron, wrought, 90
Iron-carbide, 66, 282

Joint, brazing, 28, 29
Joint design, 179, 190–98
Joule, 282

Kerf, 282
Keyholing, 282
KIP, 282
Kirchhoff's Law, 4

Lacing, 16, 20
Land, 282
Lead, welding, 155
Lead burning, 282 (*see* Lead welding)
Leaded brasses, 152
Liquidus, 59, 61, 64, 67, 71
Liter (*see* Metric system)
Load, 282
Low-alloy steels, 77, 88, 89

Machine, welding:
  gas-engine driven, 10
  motor-generator, 9
  setting welding current, 13, 14
  transformer-rectifier, 7, 9
  understanding the, 8
Magnetic pinch effect, 4, 282, 284
Malleable iron (*see* Cast iron)
Malleabilizing, 113, 283
Manganese steel (*see* Austenitic manganese steels)
MAPP gas, 283
Martempering, 100, 283
Martensite, 72, 93, 95, 96, 283
Matrix, 62
Medium-carbon steels, 79–81, 88, 89
Meld, 283
Melder, 283
Melting point, 320, 321
Metal fatigue (*see* Fatigue failure)
Metal working, 199–213

Meter (see Metric system)
Metric system, 295
Micrometer (micron), 295
MIG welding (see Gas Metallic-arc welding)
MOH scale of hardness, 295
Molybdenum, 111, 320
MONEL, 143, 282
MONEL to steel, welding, 158
Mottled iron, 111, 282
M-point, 95, 96, 97, 282

Neutral flame, 282
(see also Flames)
Newton, 282
Nickel-based alloys, 138-50
  brazing, 145
  care of electrodes, 141
  characteristics, 146
  cleaning, 139, 140, 142
  crater cracks, 139
  cutting, 148
  filler metals, 145, 146, 147
  gas welding, 149
  HASTELLOY alloys, 147
  Huntington alloys, 145, 146
  INCOLOY, 144
  INCONEL, 143
  joint preparation, 139, 141
  machining, 142
  MIG welding, 142, 146
  MONEL, 143
  nickel, pure, 143
  nickel-molybdenum alloys (see HASTELLOY alloys)
  TIG welding, 141, 142
  welding techinques, 139-50
Nickel-silver alloys, 283
Nitriding, 75
Nitrogen, 40, 78, 321
Normalizing, 74, 283
Nozzle dip, 284

Ohm, 3, 284
Ohm's Law, 3
Oil-hardening steel, 101, 106
Open-circuit voltage, 284
Overlay (see Wearfacing)
Oxide, 284
Oxidizing flame 30, 284
Oxy-acetylene cutting, 199-207, 284
Oxy-acetylene welding, 21-30
  brazing, 21
  fusion welding, 21
  lighting the torch, 31
  setting guages and valves, 199, 202, 203, 204
  sweating, 29

understanding flu, 24
understanding heat, 24
understanding the torch 30
welding tips, 30, 31
(see each metal separately)
Oxygen, 40, 78, 321
Oxygen-arc cutting, 208

Pascal, 284
Pearlite, 68, 69, 72, 73, 93, 284
Peeling, 284
Peening, 284
Phosphor bronze, 153, 284
Phosphorus, 78, 321
Piercing, flame, 205
Pig iron, 284
Pinch effect, 284 (see also Magnetic pinch effect)
Pipefitting and welding, 216-55
  backstub, 245
  blind, installation of, 234
  boilermaker, 250
  branch pipe from pipe-turn, 245
  cracking a flange, 234
  double-ending, 228
  fittings, fabrication of, 239, 243-45, 249, 250
  flanges, 230, 231
  flanging a 90-degree pipe-turn, 233, 234
  flanging a spool, 231, 232
  flop weld, 280
  formulas, 226-28
  full-size tee, 241, 242
  laying out angles, 235-37
  laying out branches, 241-42
  laying out ellipse, 236
  laying out miter cuts, 237, 239
  offsets, 246, 247
  pipe test, welding, 217-19
  pipe-turn, Joining of, 229
  protractor, use of, 237, 238
  reducing Tee, 242
  rolling weld, 216
  safety precautions, 252-54
  steam line, installations of, 247, 248
  trigonometry in pipefitting, 223, 224, 255-57
  trigonometry table, 255-57
  two-holing, 231
  wrap-around, use of, 202
Plasticity, 285
Platinum, welding, 156
Potential (see Voltage)
Precipitation, 63, 65, 285
Precipitation hardening, 285
  (see also Age hardening)
Preheating, 81, 82, 88, 89, 105

Puddle, the weld, 6, 7
Pulsed arc, 48
Pure metals, 53

Quench cracks, 285
Quenching, 74

Radial cracks, 84, 285
Reactive metals, 154, 155
Red hardness, 285
Red shortness, 285
Reducing flame (*see* Flames, carburizing)
Refractory metals, 154, 155, 285
Resistance:
  in the arc, 4
  in the electrode, 4
  in the stick-out, 46
  in the weld cable, 4, 5
Reverse polarity, 285
Rockwell hardness, 285
Rolling weld, 216
Root cracks, 83
Root opening, 197
Root pass, 285
Roping (*see* Stubbing)
Rosebud, 285

Scarfing, 207, 285
Scleroscope, 285
Seizing, 285
Self-hardening steel (*see* Air-hardening steels)
Semi-automatic welding, 285 (*see* MIG)
Shielding gases, 39, 40, 48, 49
Short-arc, 47
Short-circuit arc (same as short-arc)
Shuler weave, 285
Sideswing, 286
Silver-brazing, 286
  (*see* Brazing under specific metal)
Silver, welding, 156
Skipwelding, 286
Slip interference theory, 56
Slip lines, 54, 55, 286
Soaking, 286
Soldering (*see* each metal separately)
Solid solution, 60, 66, 68, 69, 286
Solidus, 60, 66, 68, 69, 286
Sorbite, 73, 93, 286
Space lattice, 53, 286
Spall, 286
Spatter, arc, 50
Spheroidizing, 74, 286
Spool, 232, 286
Spray transfer, 47, 286
Squirt-through, 49, 287

Stack-cutting steel, 207
Staggered intermittent, 187, 287
Stainless steel, 128-37, 287
  chemical composition and characteristics, 308, 309
  class 2xx, 129
  class 3xx, 129
  class 4xx, 131
  classification, 129, 136
  heat treatment, 136, 137
  understanding of, 128
  wearfacing applications, 262
Stainless steel, welding, 131-37
  brazing, 134
  electrodes, 135
  oxy-acetylene, 132
  soldering, 135
  stick (SMAW), 131-33
  TIG and MIG, 133, 134
  to cast iron, 157
  to INCONEL, 158
  to MONEL, 158
  to steel, 137
Steel, 54, 67, 68, 88, 89
  electrodes, 84, 85, 86, 88, 89
Steel, welding:
  to aluminum, 158
  high-carbon steel, 83
  low-alloy steel, 88, 89
  low-carbon steel, 77
  medium-carbon steel, 79
  to other metals, 157, 158
  stainless steel (*see* Stainless steel)
Stere (*see* Metric system)
Stick-out, 38, 50, 287
Stick welding, 287
  arc length, 4-6
  arc, mastering the, 1
  arc, understanding of, 2
  arc shielding, 6, 14
  arc transfer, 4
  current control, 10-12
  electrode classification (*see* each metal separately)
  welding machine (*see* Machine, welding)
  welding techniques, advanced, 179-212
  welding techniques, novice, 15-17
  (*see also* each metal separately)
Stitch welding (*see* Intermittent welding)
Stovepipe welding, 287
Strain, 180, 181, 287
Stress, 180, 181, 287
Stress crack, 287
Stress relief, 180, 287
Stubbing, 50, 287
Surface alloying, 22, 29
Surface tension, 5, 8, 24
Surfacing, 287
Sweating, 22, 29, 287

Index   **335**

Symbols:
  elements, 289-94, 320, 321
  welding, 189-98

Tables:
  Abrasion Resistance and Impact
    Strength, Interrelations of, 301
  Aluminum, Alloying Constituents in, 160
  Aluminum, Solderability of, 172
  Aluminum, TIG-welding with ACHF, 43
  Aluminum-Copper Equilibrium Diagram, 161
  Aluminum Filler Metal, Selection of, 168
  Arc-air Torch, Current Ranges for, 210
  Cast-Iron Welding, Expected Results in, 127
  Copper and Copper-Based Alloys, 312-15
  Cutting Specifications, Oxy-acetylene, 203
  Electrodes, Steel, 18, 88, 89
  Electrodes for Carbon Steels and Low-Alloy Steels, 88, 89
  Elements, Properties of, 320, 321
  Guages, Cross-Reference of Standard, 324, 325
  Hardness Values, Cross Reference of, 322, 323
  HASTELLOY Alloys, Characteristics and Filler Metals, 147
  HASTELLOY Alloys, Chemical Compositions of, 307
  Heat-Resisting Steels, 308-11
  High-Alloy Castings, 316, 317
  Huntington Alloys, Characteristics and Filler Metals, 146
  Huntington Alloys, Chemical Compositions of, 302
  Huntington Alloys, Specifications of, 304-6
  Huntingon Alloys Welding Products, 303
  Nickel and Nickel-Based Alloys, Filler Metals for, 145
  Nickel and Nickel-Based Alloys, General Characteristics, 146-48
  Oxy-acetylene Guage Settings, 33
  Pipe, Dimensions of Standard, 326
  Reactive and Refractory Metals, 154
  Silver-Brazing Alloys, 149
  Stainless Steel Electrodes, 136, 137
  Tempering Temperature, 100
  TIG-Welding Currents for Various Metals, 43
  Welding Tip Sizes, 32
Tantalum, 154, 321
Tempering, 74, 89-100, 287
  austempering, 100, 275
  martempering, 100, 283
  (*see* Tool-and-die steels)
Tensile strength, 287
Thermit, 287
Thermit welding, 124, 125, 287
Thorium, 321
TIG-welding (*see* Gas Tungsten-Arc Welding)
Tin brasses, the, 153, 312, 313
Titanium, 154, 321
Tool-and-die steels:
  air-hardening steels, 102, 106
  appealing, 106, 107
  drawing (*see* Tempering)
  electrodes, welding, 106, 107
  hardening temperature, 96, 106
  high-speed steel (HSS), 102, 106
  hot-working dies, 103, 106
  maintenance welder's view of, 104
  oil-hardening steel, 101, 106
  preheating, 105, 106
  prevention of cracked welds, 104
  quenching, 95-97, 106, 107 (*see also* eacl metal separately)
  self-hardening (*see* Air-hardening)
  silver-brazing of, 107
  stick welding, 106
  tempering, 98, 100, 105, 106 (*see* each type)
  water-hardening steel, 101
  welding, general, 104, 106
Transfer, arc:
  buried arc, 48
  globular, 47
  pulsed, 48
  short circuit, 48
  spray, 47
  submerged (*see* buried arc, above)
Transformation range, 67, 70, 287
Transformer-rectifier welding machine, 8-12
Travel speed, 50
Trigonometric functions, 256-58
Troostite, 72, 93, 288
Tube-Turn, 288
Tungsten, 42, 43, 321
Tungsten inert-gas welding (*see* Gas Tungsten-Arc Welding)

Underbead cracking, 84
Undercut, 288 (*see* Puddle)
Upset, 288
Upset joint, 195

Vanadium, 154, 321
Vertical-down welding, 288
Vertical-up welding, 16, 288

Voltage:
  closed circuit, 10
  constant, 4
  load, 10
  open circuit, 10

Wagon tracks, 288
Warping (see Distortion and Flame straightening)
Water-hardening steel, 101, 106, 288
Wearfacing, 258–71, 288
  materials, 262–64
  types of wear, 259
  understanding of, 258
Wearfacing applications, automotive:
  engine valves, cams, rocker arms, 272
  tappets, valve-stem tips, 272
  valve rings and valve disks, 272
Wearfacing applications, earthmoving equipment:
  asphalt mixer paddles, 269
  cable sheaves, 268
  ditcher-drive segments, 269
  ditcher-drive sprockets, 269
  ditcher rolls, 269
  dozer blades, 268
  dozer end bits, 267
  dragline buckets, pins, chains, and clevis, 267, 268
  drive sprockets, 268
  paving screw conveyor, 269
  ripper teeth, 267
  shovel-boom heels, 267
  shovel-bucket tooth adaptors, 265, 267
  shovel-drive tumblers, 267
  shovel-house rolls, 268
  shovel idlers, 267
  shovel rollers, 267
  shovel teeth, 265, 266
  shovel track pads, 268
  top carrier rolls, 268
  track rollers, 267
  tractor grousers, 267
  tractor idlers, 267
  tractor rails, 268

Wearfacing applications, logging and lumber:
  anvil knives, 269
  bark-conveyor trunnions, 270
  bed plates, 270
  chain-drive tumblers, 270
  chipper chutes and disks, 270
  clutch fingers and jaws, 270
  conveyor chain links, 270
  conveyor chains, 270
  debarking hammers, rotors, knives, and chain links, 270
  hog rotors, 269
  knife keepers, 269
  log escalators, 270
  log-haul chairs, 270
  rotary-head chipper blades, 270
  saw-carriage wheels, 269
Wearfacing applications, mining:
  auger bits, 271
  ball-mill scoop lips, 271
  ball-mill scoops, 271
  clutch lugs, 271
  coal-recovery augers, 271
  collar puller cutters, 271
  core barrels (coal recovery), 271
  cutter chain lets, 271
  digging arms, 271
  duck bills, 271
  grizzlies, 271
  skip hoists, 272
  sprocket drum and traveling sprockets, 271
Wearfacing well-drilling core heads, 272
Welding symbols, 189–98
Welding techniques, general, 179–213
Weldment, 288
Weld-neck flange, 231, 288
Wetting out, 288 (see Capillary action)
Whipping, 288
Whiskers, 49, 288
White iron, 288 (see Cast iron)
Work hardening, 175, 176, 288
Wrought iron, 90, 91

Zinc die-cast, welding of, 156
Zirconium, 154, 321